Signal Processing in Electronic Communications

The direction in which education starts a man will determine his future life
Plato: *The Republic, Book* (427-347 BC)

"Talking of education, people have now a-days" (said he) "got a strange opinion that every thing should be taught by lectures. Now, I cannot see that lectures can do so much good as reading the books from which the lectures are taken. I know nothing that can be best taught by lectures, expect where experiments are to be shewn. You may teach chymestry by lectures. — You might teach making of shoes by lectures!"
James Boswell: *Life of Samuel Johnson, 1766* (1709-1784 AD)

Signal Processing in Electronic Communications

MICHAEL J. CHAPMAN
DAVID P. GOODALL
and NIGEL C. STEELE
School of Mathematics and Information Sciences
University of Coventry

Horwood Publishing
Chichester

First published in 1997 by
HORWOOD PUBLISHING LIMITED
International Publishers
Coll House, Westergate, Chichester, West Sussex, PO20 6QL
England

British Library Cataloguing in Publication Data
A catalogue record of this book is available from the British Library

ISBN 1-898563-23-3

Printed in Great Britain by Hartnolls, Bodmin, Cornwall

Preface

Communication is the process by which information is exchanged between human beings, between machines, or between human beings and machines. Communication Theory is the theory of the transmission process, and the language of this theory is mathematics. This book sets out to explain some of the mathematical concepts and techniques which form the elements and syntax of that language, and thus enable the reader to appreciate some of the results from the theory.

In some ways, the degree of evolution of a nation or state may be measured by the sophistication of its communication processes, particularly those based on electronic means. Within the lifetime of one of the authors, the telephone has become an everyday means of communication, and television has moved from a rarely seen novelty to the means of mass entertainment. Instantaneous communication over long distances, across continents or oceans, by voice or text has become an everyday requirement in many walks of life. More recently, the internet has provided a new dimension to the way many of us work. Electronic mail is not only an indespensible tool for collaborative research, it is a means by which colleagues, perhaps in different countries, communicate on a day-to-day basis. The wider resource of the so-called world-wide-web gives access to a mass of data. Perhaps the major problem which faces us at the time of writing is how these data can be turned into information efficiently, but that is a debate for a different forum!

All these forms of communication are conducted by the transmission of electronic signals using an appropriate method and this book concentrates on the description of signals and the systems which may be used to process them. Such processing is for the purpose of enabling the transmission of information by, and the extraction of information from, signals. Over the last few years, the unifying concepts of signals and linear systems have come to be recognised as a particularly convenient way of formulating and discussing those branches of applied mathematics concerned with the design and control of 'processes'. The process under discussion may be mechanical, electrical, biological, economic or sociological. In this text, we consider only a restricted subset of such processes, those related to communication by electronic means. There are several first class treatments of the general field of signals and linear systems and indeed, some specifically related to communication theory. However, in many cases, the haste to discuss the engineering applications of the material means that the mathematical development takes second place. Such a treatment may not appeal to, or even be readily accessible to those whose first subject, or interest, is mathematics, and who thus may not be able to supply the necessary engineering insight to follow the discussion easily. This book is aimed in part at such readers who may wish to gain some understanding of this fascinating application area of mathematics.

We are also aware from our teaching experience that many of our engineering students (possibly more than is commonly acknowledged!) also appreciate such a development to complement and support their engineering studies. This book is also written for this readership. It is interesting to recall that, whilst the need for engineers to become competent applied mathematicians was widely recognised in the recent past, this need is not so often expressed, at least in some countries, today. It will be interesting to compare future performance in the field of design and innovation, and thus in economic performance, between countries which adopt different educational strategies.

The subject matter which is included within the book is largely self-contained, although we assume that the reader will have completed a first course in mathematical methods, as given for engineering and computer science students in most UK Universities. The style adopted is an attempt to capture that established for textbooks in other areas of applied mathematics, with an appropriate, but not overwhelming, level of mathematical rigour. We have derived results, but avoided theorems almost everywhere!

Many books on applied mathematics seem to concentrate almost exclusively on the analysis of 'given' systems or configurations. In producing this text, we have attempted to demonstrate the use of mathematics as a design or synthesis tool. Before such a task may be undertaken, it is necessary for the user or designer to achieve a considerable degree of experience of the field by the careful analysis of the relevant types of system or structure, and we have attempted to provide a suitable vehicle for this experience to be gained. Nevertheless, it has been at the forefront of our thinking as we have approached our task, that the aim of many readers will eventually be the production of network or system designs of their own. We cannot, in a book such as this, hope to give a sufficiently full treatment of any one topic area to satisfy this aim entirely. However, by focusing on the design task, we hope to demonstrate to the reader that an understanding of the underlying mathematics is an essential pre-requisite for such work.

Most of the material has been taught as a single course to students of Mathematics at Coventry University, where student reaction has been favourable. The material, to a suitable engineering interface, has been given to students of Engineering, again with encouraging results. Engineering students have generally acknowledged that the course provided an essential complement to their Engineering studies.

Many of the concepts considered within the book can be demonstrated on a PC using the MATLAB package, together with the various toolboxes. In Appendix B, we give some 'm' files, with application to the processing of speech signals.

OUTLINE OF THE BOOK.

Chapter 1 introduces the concepts of signals and linear systems. Mathematical models of some simple circuits are constructed, and the Laplace transform is introduced as a method of describing the input/output relationship. Simulation diagrams are also discussed, and a brief introduction to generalized functions is presented.

Chapters 2 and 3 provide much of the technique and analytical experience necessary for our later work. Chapter 2 is concerned with system responses. By examining the type of response which can be obtained from a linear, time invariant system, concepts of stability are developed. An introduction to signal decomposition and the convolution operation precedes a discussion of the frequency response. Chapter 3 is devoted to the harmonic decomposition of signals by use of the Fourier transform, leading to the idea of the amplitude and phase spectra of a signal. The effect of sampling a continuous-time signal on these spectra is first discussed here.

Chapter 4 deals with the design of analogue filters. Based on the analytical experience gained in the previous three chapters, the task of designing a low-pass filter is addressed first. Butterworth filters emerge as one solution to the design task, and the question of their realization using elementary circuits is considered. Transformations which produce band-pass, band-reject and high-pass filters are investigated, and a brief introduction is given to Chebyshev designs.

Chapters 5–7 provide a discussion of discrete-time signals and systems. Difference equations, the z-transform and the extension of Fourier techniques to discrete time are all discussed in some detail. The need for a fast, computationally efficient algorithm for Fourier analysis in discrete time rapidly emerges, and a Fast Fourier Transform algorithm is developed here. This chapter presents several opportunities for those readers with access to a personal computer to conduct their own investigations into the subject area.

Chapter 8 returns to the theme of design. Building on the material in the earlier chapters, it is now possible to see how digital filters can be designed either to emulate the analogue designs of Chapter 4, or from an *ab initio* basis. Infinite-impulse and finite-impulse response designs are developed, together with their realizations as difference equations. Here again, the reader with access to a personal computer, together with minimal coding skill, will find their study considerably enhanced.

Finally, in Chapter 9, we apply some of the work in the previous chapters to the processing of speech signals: speech processing, as well as image processing, being an important part of Communication Theory. There are many aspects of speech processing such as, for example, speech production, modelling of speech, speech analysis, speech recognition, speech enhancement, synthesis of speech, etc. All these aspects have numerous applications, including speech coding for communications, speech recognition systems in the robotic industry or world of finance (to name but a few), and speech synthesis, the technology for which has been incorporated in a number of modern educational toys for example. One important analysis technique is that of Linear Predictive Coding. Using this technique, a speech signal can be processed to extract salient features of the signal which can be used in various applications. For example, in the synthesis problem, these features can be used to synthesize electronically an approximation to the original speech sound. In this chapter, some specific topics considered are: a speech production model, linear predictive filters, lattice filters, and cepstral analysis, with application to recognition of non-nasal voiced speech and formant estimation.

PREREQUISITES

Chapter 1 assumes a knowledge of elementary calculus, and a familiarity with the Laplace transform would be helpful, but is not essential. Also in Chapter 1, we delve into a little analysis in connection with the discussion on generalized functions. The interested reader who wishes to develop understanding further should consult the excellent text by Hoskins, cited in the references, for a full and clearly presented account.

EXERCISES

These are designed principally to test the reader's understanding of the material but there are, in addition, some more testing exercises. In addition to stressing applications of the material, answers to Exercises are given in an Appendix.

ACKNOWLEDGEMENTS

We have benefitted from discussions with a number of our colleagues at Coventry, in particular with Peter Lockett, Robert Low, and Ann Round. The book has been typeset by the authors using LaTeX, the fact that this is the case is a tribute to the skills of its inventors. In undertaking a project of this nature, widespread encouragement is needed and, as in the earlier version of the work, there has been considerable feline support, albeit from a changed workforce! David Goodall acknowledges support given by his wife, Alison, his three daughters, Tanja, Emma and Fiona, and, in addition, helpful discussions with Mike Herring of Cheltenham and Gloucester College of Higher Education (U.K.) during initial exploratory excursions into Speech Processing. Others have helped too, and we would wish to express our thanks to Renate Fechter of Siemens (Austria), Les Balmer, Sandra Prince and Julie Clayton, all at Coventry.

M J Chapman, D P Goodall and N C Steele

Coventry University

July 1997.

	Notation
$\in$	'is a member of'
$\subset$	'is contained in'
$\mathbb{Z}$	the integers: $0,\ \pm 1,\ \pm 2, \ldots$
$\mathbb{N}$	the natural numbers: 1, 2, 3, ...
$\mathbb{R}$	the real numbers
$\mathbb{C}$	the complex numbers
$\mathcal{L}$	the Laplace transform operator
$\mathcal{Z}$	the z-transform operator
$\mathcal{F}$	the Fourier transform operator
$\Re$	'the real part of'
$\Im$	'the imaginary part of'
$\sim$	'is represented by'
$\leftrightarrow$	transform pairs
j	$\sqrt{-1}$
$\delta(t)$	the unit impulse (Dirac δ-) function
$\zeta(t)$	the Heaviside unit step function
$\{\delta_k\},\ \{\delta(k)\}$	the unit impulse sequence
$\{\zeta_k\},\ \{\zeta(k)\}$	the unit step sequence
δ_{ij}	the Kronecker delta, $(\delta_{ij} = \delta_{i-j} = \delta(i-j))$
$\mathbf{A}^{\mathrm{T}}$	the transpose of a matrix $\mathbf{A}$
$F(s)$	the Laplace transform of a signal, $f(t)$
$U(z)$	the z-transform of a sequence, $\{u_k\}$
$F(j\omega)$	the Fourier transform of a signal, $f(t)$
P	the period of a periodic signal
ω_0	the fundamental frequency in rads./sec. $\omega_0 = 2\pi/P$
T	sample period
$*$	convolution

Table of contents

Table of contents

Table of contents

List of Tables

1

Signals and linear system fundamentals

1.1 INTRODUCTION

In this first chapter we concern ourselves with **signals** and operations on signals. The first task is to define what is meant by a signal and then to classify signals into different types according to their nature. When this has been achieved, the concept of a **system** is introduced by the consideration of some elementary ideas from the theory of electrical circuits. Useful ideas and insight can be obtained from **simulation diagrams** representing circuits (or systems) and these are discussed for the time domain. **Laplace transforms** are reviewed, with their rôle seen as that of system representation rather than as a method for the solution of differential equations.

1.2 SIGNALS AND SYSTEMS

A signal is a time-varying quantity, used to cause some effect or produce some action. Mathematically, we describe a signal as a function of time used to represent a variable of interest associated with a system, and we classify signals according to both the way in which they vary with time and the manner of that variation. The classification which we use discriminates between **continuous-time** and **discrete-time signals** and, thereafter, between **deterministic** and **stochastic signals**, although we concern ourselves only with deterministic signals. Deterministic signals can be modelled or represented using completely specified functions of time, for example:

1. $f_1(t) = a\sin(\omega t)$, with a and ω constant and $-\infty < t < \infty$,

2. $f_2(t) = ce^{-dt}$, with c and d constant and $t \geq 0$,

3. $f_3(t) = \begin{cases} 1, & |t| \leq A, \\ 0, & |t| > A, \end{cases}$ with A constant and $-\infty < t < \infty$.

Each signal $f_1(t)$, $f_2(t)$ and $f_3(t)$ above is a function of the continuous-time variable t, and thus are continuous-time signals. Notice that although $f_3(t)$ is a continuous-time signal, it is not a continuous function of time.

In some applications, notably in connection with digital computers, signals are represented at discrete (or separated) values of the time variable (or index). Between these discrete-time instants the signal may take the value zero, be undefined or be simply of no interest. Examples of discrete-time signals are

1. $f_4(nT) = a\sin(nT)$, with a and T constant and $n \in \mathbb{Z}$, i.e. n is an integer.

2. $f_5(k) = bk + c$, with b and c constants and $k = 0, 1, 2, \ldots$, i.e. k is a non-negative integer.

The signals $f_4(nT)$ and $f_5(k)$ are functions of nT and k, respectively, where n and k may take only specified integer values. Thus, values of the signal are only defined at discrete points. The two notations have been used for a purpose: the origin of many discrete-time signals is in the **sampling** of a continuous-time signal $f(t)$ at (usually) equal intervals, T. If n represents a sampling index or counter, taking values from a set $I \subset \mathbb{Z}$, then this process generates the sequence of values $\{f(nT); \ n \in I\}$, with each term $f(nT)$ generated by a formula as in, for example, $f_4(nT)$ above. Using the notation as in $f_5(k) = bk + c$ merely suppresses the information on the sampling interval and uses instead the index of position, k, in the sequence $\{f(k); \ k \in I\}$ as the independent variable. Figures 1.1a–e exhibits the graphs of some of these signals. Stochastic signals, either continuous-time or discrete-time, cannot be

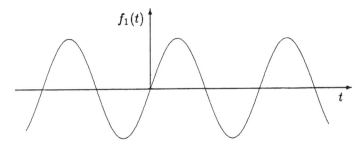

Figure 1.1a: Graph of $f_1(t)$.

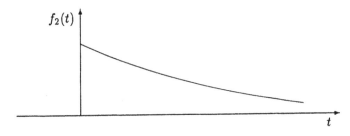

Figure 1.1b: Graph of $f_2(t)$.

so represented and their description has to be in terms of statistical properties. The analysis and processing of such signals is an important part of communication theory, but depends on an understanding of deterministic signal processing. This book concentrates on deterministic signals, and thus may serve as an introduction to the more advanced texts in this area.

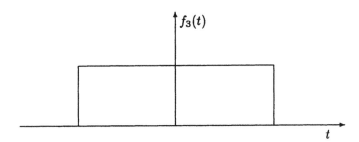

Figure 1.1c: Graph of $f_3(t)$.

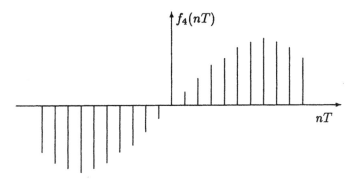

Figure 1.1d: Graph of $f_4(nT)$.

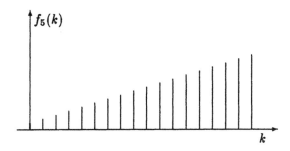

Figure 1.1e: Graph of $f_5(k)$.

A dictionary definition of a system is a set of things considered as a collective whole. Restricting the scope somewhat, we consider systems to be processors which operate on **input signals** to produce **output signals**. Since we are working with representations or mathematical models of signals, our systems are mathematical models of real systems. Input and output signals may be of any of the types discussed previously and we think of our mathematical models as *models of systems which have the purpose of processing signals*. The continuous-time systems we consider have their origin in simple electrical circuits which can be modelled quite simply using differential equations.

1.3   L–C–R CIRCUITS

We consider circuits which are configurations of three basic elements and which are driven by a voltage source. We discover that, from these basic building blocks, systems can be constructed which are capable of performing useful tasks in the field of signal processing. First, we must define the circuit elements and the manner of their interaction. The three elements are:

1. a **resistor**;

2. an **inductor**;

3. a **capacitor**.

The effect of each of these devices in a loop of a circuit in which a current $i(t)$ flows is expressed in terms of the voltage drop measured across each device. This is illustrated for each case in Figures 1.2–1.4.

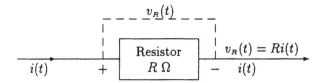

Figure 1.2: Resistor, with constant resistance R ohms.

Figure 1.3: Inductor, with constant inductance L Henrys.

$$v_R(t) = Ri(t) \ . \tag{1.1}$$

$$v_L(t) = L\frac{\mathrm{d}i}{\mathrm{d}t}(t) \ . \tag{1.2}$$

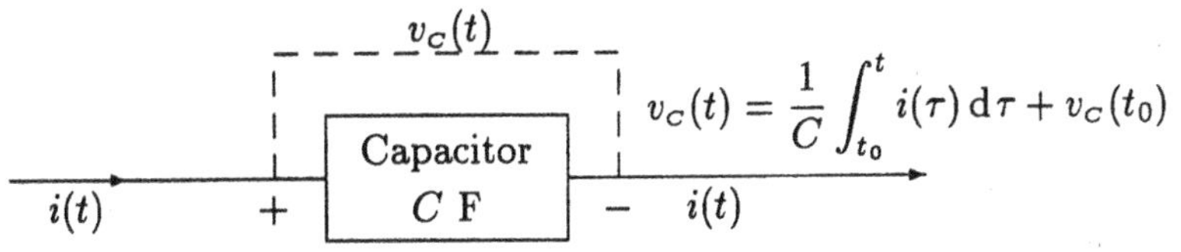

Figure 1.4: Capacitor, with constant capacitance C Farads.

$$v_C(t) = \frac{1}{C} \int_{t_0}^{t} i(\tau) \, d\tau + v_C(t_0) \; . \tag{1.3}$$

An alternative form for (1.3) is obtained by differentiation with respect to time as

$$i(t) = C \frac{d v_C}{dt}(t) \; . \tag{1.4}$$

The relations (1.1)–(1.3) are the component constitutive equations and we note that the loop current measured on each side of each device is the same. Circuits containing the above components are driven by a voltage source, delivering a voltage $e(t)$ independent of the current $i(t)$. Such a device is shown at the left-hand side of Figure 1.5 in Example 1.1.

Equations describing circuit behaviour are obtained by using **Kirchhoff's laws**. These are simply stated as:

(A) **Kirchhoff's voltage law**: *the sum of the voltage drops around a closed loop of a circuit is zero.*

(B) **Kirchhoff's current law**: *the algebraic sum of the currents at a junction is zero.*

The use of these laws and the formation of circuit equations for simple cases are demonstated through examples. The interested reader seeking a more advanced discussion of circuits is referred to Papoulis [24] and Adby [1].

Example 1.1 The L–R circuit
Figure 1.5 shows a circuit containing a resistor with constant resistance R and an inductor with constant inductance L, in future referred to as a resistor R and an inductor L. Find an equation governing circuit behaviour.

There are no junctions in this circuit and we have only Kirchhoff's voltage law available to determine the behaviour of this circuit when driven by the voltage source $e(t)$ and with an assumed current $i(t)$ flowing.

Taking as our starting point P as shown in Figure 1.5, we must have, progressing in a clockwise sense,

$$v_R(t) + v_L(t) - e(t) = 0 \; . \tag{1.5}$$

The last term has the form shown, $-e(t)$, since, with the given conditions and passing through the source $e(t)$ in the sense described, there is a presumed voltage rise. In fact, whether there is actually a voltage rise or a voltage drop at this stage, (1.5) still represents the given conditions correctly because the system is **linear**. It

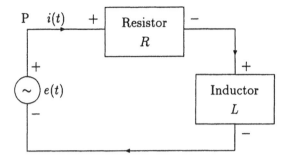

Figure 1.5: The L–R circuit of Example 1.1.

now remains to rearrange (1.5) and use the appropriate constitutive relationships (1.1) and (1.2) to obtain

$$Ri(t) + L\frac{\mathrm{d}i}{\mathrm{d}t}(t) = e(t)$$

or

$$\frac{\mathrm{d}i}{\mathrm{d}t}(t) + \frac{R}{L}i(t) = \frac{1}{L}e(t) \ . \tag{1.6}$$

The solution of this differential equation will produce $i(t)$, the current in the circuit, given the input $e(t)$ and an initial condition, $i(t_0)$ say. It may not be that $i(t)$ is the desired output for the system: other possibilities might be $v_R(t)$ or $v_L(t)$, the voltages measured across R and L respectively. However, a knowledge of $i(t)$ is easily seen to be sufficient to obtain either of these quantities.

□

Example 1.2 The C–R circuit
Repeat the procedure of Example 1.1 for the circuit of Figure 1.6.

We obtain from Figure 1.6,

$$v_R(t) + v_C(t) - e(t) = 0 \tag{1.7}$$

and, using (1.1) and (1.3), we have

$$Ri(t) + \frac{1}{C}\int_{t_0}^{t} i(\tau)\,\mathrm{d}\tau + v_C(t_0) = e(t) \ . \tag{1.8}$$

To remove the integral in (1.8) we could differentiate with respect to time to produce

$$R\frac{\mathrm{d}i}{\mathrm{d}t}(t) + \frac{1}{C}i(t) = \frac{\mathrm{d}e}{\mathrm{d}t}(t) \ . \tag{1.9}$$

Equation (1.9) is an acceptable form for the representation of the circuit, but the presence of the time derivative of the input, $\frac{\mathrm{d}e}{\mathrm{d}t}(t)$, could produce difficulties.

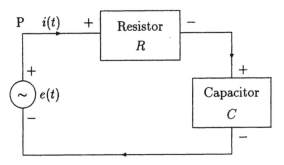

Figure 1.6: The C–R circuit.

An alternative approach is to return to Equation (1.7) and make use of the alternative form of the constitutive equation for the capacitor, (1.4), i.e.

$$i(t) = C\frac{dv_C}{dt}(t) \ .$$

Since $v_R(t) = Ri(t)$, this means that

$$v_R(t) = RC\frac{dv_C}{dt}(t)$$

and (1.7) becomes

$$RC\frac{dv_C}{dt}(t) + v_C(t) = e(t)$$

or

$$\frac{dv_C}{dt}(t) + \frac{1}{RC}v_C(t) = \frac{1}{RC}e(t) \ . \tag{1.10}$$

Thus, by choice of quantity $v_C(t)$ as the variable to describe the behaviour of this circuit, we have avoided the occurrence of the term $\dfrac{de}{dt}(t)$. The solution of (1.10) will produce $v_C(t)$, the voltage drop measured across the capacitor, given the input $e(t)$ and the initial condition $v_C(t_0)$, say.

□

It is interesting to observe the similar forms of (1.6) and (1.10) and to note that the form in (1.10) was obtained after the choice of $v_C(t)$ as dependent variable. There is a physical reason for this; and an account, using the concept of **state variables**, (which is discussed below), may be found in Gabel and Roberts [6]. Our next step is to ask what happens in a circuit which contains all three elements, namely resistor, inductor and capacitor, connected in series. Such a circuit is illustrated in Figure 1.7 and, in this case, Kirchhoff's voltage law yields

$$v_R(t) + v_L(t) + v_C(t) = e(t) \ . \tag{1.11}$$

The use of the constitutive laws (1.1)–(1.3) would yield a circuit equation in the manner of Examples 1.1 and 1.2. Initially, we do not perform this technique: instead, we observe that if we use the constitutive relationships (1.1) and (1.2), then

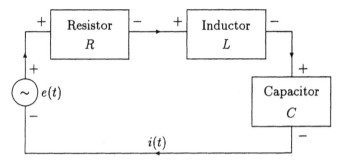

Figure 1.7: The L–C–R circuit.

we obtain

$$Ri(t) + L\frac{\mathrm{d}i}{\mathrm{d}t}(t) + v_C(t) = e(t) \ . \tag{1.12}$$

Writing the constitutive equation (1.4) for the capacitor C as

$$\frac{\mathrm{d}v_C}{\mathrm{d}t}(t) - \frac{1}{C}i(t) = 0 \ , \tag{1.13}$$

we see that (1.12) and (1.13) form a pair of **coupled differential equations** which describe the behaviour of the circuit in Figure 1.7. This pair of equations determines the behaviour of the two quantities $i(t)$ and $v_C(t)$ which emerged as 'natural' when discussing circuits with either an inductor or a capacitor, respectively. These two quantities are called state variables for the L–C–R circuit, and when we know the value of both of these quantities for all time t, we say we know the **state** of the circuit at any time, implying a complete knowledge of its behaviour.

It is natural, since the values of L, C and R are constant, to consider a matrix representation of the system of equations (1.12) and (1.13). To do this, define a **state vector** $\boldsymbol{x}(t) = [v_C(t) \ i(t)]^{\mathrm{T}}$ and rearrange (1.12) to give

$$\frac{\mathrm{d}i}{\mathrm{d}t}(t) = -\frac{1}{L}v_C(t) - \frac{R}{L}i(t) + \frac{1}{L}e(t) \ .$$

This leads to the matrix equation

$$\begin{bmatrix} \dfrac{\mathrm{d}v_C}{\mathrm{d}t}(t) \\[2ex] \dfrac{\mathrm{d}i}{\mathrm{d}t}(t) \end{bmatrix} = \begin{bmatrix} 0 & \dfrac{1}{C} \\[2ex] -\dfrac{1}{L} & -\dfrac{R}{L} \end{bmatrix} \begin{bmatrix} v_C(t) \\[2ex] i(t) \end{bmatrix} + \begin{bmatrix} 0 \\[2ex] \dfrac{1}{L} \end{bmatrix} e(t)$$

or

$$\dot{\boldsymbol{x}}(t) = \boldsymbol{A}\boldsymbol{x}(t) + \boldsymbol{b}u(t) \ , \tag{1.14}$$

where

$$\boldsymbol{A} = \begin{bmatrix} 0 & \dfrac{1}{C} \\[2ex] -\dfrac{1}{L} & -\dfrac{R}{L} \end{bmatrix} \ , \quad \boldsymbol{b} = \begin{bmatrix} 0 \\[2ex] \dfrac{1}{L} \end{bmatrix} \ , \quad u(t) = e(t) \ .$$

If the output is to be $y(t) = i(t)$, then we can write $y(t) = [0 \ 1]\boldsymbol{x}(t)$.

The form (1.14) is known as the **state-space representation** of the circuit and generalizes to

$$\dot{\boldsymbol{x}}(t) = \boldsymbol{A}\boldsymbol{x}(t) + \boldsymbol{B}\boldsymbol{u}(t)$$
$$\boldsymbol{y}(t) = \boldsymbol{C}\boldsymbol{x}(t) + \boldsymbol{D}\boldsymbol{u}(t) \ ,$$

where $\boldsymbol{B}$ is now a matrix and $\boldsymbol{u}(t) = [u_1(t) \ldots u_m(t)]^{\mathrm{T}}$ is a vector of inputs. Here, $\boldsymbol{y}(t) = [y_1(t) \ldots y_p(t)]^{\mathrm{T}}$ is a vector of outputs and $\boldsymbol{C}$ and $\boldsymbol{D}$ are matrices of appropriate dimension. This generalization provides a useful method for modelling systems with more than one input and is exploited particularly in the field of control theory. For more details, see Burghes and Graham [3].

1.4   LINEAR SYSTEMS

So far in our development we have identified the concept of a system and looked in some detail at systems which take the form of simple electric circuits. **Linear** was only mentioned in passing during the derivation of a particular circuit equation and, as yet, the concept has not been explained. To rectify this, we consider first an ordinary differential equation of the form

$$\frac{\mathrm{d}^n y}{\mathrm{d}t^n}(t) + a_{n-1}(t)\frac{\mathrm{d}^{n-1} y}{\mathrm{d}t^{n-1}}(t) + \ldots + a_0(t)y(t) =$$
$$b_m(t)\frac{\mathrm{d}^m u}{\mathrm{d}t^m}(t) + b_{m-1}(t)\frac{\mathrm{d}^{m-1} u}{\mathrm{d}t^{m-1}}(t) + \ldots + b_0(t)u(t) \quad (1.15)$$

which relates the input $u(t)$ to the output $y(t)$ and where the coefficients $a_i(t)$, $i = 0, \ldots n-1$ and $b_j(t)$, $j = 0, \ldots m$ are functions of time t. We write (1.15) in the compact form

$$L_1[y(t)] = L_2[u(t)] \ , \tag{1.16}$$

where L_1 and L_2 are operators defined by

$$L_1 \equiv \left[\frac{\mathrm{d}^n}{\mathrm{d}t^n} + a_{n-1}(t)\frac{\mathrm{d}^{n-1}}{\mathrm{d}t^{n-1}} + \ldots + a_0(t) \right] \ ,$$

$$L_2 \equiv \left[b_m(t)\frac{\mathrm{d}^m}{\mathrm{d}t^m} + b_{m-1}(t)\frac{\mathrm{d}^{m-1}}{\mathrm{d}t^{m-1}} + \ldots + b_0(t) \right] \ .$$

An operator L is said to be **linear** if

$$L[\alpha u(t) + \beta v(t)] = \alpha L[u(t)] + \beta L[v(t)] \ ,$$

where $u(t)$ and $v(t)$ are two functions with derivatives to a sufficient order and α and β are any constants. It is easy to see that L_1 and L_2, as defined above, are linear operators under this definition.

The differential equation (1.16), which contains only linear operators, is therefore known as a **linear ordinary differential equation** and, by association, the system which models it is called a **linear system**. We note that the differential equations (1.6) and (1.10), which represent the L–R circuit and the C–R circuit respectively, are of the form (1.15) and, hence, that of (1.16). In fact, each equation has the further property that all the coefficients are constant, that is non-time-varying. Such linear systems are called **linear time-invariant systems**, indicating physically that the circuit parameters are fixed for all time.

Example 1.3
Show that the system represented by the differential equation

$$\frac{d^2 y}{dt^2}(t) + 3t\frac{dy}{dt}(t) + 4e^{-t}y(t) = \frac{du}{dt}(t) , \qquad (1.17)$$

where $u(t)$ is the system input and $y(t)$ is the system output, is linear.

We write the differential equation as

$$L_1[y(t)] = L_2[u(t)] ,$$

where

$$L_1 \equiv \left[\frac{d^2}{dt^2} + 3t\frac{d}{dt} + 4e^{-t}\right] \quad \text{and} \quad L_2 \equiv \left[\frac{d}{dt}\right] .$$

Now

$$L_1[\alpha y_1(t) + \beta y_2(t)] = \left[\frac{d^2}{dt^2} + 3t\frac{d}{dt} + 4e^{-t}\right][\alpha y_1(t) + \beta y_2(t)]$$

$$= \frac{d^2}{dt^2}(\alpha y_1(t) + \beta y_2(t)) + 3t\frac{d}{dt}(\alpha y_1(t) + \beta y_2(t))$$

$$+ 4e^{-t}(\alpha y_1(t) + \beta y_2(t))$$

$$= \alpha\left(\frac{d^2 y_1}{dt^2}(t) + 3t\frac{dy_1}{dt}(t) + 4e^{-t}y_1(t)\right)$$

$$+ \beta\left(\frac{d^2 y_2}{dt^2}(t) + 3t\frac{dy_2}{dt}(t) + 4e^{-t}y_2(t)\right)$$

$$= \alpha L_1[y_1(t)] + \beta L_1[y_2(t)] .$$

Therefore, L_1 is a linear operator. Similarly,

$$L_2[\alpha u_1(t) + \beta u_2(t)] = \left[\frac{d}{dt}\right][\alpha u_1(t) + \beta u_2(t)]$$

$$= \alpha\frac{du_1}{dt}(t) + \beta\frac{du_2}{dt}(t)$$

$$= \alpha L_2[u_1(t)] + \beta L_2[u_2(t)]$$

and, thus, L_2 is a linear operator. Hence, (1.17) is a linear differential equation. Although the system represented by (1.17) is linear, *it is not time-invariant*.

□

We can now infer a most useful result for linear systems, known as **the principle of superposition**, which finds considerable application. Suppose we have a linear system modelled by

$$L_1[y(t)] = L_2[u(t)] ,$$

where L_1 and L_2 are linear operators and $y(t)$ is the output or response corresponding to the input $u(t)$, with zero initial conditions. Suppose that a second input $v(t)$
produces the response $x(t)$ when also subject to zero initial conditions. That is

$$L_1[x(t)] = L_2[v(t)] .$$

The response or output, $z(t)$, corresponding to the **linear combination** of these
two inputs

$$w(t) = \alpha u(t) + \beta v(t), \quad \text{where } \alpha \text{ and } \beta \text{ are constant,}$$

with zero initial conditions, is the solution of

$$L_1[z(t)] = L_2[w(t)] . \tag{1.18}$$

However,

$$\begin{aligned}
L_2[w(t)] &= L_2[\alpha u(t) + \beta v(t)] \\
&= \alpha L_2[u(t)] + \beta L_2[v(t)] \\
&= \alpha L_1[y(t)] + \beta L_1[x(t)] \\
&= L_1[\alpha y(t) + \beta x(t)] .
\end{aligned}$$

Therefore, $L_1[z(t)] = L_1[\alpha y(t) + \beta x(t)]$; in other words, $z(t) = \alpha y(t) + \beta x(t)$ is a
solution of (1.18) and the fact that $z(t)$ satisfies the requirement of zero initial conditions ensures uniqueness of the solution. These ideas are clarified in the following
examples.

Example 1.4
Suppose a linear time-invariant system, with input u and corresponding output y,
is modelled as

$$\frac{dy}{dt}(t) + 2y(t) = u(t), \quad t \geq 0 .$$

Demonstrate the principle of superposition using the two inputs $u_1(t) = 1$ and
$u_2(t) = e^{-t}$.

   Let us find the outputs or responses when $y(0) = 0$. These are so-called **zero-
state responses** of the system. When $u(t) = u_1(t) = 1$, we determine the response
$y_1(t)$ that satisfies

$$\frac{dy_1}{dt}(t) + 2y_1(t) = 1.$$

By inspection (or otherwise), we see that

$$y_1(t) = Ae^{-2t} + \frac{1}{2} ,$$

where A is an arbitrary constant. However, $y_1(0) = 0$ and so $A = -\frac{1}{2}$. Thus,
$y_1(t) = \frac{1}{2}(1 - e^{-2t})$. On the other hand, with $u(t) = u_2(t) = e^{-t}$, we find that

$$\frac{dy_2}{dt}(t) + 2y_2(t) = e^{-t} ,$$

which implies that

$$y_2(t) = Be^{-2t} + e^{-t} ,$$

with B determined from $y_2(0) = 0$ as $B = -1$, so that

$$y_2(t) = e^{-t} - e^{-2t} .$$

With the composite input $v(t) = \alpha u_1(t) + \beta u_2(t) = \alpha + \beta e^{-t}$, where α and β are constant, we must solve for the output $z(t)$ which satisfies

$$\frac{\mathrm{d}z}{\mathrm{d}t}(t) + 2z(t) = \alpha + \beta e^{-t} .$$

We find that $z(t) = Ce^{-2t} + \frac{1}{2}\alpha + \beta e^{-t}$ and $z(0) = 0$ determines the arbitrary constant C as $C = -\frac{1}{2}\alpha - \beta$ and, thus,

$$
\begin{aligned}
z(t) &= \tfrac{1}{2}\alpha(1 - e^{-2t}) + \beta(e^{-t} - e^{-2t}) \\
&= \alpha[\tfrac{1}{2}(1 - e^{-2t})] + \beta[e^{-t} - e^{-2t}] \\
&= \alpha y_1(t) + \beta y_2(t) .
\end{aligned}
$$

$\square$

As we have noted, the response of a linear system to a particular input when all initial conditions are zero, is called the zero-state response. We have shown that, if the zero-state responses to two different inputs are known, then *the zero-state response to a linear combination of these inputs is the same linear combination of the individual responses.* Clearly this can be generalized to any number of different inputs and their zero-state responses.

1.5 SIMULATION DIAGRAMS

Before dealing with this topic, we consider an alternative form of the circuit diagram which highlights the concept of the circuit processing an input signal to produce an output signal. Consider, again, the C–R circuit in Example 1.2. The input to this circuit is $e(t)$ and, if the desired output, $y(t)$, is the voltage $v_C(t)$ measured across the capacitor, we can represent this situation in the re-drawn form illustrated in Figure 1.8. It is assumed that the readings obtained for $y(t)$ are exactly the values of $v_C(t)$, implying that an ideal measuring device is used, taking no current. In fact, this is not physically possible. However, the current through the measuring device is assumed to be so small that the presence of the device may be ignored in the analysis of the circuit.

Although the representation of the circuit in Figure 1.8 is helpful, simulation diagrams can give further insight. Simulation diagrams focus on the mathematical model of the circuit or system and are formed from four basic building blocks. For continuous-time systems, these blocks represent the operations of:

1. **multiplication by a constant**,

2. **time differentiation**,

3. **time integration**, and

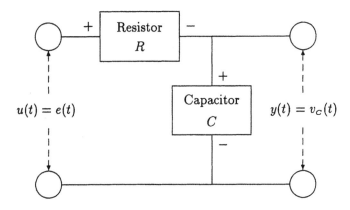

Figure 1.8: The C–R circuit.

4. **summation**, including the possibility of **sign reversal**.

Diagrammatically, these are represented in Figure 1.9. In fact, time differentiation rarely appears and, frequently, initial conditions are zero with the result that the vertical arrow in time integration, representing the input of the initial condition, is absent. Simulation diagrams are sometimes referred to as **block diagrams**. This terminology is commonly used for diagrams that may contain a wider scope for the blocks than the four basic building blocks described above.

In Example 1.2, we saw that the C–R circuit of Figure 1.8 could be modelled by the differential equation

$$\frac{dy}{dt}(t) + \frac{1}{RC}y(t) = \frac{1}{RC}u(t) ,$$

where $y(t) = v_C(t)$, the voltage drop measured across C, is the output and $u(t) = e(t)$ is the input. We solve this equation for $\frac{dy}{dt}(t)$, the highest occurring derivative, to obtain

$$\frac{dy}{dt}(t) = \frac{1}{RC}(u(t) - y(t)) . \qquad (1.19)$$

Suppose, for a moment, that we have somehow formed a signal to represent $\frac{dy}{dt}(t)$, then one integration together with a specified initial condition would generate $y(t)$. If $y(t_0) = 0$, this can be illustrated as in Figure 1.10. We now see (1.19) as a 'prescription' for generating the signal $\frac{dy}{dt}(t)$ as a scaled sum of the input $u(t)$ and $-y(t)$, and this leads to the system simulation diagram of Figure 1.11. Figure 1.11 gives us more insight into how the system processes the input signal. The input $u(t)$ is combined with a negative feedback of the system output: the composite signal is then scaled (amplified if $RC < 1$, or attenuated if $RC > 1$) and integrated to form the output. The simulation diagram, as built up for this example, illustrates the processing effect of the circuit rather more clearly by showing the path (and the construction) of signals within the system. This visual representation in blocks is utilized in later sections.

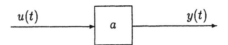

(1) Multiplication, $\;y(t) = au(t)$.

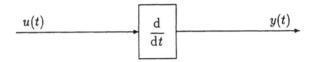

(2) Time differentiation, $\;y(t) = \dfrac{\mathrm{d}u}{\mathrm{d}t}(t)$.

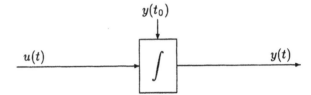

(3) Time integration, $\;y(t) = \displaystyle\int_{t_0}^{t} u(\tau)\,\mathrm{d}\tau + y(t_0)$.

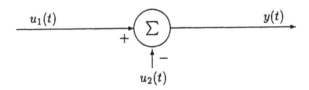

(4) Summation, $\;y(t) = u_1(t) - u_2(t)$.

Figure 1.9: Simulation diagram operations.

$\dfrac{\mathrm{d}y}{\mathrm{d}t}(t)$ $\displaystyle\int$ $y(t)$

Figure 1.10: Integration of $\dfrac{\mathrm{d}y}{\mathrm{d}t}(t)$ to produce $y(t)$ with $y(t_0) = 0$.

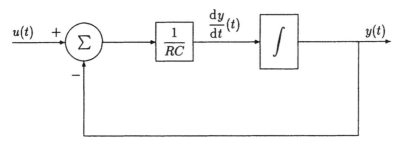

Figure 1.11: Simulation diagram for the C–R circuit.

1.6   THE LAPLACE TRANSFORM

Readers are assumed to be familiar with an elementary treatment of the Laplace transform. It is likely that such a treatment will have focused on the transform as a solution technique for linear ordinary differential equations with constant coefficients. We show that there is a wider rôle to be played by the transform and that we can generate a useful tool for the analysis of linear time-invariant systems. Initially, we take a slightly wider view of the topic and, first, define the **two-sided** or **bilateral Laplace transform** of a continuous function of time, $f(t)$, defined on $\mathbb{R}$ (i.e. t satisfies $-\infty < t < \infty$). The bilateral Laplace transform of $f(t)$ is defined as

$$F(s) \stackrel{\text{def}}{=} \int_{-\infty}^{\infty} f(t)e^{-st}\,dt \tag{1.20}$$
$$= \mathcal{L}_\mathrm{B}\{f(t)\}\,,$$

where s is a complex variable. The transform is said to exist whenever (1.20) exists, that is when the integral takes a finite value. Most of the time, the functions we are concerned with represent time-varying signals which are **causal**, that is $f(t) = 0$ for $t < 0$. Under these circumstances, (1.20) reduces to

$$F(s) = \int_{0}^{\infty} f(t)e^{-st}\,dt\,, \tag{1.21}$$

and $F(s)$ is called the **one-sided** or **unilateral Laplace transform** of the causal function $f(t)$. It is possible to take the unilateral transform of non-causal signals, but it is important to note that all information concerning the behaviour for $t < 0$ is then lost under transformation.

Example 1.5
Find the unilateral Laplace transform of $f(t)$ when $f(t) = e^{-at}\zeta(t)$, where

$$\zeta(t) = \begin{cases} 0, & t < 0 \\ 1, & t \geq 0 \end{cases}$$

is the **Heaviside unit step function** and $a > 0$.

Here, we observe that $f(t)$ is causal because

$$f(t) = \begin{cases} 0, & t < 0 \\ e^{-at}, & t \geq 0 \end{cases}.$$

Using (1.21),

$$F(s) = \int_0^\infty e^{-at} e^{-st}\, dt = \int_0^\infty e^{-(s+a)t}\, dt$$

$$= \frac{1}{s+a}, \quad \text{if } \Re(s) > -a, \tag{1.22}$$

where $\Re(\cdot)$ denotes 'real part'.

$\square$

We note that $f(t)$ is causal in view of the presence of $\zeta(t)$, and it is easy to see, from (1.20), that $g(t) = e^{-at}$, $-\infty < t < \infty$, does not have a bilateral transform. Table 1.1 lists frequently occurring causal functions and their Laplace transforms.

$f(t)$	$F(s)$
$a\zeta(t)$	$\dfrac{a}{s}$
$t\zeta(t)$	$\dfrac{1}{s^2}$
$t^n\zeta(t)$	$\dfrac{n!}{s^{n+1}}$
$\sin(at)\zeta(t)$	$\dfrac{a}{s^2+a^2}$
$\cos(at)\zeta(t)$	$\dfrac{s}{s^2+a^2}$
$e^{-at}\zeta(t)$	$\dfrac{1}{s+a}$
$e^{-at}\cos(bt)\zeta(t)$	$\dfrac{s+a}{(s+a)^2+b^2}$
$e^{-at}\sin(bt)\zeta(t)$	$\dfrac{b}{(s+a)^2+b^2}$

Table 1.1: Laplace transforms of elementary causal functions.

It is common practice to drop $\zeta(t)$ from the left hand side terms in Table 1.1 and, therefore, to consider it as a table of unilateral Laplace transforms. Henceforth, unless specified otherwise, all Laplace transforms will be one-sided and we shall accordingly drop the word 'unilateral'.

1.7 INTRODUCTION TO GENERALIZED FUNCTIONS

As seen in §1.6, all causal functions can be defined in terms of the unit step function, $\zeta(t)$. Suppose we require to differentiate a causal function; then, consequently, this would entail differentiating $\zeta(t)$. However, since $\zeta(t)$ has a discontinuity at $t = 0$, the derivative of $\zeta(t)$ does not exist in the 'ordinary' sense. To enable us to find a 'function' which is the derivative of the unit step function, we need to generalize our idea of a function. This can be achieved by utilizing **distribution theory** in which the concept of **a generalized function** can be made rigorous. In this text, we do not discuss details of distribution theory, although rigorous arguments are used in the applications of generalized functions. For those readers who may wish to consult texts on distribution theory, the book by Kanwal [13], for example, may be useful.

1.7.1   The Unit Impulse Function

We define the **(generalized) unit impulse function**, $\delta(t)$, also known as the **Dirac δ-function**, in terms of the following property (sometimes referred to as the **sifting property**).

Sifting property Assuming that $f(t)$ is continuous at $t = a$,

$$\int_c^d f(t)\delta(t-a)\,\mathrm{d}t = \begin{cases} f(a)\,, & \text{if } c < a < d \\ 0\,, & \text{if } a < c \text{ or } a > d \\ \text{undefined}\,, & \text{if } a = c \text{ or } a = d\,. \end{cases} \tag{1.23}$$

From the sifting property (1.23), it follows that

(1) $\delta(t) = 0 \quad t \neq 0$,   (1.24a)

(2) $\delta(t)$ is undefined at $t = 0$,   (1.24b)

(3) $\displaystyle\int_{-\infty}^{\infty} \delta(t)\,\mathrm{d}t = 1$.   (1.24c)

We see from (1.24a–b) that we cannot define $\delta(t)$ point by point as we do for 'ordinary' functions. However, we can give an interpretation which may help in the understanding of a generalized function. For a fuller treatment, the reader should consult Hoskins [9], who uses a similar development.

Consider the sequence of functions $\{p_n(t) : n \in \mathbb{N}\}$, where $\mathbb{N}$ denotes the set of natural numbers $\{1, 2, 3, \ldots\}$, defined by

$$p_n(t) = \begin{cases} 0\,, & |t| > \dfrac{1}{2n}\,, \\[2mm] n\,, & |t| < \dfrac{1}{2n}\,. \end{cases} \tag{1.25}$$

A typical member of this sequence is illustrated in Figure 1.12. From (1.25), taking the limit as $n \to \infty$, we see that

$$\lim_{n \to \infty} p_n(t) = \begin{cases} \text{undefined when } t = 0\,, \\ 0 \qquad \text{when } t \neq 0 \end{cases}$$

and the area under the graph of $p_n(t)$ is 1, for all n. Thus, the limit of the sequence of functions $p_n(t)$ as $n \to \infty$ has just those properties which were ascribed to $\delta(t)$

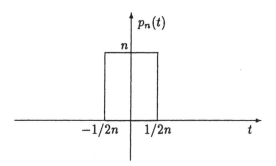

Figure 1.12: Graph of $p_n(t)$.

in (1.24a–c). Now, if $f(t)$ is continuous in some region including $t = 0$, we have, by the First Mean Value Theorem for integrals[1]

$$\lim_{n \to \infty} \int_{-\infty}^{\infty} f(t)p_n(t)\,\mathrm{d}t = \lim_{n \to \infty} \int_{-\frac{1}{2n}}^{\frac{1}{2n}} f(t)p_n(t)\,\mathrm{d}t$$

$$= \lim_{n \to \infty} f(\varepsilon_n)\ , \quad \text{for some } \varepsilon_n \text{ satisfying } -\frac{1}{2n} \le \varepsilon_n \le \frac{1}{2n}$$

$$= f(0)\ .$$

Assuming that we can interchange the processes of taking the limit and integration, formally we have

$$\int_{-\infty}^{\infty} f(t) \left[\lim_{n \to \infty} p_n(t) \right]\,\mathrm{d}t = f(0)\ ,$$

which can be compared to (1.23) with $a = 0$, $c = -\infty$ and $d = \infty$. From this, we may infer that $\delta(t)$ may be interpreted as the limit of a sequence of 'ordinary' functions $\{p_n(t)\}$. The process of generalization of the concept of a function, then, means that we admit such 'generalized functions', defined as limits of sequences of ordinary functions, to our function (or signal) space. The reader who is familiar with the proof of the existence of the real number $\sqrt{2}$ may see a similarity of concept here.

In describing $\delta(t)$ as the limit of a sequence of ordinary functions, it is apparent that there may be more than one suitable sequence of functions whose limit is $\delta(t)$. Thus, we need to give a definition of equivalence for such generalized functions. Such a definition is possible if we introduce the concept of a **testing function**. Testing functions are continuous, have continuous derivatives of all orders, and are zero outside a finite interval. An example of a class of testing functions is

$$\theta_a(t) = \begin{cases} e^{-a^2/(a^2 - t^2)}\ , & |t| < a \\ 0\ , & |t| \ge a\ , \end{cases}$$

where a is a positive, real parameter.

---

[1] If f is continuous for $a \le t \le b$ then for some value ε, $a \le \varepsilon \le b$, we have
$$\int_{a}^{b} f(t)\,\mathrm{d}t = f(\varepsilon)(b - a).$$

Suppose $g_1(t)$ and $g_2(t)$ are two generalized functions (or combinations of 'ordinary' functions and generalized functions), then $g_1(t)$ is **equivalent** to $g_2(t)$ if

$$\int_{-\infty}^{\infty} g_1(t)\theta(t)\,\mathrm{d}t = \int_{-\infty}^{\infty} g_2(t)\theta(t)\,\mathrm{d}t \qquad (1.26)$$

for all testing functions $\theta(t)$. A helpful interpretation of the relationship (1.26) is given in Gabel and Roberts [6], where the process of 'examining' $g_1(t)$ and $g_2(t)$ by multiplication by $\theta(t)$ and integration is likened to the use of a measuring instrument. Essentially, the definition of equivalence means that two generalized functions are equivalent if the measuring instrument cannot detect any differences between them. One is quite used to this idea elsewhere: two electrical supplies are deemed equivalent if measurements of the voltage, current and frequency yield equal values, irrespective of their source.

Example 1.6

Show that $\delta(t) = \dfrac{\mathrm{d}\zeta}{\mathrm{d}t}(t)$, where $\zeta(t) = \begin{cases} 1\,, & t \geq 0 \\ 0\,, & t < 0 \end{cases}$ is the Heaviside unit step

function and $\dfrac{\mathrm{d}\zeta}{\mathrm{d}t}(t)$ is the generalized derivative.

Before demonstrating the equivalence of $\delta(t)$ and $\dfrac{\mathrm{d}\zeta}{\mathrm{d}t}(t)$ under the definition (1.26), we demonstrate the plausibility of the result. First of all, note that in the pointwise sense,

$$\frac{\mathrm{d}\zeta}{\mathrm{d}t}(t) = \begin{cases} 0\,, & t \neq 0 \\ \text{undefined}\,, & t = 0\,. \end{cases}$$

Also, if $f(t)$ is any function continuous in a region containing the origin, $-\alpha < t < \alpha$ say,

$$\int_{-\infty}^{\infty} f(t)\frac{\mathrm{d}\zeta}{\mathrm{d}t}(t)\,\mathrm{d}t = \int_{-\alpha}^{\alpha} f(t)\frac{\mathrm{d}\zeta}{\mathrm{d}t}(t)\,\mathrm{d}t$$

$$= \int_{-\alpha}^{\alpha} f(t)\lim_{\Delta t \to 0} \frac{\zeta(t+\Delta t) - \zeta(t)}{\Delta t}\,\mathrm{d}t$$

$$= \lim_{\Delta t \to 0} \frac{1}{\Delta t}\int_{-\Delta t}^{0} f(t)\,\mathrm{d}t\,,$$

assuming the interchange of order is permissible. As a consequence of the First Mean Value Theorem for integrals,

$$\int_{-\infty}^{\infty} f(t)\frac{\mathrm{d}\zeta}{\mathrm{d}t}(t)\,\mathrm{d}t = \lim_{\Delta t \to 0} \frac{1}{\Delta t}\Delta t\, f(\varepsilon_n)\,, \qquad -\Delta t \leq \varepsilon_n \leq 0\,,$$

$$= f(0)\,,$$

from the continuity of $f(t)$ on $(-\alpha,\ \alpha)$. Thus, we see that $\dfrac{\mathrm{d}\zeta}{\mathrm{d}t}(t)$ has the property required of $\delta(t)$ and we now establish the equivalence formally.

Using (1.26), we must show that

$$\int_{-\infty}^{\infty} \frac{d\zeta}{dt}(t)\theta(t)\,dt = \int_{-\infty}^{\infty} \delta(t)\theta(t)\,dt \qquad (1.27)$$

for all testing functions $\theta(t)$. From (1.23), with $a = 0$ $c = -\infty$ and $d = \infty$, the right-hand side of (1.27) is

$$\int_{-\infty}^{\infty} \delta(t)\theta(t)\,dt = \theta(0) \ .$$

On the other hand, the left-hand side can be integrated by parts, that is

$$\int_{-\infty}^{\infty} \frac{d\zeta}{dt}(t)\theta(t)\,dt = [\zeta(t)\theta(t)]\Big|_{-\infty}^{\infty} - \int_{-\infty}^{\infty} \zeta(t)\frac{d\theta}{dt}(t)\,dt \ .$$

(This is effectively the means by which generalized differentiation is defined.) The first term vanishes by the properties of testing functions (zero outside a finite interval, $(-\alpha, \alpha)$ say) and we are similarly guaranteed the existence of $\frac{d\theta}{dt}(t)$ as a continuous function. Therefore,

$$\int_{-\infty}^{\infty} \frac{d\zeta}{dt}(t)\theta(t)\,dt = -\int_{-\infty}^{\infty} \zeta(t)\frac{d\theta}{dt}(t)\,dt$$

$$= -\int_{0}^{\infty} \frac{d\theta}{dt}(t)\,dt$$

$$= -[\theta(t)]\Big|_{0}^{\infty}$$

$$= \theta(0) \ ,$$

since $\theta(t) = 0$, $|t| > \alpha$. Thus, both left- and right-hand sides of (1.27) yield the same number $\theta(0)$ for every testing function $\theta(t)$. Hence, we have established the equivalence of $\delta(t)$ and $\frac{d\zeta}{dt}(t)$, where $\frac{d\zeta}{dt}(t)$ denotes the generalized derivative of $\zeta(t)$.

□

Example 1.7
Find the (generalized) derivative of the causal function $f(t)$ given by

$$f(t) = e^{-at}\zeta(t) \ , \quad a > 0 \ .$$

Differentiating,

$$\frac{df}{dt}(t) = -ae^{-at}\zeta(t) + e^{-at}\frac{d\zeta}{dt}(t)$$

$$= -ae^{-at}\zeta(t) + e^{-at}\delta(t) \ .$$

As shown in Figure 1.13, $f(t)$ has a jump discontinuity at $t = 0$. This behaviour is

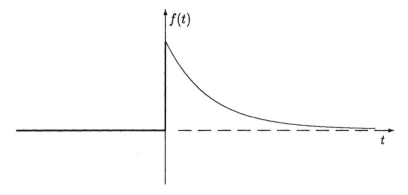

Figure 1.13: Graph of $f_1(t) = e^{-at}\zeta(t), \quad a > 0.$

contained in the generalized derivative with the appearance of a term in $\delta(t)$. Note that we cannot associate a value with $\dfrac{\mathrm{d}f}{\mathrm{d}t}(t)$ at $t = 0$; however, we shall find that we only need to make use of such expressions as *integrands* and the difficulty does not arise.

□

Example 1.8
Show, for $T \in \mathbb{R}$, that the frequently used equivalence

$$f(t)\delta(t - T) = f(T)\delta(t - T) \tag{1.28}$$

holds when $f(t)$ is continuous at $t = T$.

This follows at once, since

$$\int_{-\infty}^{\infty} f(t)\delta(t - T)\theta(t)\,\mathrm{d}t = f(T)\theta(T)$$

and

$$\int_{-\infty}^{\infty} f(T)\delta(t - T)\theta(t)\,\mathrm{d}t = f(T)\theta(T) ,$$

for all testing functions $\theta(t)$.

□

Example 1.9
Differentiate the functions defined by

$$f_1(t) = \sin(\omega t)\zeta(t) \quad \text{and} \quad f_2(t) = \cos(\omega t)\zeta(t) .$$

Now

$$\frac{\mathrm{d}f_1}{\mathrm{d}t}(t) = \omega\cos(\omega t)\zeta(t) + \sin(\omega t)\delta(t) = \omega\cos(\omega t)\zeta(t) , \quad \text{by (1.28)} .$$

Also,

$$\frac{\mathrm{d}f_2}{\mathrm{d}t}(t) = -\omega\sin(\omega t)\zeta(t) + \cos(\omega t)\delta(t)$$
$$= -\omega\sin(\omega t)\zeta(t) + \delta(t) , \quad \text{again by (1.28)} .$$

We see that $\dfrac{\mathrm{d}f_1}{\mathrm{d}t}(t)$ does not, in fact, contain a term in $\delta(t)$. This is not surprising, since $f_1(t) = \sin(\omega t)\zeta(t)$ is continuous at $t = 0$.

□

1.7.2   Laplace Transform of the Impulse Function

Using the sifting property (1.23), the bilateral Laplace transform of $\delta(t-T)$ is given by

$$\mathcal{L}_{\mathrm{B}}\{\delta(t-T)\} = \int_{-\infty}^{\infty} \delta(t-T)e^{-st}\,\mathrm{d}t = e^{-sT} .$$

For the special case $T = 0$, we obtain $\mathcal{L}_{\mathrm{B}}\{\delta(t)\} = 1$.

The above result still remains valid when we take the usual (unilateral) Laplace transform, provided the signal is causal with $T > 0$, that is

$$\mathcal{L}\{\delta(t-T)\} = \int_{0}^{\infty} \delta(t-T)e^{-st}\,\mathrm{d}t = e^{-sT} .$$

The case $T = 0$, however, introduces a minor problem, since we then have

$$\mathcal{L}\{\delta(t)\} = \int_{0}^{\infty} \delta(t)e^{-st}\,\mathrm{d}t$$

and, according to the sifting property of Equation (1.23), this is undefined. In order to circumvent this problem, the definition of the Laplace transform is modified so that the lower limit of the integral is given by 0^- rather than 0, where 0^- is considered to be a number that is infinitesimally less than zero. More precisely, we redefine

$$\mathcal{L}\{f(t)\} = F(s) = \int_{0^-}^{\infty} f(t)\,e^{-st}\,\mathrm{d}t \stackrel{\mathrm{def}}{=} \lim_{\varepsilon\downarrow 0}\int_{-\varepsilon}^{\infty} f(t)e^{-st}\,\mathrm{d}t .$$

This new definition of the Laplace transform does not effect functions in the usual sense, but allows the unit impulse at time $t = 0$ to have a well defined Laplace transform, namely

$$\mathcal{L}\{\delta(t)\} = 1 .$$

1.8   SOME PROPERTIES OF THE LAPLACE TRANSFORM

We summarize, here, the most immediately applicable results from the theory of Laplace transforms: for a fuller discussion, see Gabel and Roberts [6], or O'Neil [20].

1. **Linearity** : If the Laplace transforms of two continuous-time signals $f_1(t)$ and $f_2(t)$ are $F_1(s)$ and $F_2(s)$, respectively, then

$$\mathcal{L}\{\alpha f_1(t) + \beta f_2(t)\} = \alpha F_1(s) + \beta F_2(s) \ ,$$

where α and β are any constants. The proof of this result is immediate :

$$\mathcal{L}\{\alpha f_1(t) + \beta f_2(t)\} = \int_0^\infty (\alpha f_1(t) + \beta f_2(t)) e^{-st} \, dt$$

$$= \alpha \int_0^\infty f_1(t) e^{-st} \, dt + \beta \int_0^\infty f_2(t) e^{-st} \, dt$$

$$= \alpha F_1(s) + \beta F_2(s) \ .$$

2. **Time differentiation** : If the Laplace transform of $f(t)$ is $F(s)$, then

$$\mathcal{L}\left\{\frac{df}{dt}(t)\right\} = sF(s) - f(0) \ . \tag{1.29}$$

N.B. If $f(t)$ has a discontinuity at $t = 0$, then we need to take this into account in defining the derivative function $\frac{df}{dt}(t)$, as detailed in the previous section. In this case, we must replace $f(0)$ in (1.29) with $f(0^-)$. An alternative approach is to ignore the discontinuity at $t = 0$ and to use $f(0^+)$, where, loosely speaking, 0^+ is a number that is infinitesimally greater than zero. This latter approach can be seen to correspond to redefining the Laplace transform so that the lower limit of integration is 0^+. For consistency, we shall always use 0^- for this lower limit and will assume $f(0^-) = f(0) = f(0^+)$, i.e. f is continuous at $t = 0$, whenever the behaviour of $f(t)$ for $0^- < t < 0^+$ is unknown.

The proof is by direct calculation; by definition

$$\mathcal{L}\left\{\frac{df}{dt}(t)\right\} = \int_0^\infty e^{-st} \frac{df}{dt}(t) \, dt \qquad .$$

$$= f(t) e^{-st} \Big|_0^\infty + s \int_0^\infty f(t) e^{-st} \, dt \ ,$$

$$= sF(s) - f(0) \ ,$$

since from the existence of $F(s)$, $f(t) e^{-st} \to 0$ as $t \to \infty$.

Repeated integration by parts, or an inductive proof, shows that

$$\mathcal{L}\left\{\frac{d^n f}{dt^n}(t)\right\} = s^n F(s) - s^{n-1} f(0) - s^{n-2} f^{(1)}(0) - \ldots - f^{(n-1)}(0) \ , \tag{1.30}$$

where

$$f^{(n)}(0) = \frac{d^n f}{dt^n}(t)\Big|_{t=0} \ .$$

Again, any discontinuities must be taken into account using appropriate generalized differentiation.

3. **Time integration** : If $\mathcal{L}\{f(t)\} = F(s)$, then

$$\mathcal{L}\left\{\int_0^t f(\tau)\,\mathrm{d}\tau\right\} = \frac{1}{s}F(s) \ .$$

This can be shown using the time differentiation property. Let

$$g(t) = \int_0^t f(\tau)\,\mathrm{d}\tau$$

then $\dot{g}(t) = f(t)$ and $g(0) = 0$. It follows that $sG(s) = F(s)$ and so

$$G(s) = \frac{1}{s}F(s)$$

as desired.

4. **Convolution** : If $\mathcal{L}\{f(t)\} = F(s)$ and $\mathcal{L}\{g(t)\} = G(s)$, then

$$\mathcal{L}\{(f * g)(t)\} = F(s)G(s) \ ,$$

where $(f * g)(t)$ is the **time-convolution** of $f(t)$ and $g(t)$, defined by the convolution integrals

$$(f * g)(t) \stackrel{\text{def}}{=} \int_0^t f(\tau)g(t - \tau)\,\mathrm{d}\tau = \int_0^t g(\tau)f(t - \tau)\,\mathrm{d}\tau \ .$$

This result can be shown as follows. We have

$$\mathcal{L}\{(f * g)(t)\} = \int_0^\infty e^{-st} \int_0^t f(\tau)g(t - \tau)\,\mathrm{d}\tau\,\mathrm{d}t$$

$$= \int_0^\infty \int_\tau^\infty e^{-st}g(t - \tau)\,\mathrm{d}t\,f(\tau)\,\mathrm{d}\tau$$

where we have swapped the order of integration. Substituting $z = t - \tau$ into the inner integral gives

$$\mathcal{L}\{(f * g)(t)\} = \int_0^\infty \int_0^\infty e^{-s(z+\tau)}g(z)\,\mathrm{d}z\,f(\tau)\,\mathrm{d}\tau$$

$$= \int_0^\infty e^{-s\tau}f(\tau)\,\mathrm{d}\tau \int_0^\infty e^{-sz}g(z)\,\mathrm{d}z$$

$$= F(s)G(s)$$

as desired.

5. **First shift theorem** : If $\mathcal{L}\{f(t)\} = F(s)$, then $\mathcal{L}\{e^{at}f(t)\} = F(s-a)$. (The proof of this is essentially straightforward and is left as an exercise for the reader.) Note that $F(s - a) = F(s)|_{s \to s-a}$, i.e. every occurrence of s in $F(s)$ is replaced by $s - a$. For example, to find the Laplace transform of $e^{-t}\cos(2t)$, we have

$$\mathcal{L}\{e^{-t}\cos(2t)\} = \left.\frac{s}{s^2 + 4}\right|_{s \to s+1} = \frac{s+1}{(s+1)^2 + 4} = \frac{s+1}{s^2 + 2s + 5} \ .$$

6. **Second shift theorem** : If $\mathcal{L}\{f(t)\} = F(s)$, then

$$\mathcal{L}\{f(t-\tau)\zeta(t-\tau)\} = e^{-\tau s}F(s).$$

(As for the first shift theorem, the proof is left as a simple exercise for the reader.) Note that, if we let $g(t) = f(t-\tau)\zeta(t-\tau)$, then $g(t+\tau) = f(t)\zeta(t)$, and so $F(s) = \mathcal{L}\{g(t+\tau)\}$. For example, to find the Laplace transform of $g(t) = (t-2)\zeta(t-3)$, we first find $g(t+3) = (t+1)\zeta(t)$. It follows that

$$\mathcal{L}\{g(t+3)\} = \frac{1}{s^2} + \frac{1}{s}$$

and hence

$$\mathcal{L}\{(t-2)\zeta(t-3)\} = e^{-3s}\left(\frac{1}{s^2} + \frac{1}{s}\right).$$

7. **Multiplication by t property** : If $\mathcal{L}\{f(t)\} = F(s)$, then $\mathcal{L}\{tf(t)\} = -\dfrac{\mathrm{d}F}{\mathrm{d}s}(s)$. By repeated use of this result, we can obtain

$$\mathcal{L}\{t^n f(t)\} = (-1)^n \frac{\mathrm{d}^n F}{\mathrm{d}s^n}(s).$$

To see this, let us note that

$$\frac{\mathrm{d}F}{\mathrm{d}s}(s) = \frac{\mathrm{d}}{\mathrm{d}s}\int_0^\infty f(t)e^{-st}\,\mathrm{d}t$$

$$= \int_0^\infty \frac{\mathrm{d}}{\mathrm{d}s}\left[f(t)e^{-st}\right]\mathrm{d}t$$

$$= \int_0^\infty -tf(t)e^{-st}\,\mathrm{d}t$$

$$= \mathcal{L}\{-tf(t)\},$$

where we have assumed we can interchange differentiation and integration. For example, the Laplace transform of $t\cos(3t)$ is given by

$$\mathcal{L}\{t\cos(3t)\} = -\frac{\mathrm{d}}{\mathrm{d}s}\left[\frac{s}{s^2+9}\right] = \frac{s^2-9}{(s^2+9)^2}.$$

1.9 APPLICATION TO TIME-INVARIANT LINEAR SYSTEMS

Let us return to our $C-R$ circuit of Example 1.2, for which the output $y(t)$ is given by

$$\frac{\mathrm{d}y}{\mathrm{d}t}(t) + \frac{1}{RC}y(t) = \frac{1}{RC}u(t).$$

Taking transforms with $\mathcal{L}\{u(t)\} = U(s)$, $\mathcal{L}\{y(t)\} = Y(s)$ and $y(0) = 0$, that is an initially relaxed system, we have

$$sY(s) + \frac{1}{RC}Y(s) = \frac{1}{RC}U(s)$$

or

$$(1 + RCs)Y(s) = U(s) \ . \tag{1.31}$$

Using (1.31) and solving for the term containing the highest power of s, we obtain

$$sY(s) = \frac{1}{RC}(U(s) - Y(s)) \ . \tag{1.32}$$

Proceeding as for (1.19), we suppose that in the transform domain we have generated the transformed signal $sY(s)$. Then, multiplication by $1/s$ will generate $Y(s)$, the transform of the output. Finally, we recognize (1.32) as a 'recipe' for generating $sY(s)$ and draw the **Laplace transform domain simulation diagram** representation of the C–R circuit as in Figure 1.14. The similarity with Figure 1.11 is

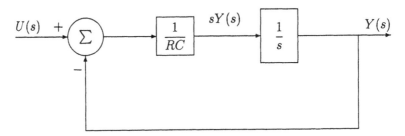

Figure 1.14: Transform domain simulation diagram for the C–R circuit.

obvious, with the '$1/s$ block' serving as the transformed integrator, clearly in line with property 3 of the Laplace transform.

An even simpler form can be deduced which suppresses structural information, but contains all the information for its reconstruction. Equation (1.31) can be represented by a single block operating on the transform of the input signal to yield the transformed output signal, as illustrated in Figure 1.15. The diagram leads natu-

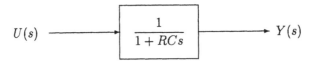

Figure 1.15: Single block representation of the C–R circuit.

rally to the idea of the system **Laplace transfer function** $H(s)$. Equation (1.31) may be rewritten in the form

$$Y(s) = H(s)U(s) \ . \tag{1.33}$$

We see that, in the transform domain, the entire action of the circuit or system, on the input signal transform $U(s)$, is achieved by multiplication by $H(s)$. The convolution theorem, see the Laplace transform property 4, allows us to invert (1.33) as

$$y(t) = (h * u)(t)$$
$$= \int_0^t h(t - \tau)u(\tau)\,d\tau\ , \tag{1.34}$$

where $h(t) = \mathcal{L}^{-1}\{H(s)\}$. Notice that (1.34) makes an important statement for linear time-invariant systems; namely, if we know $h(t)$, then we can express the zero-state system response corresponding to an arbitrary input as a convolution integral, (1.34). We discuss this point later; meanwhile we illustrate (1.34) by an example.

Example 1.10
Find the response of the system with

$$h(t) = \beta e^{-\alpha t}\zeta(t)\ , \quad \alpha,\ \beta > 0\ ,$$

to a unit step input signal, $u(t) = \zeta(t)$.

Using (1.34), the step response, which is understood to mean from the zero initial state, is $y_\zeta(t)$, where

$$y_\zeta(t) = \int_0^t \beta e^{-\alpha(t-\tau)}\,d\tau$$
$$= \int_0^t \beta e^{-\alpha\tau}\,d\tau\ .$$

Here we have used the commutative property for time convolution:

$$h * u = u * h\ .$$

Thus,

$$y_\zeta(t) = -\frac{\beta}{\alpha}e^{-\alpha\tau}\Big|_0^t = \frac{\beta}{\alpha}(1 - e^{-\alpha t})\ .$$

(See Figure 1.16.)

□

As a further demonstration of these ideas, we examine the L–C–R circuit discussed in §1.3. Our analysis produced the pair of coupled first-order, linear, constant coefficient differential equations (1.12) and (1.13), repeated here:

$$Ri(t) + L\frac{di}{dt}(t) + v_C(t) = e(t)\ ,$$
$$\frac{dv_C}{dt}(t) - \frac{1}{C}i(t) = 0\ .$$

Take the Laplace transform, with the notation $\mathcal{L}\{e(t)\} = E(s)$, $\mathcal{L}\{i(t)\} = I(s)$, $\mathcal{L}\{v_C(t)\} = V_C(s)$ and $i(0) = v_C(0) = 0$, corresponding to an initial zero-state, to

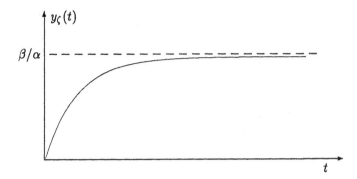

Figure 1.16: Step response for Example 1.10.

obtain

$$RI(s) + LsI(s) + V_C(s) = E(s) \ ,$$

$$sV_C(s) - \frac{1}{C}I(s) = 0 \ .$$

Proceeding as before, we solve for $sI(s)$ and $sV_C(s)$, the terms in the highest power of s in each equation, that is

$$sI(s) = \frac{1}{L}(E(s) - V_C(s) - RI(s)) \ , \tag{1.35}$$

$$sV_C(s) = \frac{1}{C}I(s) \ . \tag{1.36}$$

Considering (1.35) first, we easily determine the structure of the simulation diagram for this subsystem, illustrated in Figure 1.17. We see that we need to have available

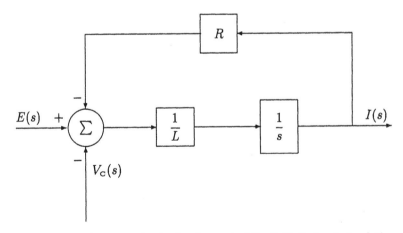

Figure 1.17: First stage in the development of the L–C–R circuit simulation diagram.

the transformed signal $V_C(s)$, but, since $sV_C(s) = I(s)/C$, scaling the signal $I(s)$ by a factor $1/C$ and then passing it through a second '1/s block' generates $V_C(s)$. We

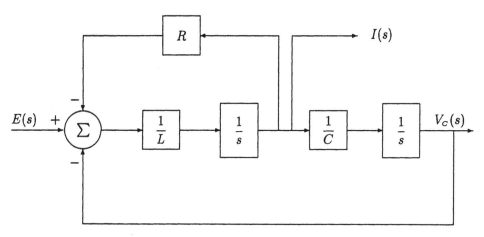

Figure 1.18: The complete L–C–R simulation diagram.

can then produce the final diagram as in Figure 1.18. Referring to the diagram, we might be interested in either $I(s)$ or $V_C(s)$, or rather the inverse of these transforms, as the output corresponding to an input voltage $e(t)$. We, thus, seek the appropriate Laplace transfer functions. Evidently, the transformed signal $I(s)$ is generated as

$$I(s) = \frac{1}{s} \times \frac{1}{L} \times (E(s) - RI(s) - V_C(s))$$

$$= \frac{1}{sL} \left\{ E(s) - RI(s) - \frac{1}{C} \times \frac{1}{s}I(s) \right\} \ .$$

Therefore,

$$I(s) \left\{ \frac{LCs^2 + RCs + 1}{LCs^2} \right\} = \frac{1}{Ls}E(s)$$

or

$$I(s) = \frac{Cs}{LCs^2 + RCs + 1}E(s) = H_1(s)E(s) \ , \tag{1.37}$$

where

$$H_1(s) \stackrel{\text{def}}{=} \frac{Cs}{LCs^2 + RCs + 1} \ .$$

Also, since

$$V_C(s) = \frac{1}{Cs}I(s) \ ,$$

$$V_C(s) = \frac{1}{LCs^2 + RCs + 1}E(s) = H_2(s)E(s) \ , \tag{1.38}$$

where

$$H_2(s) \stackrel{\text{def}}{=} \frac{1}{LCs^2 + RCs + 1} \ .$$

Thus, depending on the desired output, we can generate the two Laplace transfer functions $H_1(s)$ and $H_2(s)$. Although these are not identical, there are clearly some similarities and, as an indication of the origin of the similarities, we can construct differential equations for each output provided that the restriction to zero initial conditions is observed.

From (1.37),

$$(LCs^2 + RCs + 1)I(s) = CsE(s)$$

and, if initial conditions on $i(t)$, $\dfrac{di}{dt}(t)$ and $e(t)$ are zero, this is the Laplace transform of the differential equation

$$LC\frac{d^2i}{dt^2}(t) + RC\frac{di}{dt}(t) + i(t) = C\frac{de}{dt}(t) \ . \tag{1.39}$$

On the other hand, from (1.38), we easily obtain the differential equation for $v_C(t)$ as the output in the form

$$LC\frac{d^2v_C}{dt^2}(t) + RC\frac{dv_C}{dt}(t) + v_C(t) = e(t) \ . \tag{1.40}$$

The two equations (1.39) and (1.40) are alternative forms of the circuit equation for the L–C–R circuit, which could in fact have been inferred by elimination between (1.12) and (1.13).

1.10   TRANSFER FUNCTIONS

We conclude this chapter by reviewing our ideas on system transfer functions for linear time-invariant (LTI) systems. Such a system, including any of the previously mentioned circuits, can be modelled by a differential equation of the form

$$\frac{d^n y}{dt^n}(t) + a_{n-1}\frac{d^{n-1}y}{dt^{n-1}}(t) + \ldots + a_0 y(t) = b_m \frac{d^m u}{dt^m}(t) + \ldots + b_0 u(t) \ , \tag{1.41}$$

where $u(t)$ is the input, $y(t)$ the output, and where $a_0, \ldots, a_{n-1}, b_0, \ldots, b_m$ are constants.

LTI systems, which have been modelled using a system of n first-order equations, can be represented in the form (1.41) after elimination, which may be performed directly or using simulation diagrams, following the method in §1.9.

With an initially relaxed system, meaning that all conditions on $y(t)$ and $u(t)$ and their derivatives at $t = 0$ are zero, taking the Laplace transform of (1.41), with $\mathcal{L}\{u(t)\} = U(s)$ and $\mathcal{L}\{y(t)\} = Y(s)$, we obtain

$$(s^n + a_{n-1}s^{n-1} + \ldots + a_0)Y(s) = (b_m s^m + \ldots + b_0)U(s) \ .$$

Thus,

$$Y(s) = \frac{b_m s^m + \ldots + b_0}{s^n + a_{n-1}s^{n-1} + \ldots + a_0}U(s) \ , \tag{1.42}$$

that is

$$Y(s) = H(s)U(s) \ , \tag{1.43}$$

where

$$H(s) \overset{\text{def}}{=} \frac{b_m s^m + \ldots + b_0}{s^n + a_{n-1}s^{n-1} + \ldots + a_0} \ . \tag{1.44}$$

Here, H(s), as previously identified, is the Laplace transfer function and, if $m \leq n$ (the normal situation), $H(s)$ is said to be **proper**. The denominator of $H(s)$, *viz.*

$$s^n + a_{n-1}s^{n-1} + \ldots + a_0 \, ,$$

is called the **characteristic polynomial** of the system and we observe that for the different representations (1.39) and (1.40) of the same system we obtain the same characteristic polynomial. This is not a coincidence and a proof that the characteristic equation is independent of the particular representation of the system chosen is suggested in the exercises.

The Laplace transfer function $H(s)$, defined by (1.44), can be inverted to yield

$$h(t) = \mathcal{L}^{-1}\{H(s)\}. \tag{1.45}$$

Recalling ideas of §1.9, we can express the zero-state response of a linear time-invariant system for a given input $u(t)$ in a compact form. From (1.43) and using the convolution theorem for Laplace transforms, we have

$$y(t) = \int_0^t h(t - \tau)u(\tau)\,\mathrm{d}\tau$$

$$= \int_0^t h(\tau)u(t - \tau)\,\mathrm{d}\tau \, ,$$

where $h(t)$ is obtained from (1.45).

Finally, we demonstrate these and earlier ideas by means of an example.

Example 1.11
Consider the circuit shown in Figure 1.19. Find the Laplace transfer function between the input voltage $u(t)$ and the output voltage $y(t) = v_2(t)$, the voltage drop measured across C_2.

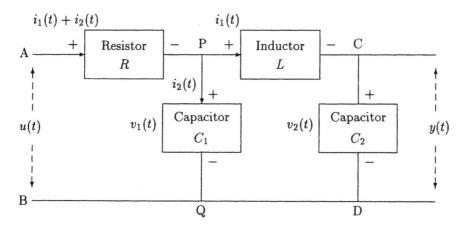

Figure 1.19: Circuit for Example 1.11.

Notice how we have chosen to denote the current in the branches of the circuit; the effect of this is to automatically satisfy Kirchhoff's current law at the junctions

P and Q. Remember that, by assumption, the measuring device used to obtain $y(t)$ is assumed to draw a negligible current, which is ignored. Thus, the output $y(t)$ is $v_2(t)$, the voltage drop across the capacitor C_2.

Using our procedure for analysing circuits, we first consider the circuit APQBA, a loop in the overall circuit, and, with the usual notation, Kirchhoff's voltage law for this loop gives

$$v_R(t) + v_1(t) - u(t) = 0 . \tag{1.46}$$

Clearly, (1.46) does not involve the output $y(t) = v_2(t)$ and we need a second 'loop' equation. There are two candidates, namely the loop APCDQBA or the loop PCDQP, and either may be chosen. Having used either of these loops, we have traversed every route in the entire network at least once and, thus, have extracted all the information available. We select for our second loop PCDQP and obtain

$$v_L(t) + v_2(t) - v_1(t) = 0 . \tag{1.47}$$

Writing down the component constitutive equations, we obtain

for R $\qquad v_R(t) = [i_1(t) + i_2(t)]R ,$ $\hfill (1.48)$

for C_1 $\qquad i_2(t) = C_1 \dfrac{dv_1}{dt}(t) ,$ $\qquad$ using the derivative form, $\hfill (1.49)$

for L $\qquad v_L(t) = L \dfrac{di_1}{dt}(t) ,$ $\hfill (1.50)$

for C_2 $\qquad i_1(t) = C_2 \dfrac{dv_2}{dt}(t) .$ $\hfill (1.51)$

Viewing the system (1.46)–(1.51) as a whole, we see that (1.49)–(1.51) is a set of three simultaneous linear differential equations with constant coefficients, whereas (1.46)–(1.48) provide a set of algebraic (or static) equations. Equations (1.49)–(1.51) imply that the variables $v_1(t)$, $v_2(t)$, $i_1(t)$ may be used to describe the circuit behaviour (in other words, provide a set of state variables), whereas the static equations (1.46)–(1.48) link the state variables, both to each other and, via the auxiliary variables $i_2(t)$, $v_L(t)$ and $v_R(t)$, to the input $u(t)$.

Our aim is to obtain the Laplace transfer function for the circuit and this task can be achieved by eliminating the auxiliary variables as follows. Using (1.49), we have

$$\frac{dv_1}{dt}(t) = \frac{1}{C_1} i_2(t) ,$$

but, from (1.48),

$$i_2(t) = \frac{1}{R} v_R(t) - i_1(t)$$
$$= \frac{1}{R}[u(t) - v_1(t)] - i_1(t) \quad \text{from (1.46).}$$

Thus, (1.49) becomes

$$\frac{dv_1}{dt}(t) = \frac{1}{RC_1} u(t) - \frac{1}{RC_1} v_1(t) - \frac{1}{C_1} i_1(t) , \tag{1.52}$$

with the right-hand side involving only state variables and the input $u(t)$. Also, (1.50) becomes, using (1.47),

$$\frac{di_1}{dt}(t) = \frac{1}{L}v_1(t) - \frac{1}{L}v_2(t) , \qquad (1.53)$$

whereas (1.51) is in the desired form when written as

$$\frac{dv_2}{dt}(t) = \frac{1}{C_2}i_1(t) . \qquad (1.54)$$

The time domain simulation diagrams for the circuit could now be built up in stages, as before, but, as our aim is the system transfer function, we now transform our equations from the time domain to the Laplace transform domain. We obtain, with zero initial conditions,

$$sV_1(s) = \frac{1}{RC_1}U(s) - \frac{1}{RC_1}V_1(s) - \frac{1}{C_1}I_1(s) , \qquad (1.55)$$

$$sI_1(s) = \frac{1}{L}V_1(s) - \frac{1}{L}V_2(s) , \qquad (1.56)$$

$$sV_2(s) = \frac{1}{C_2}I_1(s) . \qquad (1.57)$$

Obviously, algebraic elimination yields the transfer function readily at this stage. However, we first construct the transform domain simulation diagram in stages as illustrated in Figure 1.20a–d.

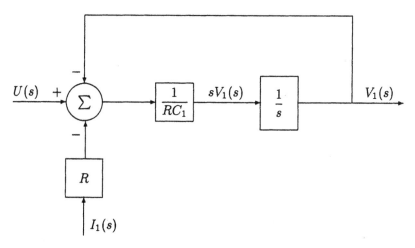

Figure 1.20a: Simulation diagram for (1.55): $I_1(s)$ is needed.

From (1.55)–(1.57), (or reading from the diagram), with $Y(s) = V_2(s)$,

$$Y(s) = \frac{1}{C_2 s}I_1(s) , \qquad (1.58)$$

$$I_1(s) = \frac{1}{Ls}(V_1(s) - Y(s)) , \qquad (1.59)$$

$$V_1(s) = \frac{1}{RC_1 s}(U(s) - V_1(s) - RI_1(s)) . \qquad (1.60)$$

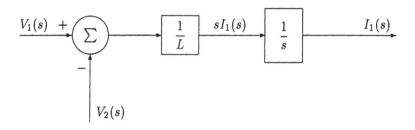

Figure 1.20b: Simulation diagram for (1.56): $V_2(s)$ is needed.

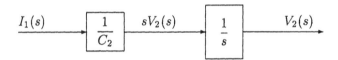

Figure 1.20c: Simulation diagram for (1.57).

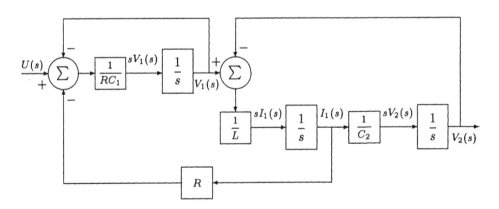

Figure 1.20d: Simulation diagram for Example 1.11.

Therefore, from (1.58) and (1.59),

$$Y(s) = \frac{1}{LC_2s^2}(V_1(s) - Y(s)) ,$$

or

$$Y(s)\left\{1 + \frac{1}{LC_2s^2}\right\} = \frac{1}{LC_2s^2}V_1(s) .$$

That is,

$$Y(s)\{1 + LC_2s^2\} = V_1(s) . \tag{1.61}$$

But

$$V_1(s)\left\{1 + \frac{1}{RC_1s}\right\} = \frac{1}{RC_1s}(U(s) - RI_1(s))$$

$$= \frac{1}{RC_1s}(U(s) - RC_2sY(s)) , \quad \text{from (1.58)}.$$

That is,

$$V_1(s)\{C_1Rs + 1\} = U(s) - RC_2sY(s) .$$

Thus, using (1.61),

$$Y(s)\{LC_2s^2 + 1\}\{C_1Rs + 1\} = U(s) - RC_2sY(s) ,$$

which may be rewritten in the form

$$Y(s)\{LC_1C_2Rs^3 + LC_2s^2 + R(C_1 + C_2)s + 1\} = U(s)$$

or

$$Y(s) = \frac{\dfrac{1}{LC_1C_2R}U(s)}{s^3 + \dfrac{1}{C_1R}s^2 + \dfrac{(C_1 + C_2)}{LC_1C_2}s + \dfrac{1}{LC_1C_2R}} \tag{1.62}$$

$$= H(s)U(s) . \tag{1.63}$$

From (1.62), the Laplace transfer function for the system is

$$H(s) = \frac{\dfrac{1}{LC_1C_2R}}{s^3 + \dfrac{1}{C_1R}s^2 + \dfrac{(C_1 + C_2)}{LC_1C_2}s + \dfrac{1}{LC_1C_2R}} \tag{1.64}$$

and we note that this is 'proper' in the sense defined.

□

Remark: We note that by inverting (1.63), using (1.64), we obtain as the time domain representation of this system

$$\frac{d^3y}{dt^3}(t) + \frac{1}{C_1R}\frac{d^2y}{dt^2}(t) + \frac{(C_1 + C_2)}{LC_1C_2}\frac{dy}{dt}(t) + \frac{1}{LC_1C_2R}y(t) = \frac{1}{LC_1C_2R}u(t) , \tag{1.65}$$

where $u(t)$ is the input voltage and $y(t) = v_2(t)$ is the output voltage measured across the capacitor C_2.

In this chapter, we have examined the concepts of signals and linear systems. We have seen that simple electrical circuits can be used to perform operations on input signals to produce output signals and we have developed a method for the analysis of such systems. The Laplace transform, with its associated simulation diagrams, has been shown to be a useful tool in such analysis and we have observed the duality between time and transform domains. In the next chapter we examine some specific responses of linear systems which help us to understand system characteristics and system behaviour. In later chapters we develop methods for the design of systems which perform specific operations on input signals.

1.11 EXERCISES

1. If the desired output of the system of Example 1.1 is v_L, obtain a corresponding differential equation representation.

2. Calculate the Laplace transform of the following causal signals using tables and transform properties :

   (a) $(5 - \cos(t/\sqrt{2}))\zeta(t)$ (b) $(3e^{-2t} - t^3)\zeta(t)$

   (c) $(2e^{-t}\sin(t) - t\cos(t))\zeta(t)$ (d) $\sinh(t)\zeta(t) + 2t\zeta(t-1)$.

3. Prove the two shift theorems for the Laplace transform given in §1.8.

4. For each of the following unilateral Laplace transforms, find the inverse transform

   (a) $\dfrac{1}{s^2 + 7s + 10}$ (b) $\dfrac{s}{(s-1)(s^2+1)}$ (c) $\dfrac{1}{(s-2)(s+2)^2}$.

5. Evaluate

   (i) $\dfrac{\mathrm{d}}{\mathrm{d}t}[e^{-3t}\zeta(t)]$ (ii) $\dfrac{\mathrm{d}}{\mathrm{d}t}[(1 - \cos(at))\zeta(t)]$,

where $\zeta(t)$ is the Heaviside step function.
Also, show that

$$\frac{d[t\zeta(t)]}{dt} = \zeta(t) .$$

6. Use the identity $|t| = -t\zeta(-t) + t\zeta(t)$ to show that

$$\frac{d(|t|)}{dt} = \zeta(t) - \zeta(-t) .$$

7. Obtain the transfer function for the system, with input $u(t)$ and output $y(t)$, represented by the differential equation

$$\frac{d^2y}{dt^2}(t) + 5\frac{dy}{dt}(t) + 6y(t) = u(t) .$$

8. What is the transfer function for the system

$$\frac{d^3y}{dt^3}(t) - 3\frac{dy}{dt}(t) + 2y(t) = 2u(t) + \frac{du}{dt}(t) ,$$

where $u(t)$ is the input and $y(t)$ is the associated output?

9. For functions defined on $(-\infty, \infty)$, the convolution operation is defined as

$$(f * g)(t) \stackrel{\text{def}}{=} \int_{-\infty}^{\infty} f(\tau)g(t - \tau)\,d\tau .$$

Show that, if $f(t)$ and $g(t)$ are both causal functions, then $f * g$ is causal and, for $t \geq 0$,

$$(f * g)(t) = \int_0^t f(\tau)g(t - \tau)\,d\tau = \int_0^t f(t - \tau)g(\tau)\,d\tau .$$

Hence, calculate $(f * g)(t)$ when

(a) $f(t) = e^t\zeta(t)$ and $g(t) = t\zeta(t)$
(b) $f(t) = e^t$ and $g(t) = t\zeta(t)$,

where $\zeta(t)$ is the unit step function.

10. Find the transfer function for a system with input $u(t)$ and output $y(t) = y_3(t)$ modelled by

$$\dot{y}_1 = y_2$$
$$\dot{y}_2 = y_3 - u$$
$$\dot{y}_3 = 2y_1 - y_2 + u.$$

11. Obtain a state-space representation in the form of Equation (1.14) for the third-order system of Example 1.11. Evaluate the **characteristic polynomial** $\det(\lambda I - A)$ for the matrix A.

12. Use the method of Example 1.3 to show that

$$e^{-2t}\frac{d^2y}{dt^2}(t) + 3e^{-t}\frac{dy}{dt}(t) + y(t) = u(t)$$

represents a linear system.

13. (a) Suppose M is a non-singular matrix and $B = M^{-1}AM$. Show that $\det(\lambda I - A) = \det(\lambda I - B)$.

(b) A system is described in state-space form as

$$\frac{d\boldsymbol{x}}{dt} = A\boldsymbol{x}(t) + \boldsymbol{b}u(t)$$
$$y(t) = \boldsymbol{c}^T\boldsymbol{x}(t)$$

and a transformation of state is defined by $\boldsymbol{z}(t) = M\boldsymbol{x}(t)$. Show that

$$\frac{d\boldsymbol{z}}{dt} = M^{-1}AM\boldsymbol{z}(t) + M^{-1}\boldsymbol{b}u(t)$$
$$y(t) = \boldsymbol{c}^TM^{-1}\boldsymbol{x}(t)$$

and, hence, deduce that the characteristic polynomial of the system is independent of any such transformation.

14. Show that

$$\int_{-\infty}^{\infty} \delta'(t) f(t) \, dt = -f'(0) \ ,$$

if $f'(t)$ is continuous at $t = 0$. Furthermore, show (by induction) that

$$\int_{-\infty}^{\infty} \delta^{(n)}(t) f(t) \, dt = (-1)^n f^{(n)}(0) \ ,$$

if $f^{(n)}(t)$ is continuous at $t = 0$.

15. Determine the solution of the differential equation

$$\frac{d^2 y}{dt^2}(t) + 7\frac{dy}{dt}(t) + 12y(t) = u(t)$$

when (i) $u(t) = \sin(t)$ and (ii) $u(t) = \sin(2t)$. What is the long-term behaviour of these solutions and does this behaviour depend on the initial conditions? What would be the long-term behaviour of the solution when $u(t) = \sin(t) + \sin(2t)$?

2

System responses

2.1 INTRODUCTION

In Chapter 1, it was shown that some elementary electrical circuits could be modelled by linear differential equations. The idea of system response was introduced, in passing, as the output of a system subjected to an input signal. In the model, the differential equation was constructed in such a way that the output was given by, (or derivable from), the solution. The first section of this chapter looks at factors which govern some important aspects of response behaviour. After developing some mathematical techniques, we consider particular responses which serve to characterize a linear system and we conclude with a discussion of response as viewed in the frequency domain.

2.2 STABILITY OF LINEAR TIME-INVARIANT SYSTEMS

Before defining what is meant by stability for linear time-invariant systems, we examine the output of two such systems which are both initially relaxed and subject to a step input. First consider the system with Laplace transfer function

$$H_1(s) = \frac{1}{(s+1)(s+2)} \ . \tag{2.1}$$

Now $Y(s) = H_1(s)U(s)$, where $Y(s) = \mathcal{L}\{y(t)\}$ is the Laplace transform of the output and $U(s) = \mathcal{L}\{u(t)\}$ is the Laplace transform of the input. With $u(t) = \zeta(t)$, the unit step function, we have $U(s) = 1/s$ and so

$$Y(s) = \frac{1}{s(s+1)(s+2)} \ .$$

Thus

$$Y(s) = \frac{\frac{1}{2}}{s} - \frac{1}{s+1} + \frac{\frac{1}{2}}{s+2}$$

and inversion gives

$$y(t) = \frac{1}{2} - e^{-t} + \frac{1}{2}e^{-2t}, \quad t \geq 0 \ . \tag{2.2}$$

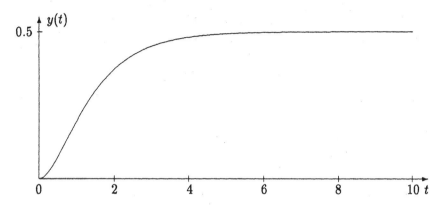

Figure 2.1: Response of system (2.1) to a step input.

Figure 2.1 shows the graph of $y(t)$, the response of the system (2.1) to the step input.

Secondly, consider the system with Laplace transfer function

$$H_2(s) = \frac{1}{(s-1)(s+2)} \tag{2.3}$$

subject to the same input. The output is then given by

$$
\begin{aligned}
y(t) &= \mathcal{L}^{-1}\left\{\frac{1}{s(s-1)(s+2)}\right\} \\
&= \mathcal{L}^{-1}\left\{\frac{\frac{1}{3}}{s-1} - \frac{\frac{1}{2}}{s} + \frac{\frac{1}{6}}{s+2}\right\} \\
&= \frac{1}{3}e^t - \frac{1}{2} + \frac{1}{6}e^{-2t}, \quad t \geq 0.
\end{aligned}
\tag{2.4}
$$

Figure 2.2 illustrates the graph of $y(t)$ in this case.

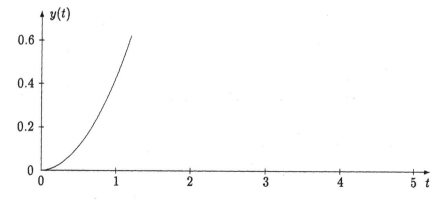

Figure 2.2: Response of system (2.3) to a step input.

We observe drastically different behaviour of the output signals as given in (2.2) and (2.4): in the former case, we see a rise to a bounded limiting value of $\frac{1}{2}$, whereas the solution in (2.4) increases without bound. In the two chosen examples, the reason for the different behaviour is easy to locate. For system (2.3), the term $(s - 1)$ in the denominator of $H_2(s)$ generates a term in e^t in the solution, which produces the unbounded behaviour of the output. We see that this behaviour is inherent in the system Laplace transfer function, and is not dependent in any way on the input signal. We say that system (2.1) is a stable system, whereas system (2.3) is unstable. More precisely, we say that a system is **BIBO, (Bounded Input, Bounded Output), stable** *if bounded inputs always give rise to bounded outputs for initially relaxed conditions.* For brevity, the term BIBO is usually dropped and we simply refer to *system stability*. Similarly, a system is said to be **unstable** if (generically) bounded inputs give rise to unbounded outputs; the precise details are given below. In between these two concepts is a condition known as **marginal stability**. It is possible to extend these concepts to take into account possibly non-zero initial conditions, but we do not explore such issues.

In the last chapter, we saw that linear time-invariant systems generated Laplace transfer functions of the form (1.44), that is

$$H(s) = \frac{b_m s^m + b_{m-1} s^{m-1} + \cdots + b_0}{s^n + a_{n-1} s^{n-1} + \cdots + a_0} .$$

Such transfer functions are said to be **strictly proper** if $m < n$ and we restrict attention to this class here. The denominator of $H(s)$, $a(s)$ given by

$$a(s) = s^n + a_{n-1} s^{n-1} + \cdots + a_0 ,$$

is the **characteristic polynomial** of the system and may be factorized into complex factors, so that

$$a(s) = \prod_{i=1}^{n} (s - \lambda_i) , \qquad (2.5)$$

where λ_i, $i = 1, 2, \ldots, n$ are the roots of the **characteristic equation**

$$a(s) = 0 ,$$

and are called the **poles** of the system.

We consider the case when the poles are distinct, i.e.

$$\lambda_i \neq \lambda_j, \qquad i \neq j ,$$

when $H(s)$ can be decomposed into partial fractions in the form

$$H(s) = \sum_{i=1}^{n} \frac{\alpha_i}{s - \lambda_i} . \qquad (2.6)$$

Inverting (2.6), we obtain

$$\mathcal{L}^{-1}\{H(s)\} = h(t)$$

$$= \sum_{i=1}^{n} \alpha_i e^{\lambda_i t} . \qquad (2.7)$$

Now λ_i, $i = 1, 2, \ldots, n$ is in general a complex number so that

$$\lambda_i = a_i + j\, b_i, \qquad \text{with } a_i,\ b_i \text{ both real,}$$

and (see Appendix A)

$$e^{\lambda_i t} = e^{(a_i + j\, b_i)t} = e^{a_i t}\big(\cos(b_i t) + j\, \sin(b_i t)\big)\ . \tag{2.8}$$

Clearly as $t \to \infty$, $|e^{\lambda_i t}| \to \infty$, (implying that $|h(t)| \to \infty$), if $a_i > 0$ for *any* i, $i = 1, 2, \ldots, n$, whereas if $a_i < 0$ for *all* i, $i = 1, 2, \ldots, n$, then $|h(t)| \to 0$ as $t \to \infty$.

It can be shown that *a system is (BIBO) stable if and only if $h(t) \to 0$ as $t \to \infty$ which in turn occurs if and only if all the system poles lie in the open left half of the complex plane, i.e. they all have strictly negative real parts.* It should be noted that this is the case whether or not the system has distinct poles.

Similarly, it can be shown that *a system is unstable if and only if $|h(t)| \to \infty$ as $t \to \infty$ which occurs if and only if there exists at least one pole in the open right half plane and/or at least one repeated pole on the imaginary axis.* We see that an unstable system subjected to a bounded input gives rise to an unbounded output unless the numerator of $U(s)$ cancels out all the unstable poles.

Finally, we say that a system is **marginally stable** if it is neither stable nor unstable. This occurs if and only if $|h(t)|$ remains bounded but does not decay to zero as $t \to \infty$, which occurs if and only if all the system poles lie in the closed left half of the complex plane and, in addition, all the poles lying on the imaginary axis, (of which there must be at least one), must be simple, i.e. not repeated.

Example 2.1
Examine the stability of the systems with transfer functions:

(i) $H_1(s) = \dfrac{1}{s^2 + \sqrt{2}\,s + 1}$,   (ii) $H_2(s) = \dfrac{1}{s^2 + 1}$,   (iii) $H_3(s) = \dfrac{1}{s^2(s + 1)}$.

For any unstable systems, determine a bounded, non-zero input signal, $u(t)$, that will in theory give rise to a bounded output signal. For any marginally stable systems, determine a bounded input signal that will give rise to an unbounded output signal. (All systems should be considered as being initially relaxed.)

For case (i), solving the characteristic equation $s^2 + \sqrt{2}\,s + 1 = 0$, we obtain the poles

$$s = \frac{1}{\sqrt{2}}(-1 \pm j)\ .$$

Since both poles lie in the open left half of the complex plane, the system given by $H_1(s)$ is stable.

For case (ii), the poles are given by $s = \pm j$ which both lie on the imaginary axis and are both simple. We therefore deduce that the system given by $H_2(s)$ is marginally stable. If we let $u(t)$ be a bounded signal for which the poles of $U(s)$ coincide with those of $H_2(s)$ then the resulting output will be unbounded. For example, if we let $u(t) = \cos(t)$ then

$$Y(s) = H_2(s)U(s) = \frac{s}{(s^2 + 1)^2}$$

which inverts to give the unbounded response, $y(t) = \frac{1}{2}t\sin(t)$, $t \geq 0$.

For case (iii), the poles are given by

$$s = 0, \text{ (twice)} \quad \text{and} \quad s = -1.$$

Although all the poles lie in the closed left half complex plane, the pole at $s = 0$ lies on the imaginary axis and is repeated. This means that the system given by $H_3(s)$ is unstable. In order to find a bounded input signal $u(t)$ that gives rise to a bounded response, we see that $U(s)$ must have a factor s in its numerator. For example, let

$$U(s) = \frac{s^2}{(s+1)^3} = \frac{1}{s+1} - \frac{2}{(s+1)^2} + \frac{1}{(s+1)^3} .$$

Inverting this gives the bounded input signal, $u(t) = (1 - 2t + \frac{1}{2}t^2)e^{-t}$, $t \geq 0$. For this input, we have

$$Y(s) = H_3(s)U(s) = \frac{1}{(s+1)^4} .$$

Inverting this gives the bounded response, $y(t) = \frac{1}{6}t^3 e^{-t}$, $t \geq 0$. It should be noted that only *very* special inputs will give bounded responses for an unstable system. In practice modelling errors and input disturbances imply that such inputs can only be implemented using some form of feedback. See textbooks on **Control Theory** for more details on this.

$\square$

2.3   THE IMPULSE RESPONSE

The **impulse response** of a system is defined to be *the output that results when the input is a unit impulse at time $t = 0$, under the assumption that the system is initially at rest.* As discussed in §1.7.1, the impulse 'function', $\delta(t)$, has an instantaneous effect at $t = 0$ and so zero initial conditions have to be interpreted as referring to time $t = 0^-$. For our purposes, it is sufficient to consider the output $y_\delta(t)$ as being identically zero for $t < 0$, i.e. $y_\delta(t)$ is causal. We recall from §1.7.2 that the unit impulse, $\delta(t)$, has the Laplace transform

$$\mathcal{L}\{\delta(t)\} = 1 . \tag{2.9}$$

Suppose we now set

$$u(t) = \delta(t)$$

as the input to the linear system with Laplace transfer function $H(s)$. Then $Y_\delta(s)$, the transform of the output, is given by

$$Y_\delta(s) = H(s)U(s) ,$$

but, from (2.9), $\mathcal{L}\{u(t)\} = U(s) = 1$ and so

$$Y_\delta(s) = H(s). \tag{2.10}$$

Inverting (2.10), we see that the impulse response can be found using

$$y_\delta(t) = \mathcal{L}^{-1}\{Y_\delta(s)\} = \mathcal{L}^{-1}\{H(s)\} ,\qquad(2.11)$$

that is, *by inverse Laplace transforming the transfer function*. For this reason, the impulse response, $y_\delta(t)$, is also often written as $h(t)$. As discussed previously, the properties of BIBO stability, marginal stability and instability can be characterized by the impulse response respectively tending to zero, remaining bounded and tending to infinity in magnitude.

Example 2.2

Find the impulse response for the following systems and state whether the systems are stable, unstable or marginally stable.

(1) $\quad \dfrac{d^2 y}{dt^2}(t) + 3\dfrac{dy}{dt}(t) + 2y(t) = u(t)$.

(2) $\quad \dfrac{d^2 y}{dt^2}(t) + \sqrt{2}\dfrac{dy}{dt}(t) + y(t) = u(t)$.

(3) $\quad \dfrac{d^2 y}{dt^2}(t) + \sqrt{2}\dfrac{dy}{dt}(t) + y(t) = u(t) + 2\dfrac{du}{dt}(t)$.

(4) $\quad \dfrac{d^4 y}{dt^4}(t) + 3\dfrac{d^3 y}{dt^3}(t) + 3\dfrac{d^2 y}{dt^2}(t) + 3\dfrac{dy}{dt}(t) + 2y(t) = u(t)$.

(5) $\quad \dfrac{d^3 y}{dt^3}(t) + \dfrac{d^2 y}{dt^2}(t) = 2u(t) + \dfrac{du}{dt}(t)$.

(1) By inspection, the Laplace transfer function is given by

$$H(s) = \frac{1}{s^2 + 3s + 2} = \frac{1}{(s+1)(s+2)} .$$

It therefore follows that

$$Y_\delta(s) = \frac{1}{(s+1)(s+2)} = \frac{1}{s+1} - \frac{1}{s+2} .$$

Thus the impulse response is

$$y_\delta(t) = \mathcal{L}^{-1}\{Y_\delta(s)\}$$
$$= e^{-t} - e^{-2t}, \quad t \geq 0.$$

Since $y_\delta(t) \to 0$ as $t \to \infty$ the system is stable. Let us also note that, if we wish to stress the causality of the impulse response, we can write

$$y_\delta(t) = (e^{-t} - e^{-2t})\zeta(t) ,$$

where $\zeta(t)$ is the unit step function.

(2) This time the transfer function can be seen to be

$$H(s) = \frac{1}{s^2 + \sqrt{2}\,s + 1} \; .$$

It follows that

$$Y_\delta(s) = \frac{1}{s^2 + \sqrt{2}\,s + 1}$$

$$= \sqrt{2}\, \frac{\frac{1}{\sqrt{2}}}{\left(s + \frac{1}{\sqrt{2}}\right)^2 + \left(\frac{1}{\sqrt{2}}\right)^2} \; ,$$

and so

$$y_\delta(t) = \left(\sqrt{2}e^{-\frac{t}{\sqrt{2}}}\sin\left(\frac{t}{\sqrt{2}}\right)\right)\zeta(t) \; .$$

We again see that $y_\delta(t) \to 0$ as $t \to \infty$ and so the system is stable, (as we had already determined in Example 2.1).

(3) The system transfer function is

$$H(s) = \frac{2s + 1}{s^2 + \sqrt{2}\,s + 1}$$

and so

$$Y_\delta(s) = \frac{2\left(s + \frac{1}{\sqrt{2}}\right)}{\left(s + \frac{1}{\sqrt{2}}\right)^2 + \left(\frac{1}{\sqrt{2}}\right)^2} - \sqrt{2}(\sqrt{2} - 1)\, \frac{\frac{1}{\sqrt{2}}}{\left(s + \frac{1}{\sqrt{2}}\right)^2 + \left(\frac{1}{\sqrt{2}}\right)^2} \; .$$

Thus

$$y_\delta(t) = e^{-\frac{t}{\sqrt{2}}}\left[2\cos\left(\frac{t}{\sqrt{2}}\right) - (2 - \sqrt{2})\sin\left(\frac{t}{\sqrt{2}}\right)\right]\zeta(t) \; .$$

In fact we can use the following argument to obtain this response from case (2).

The transform of the impulse response for case (2) was given by

$$(s^2 + \sqrt{2}\,s + 1)Y_\delta(s) = 1 \; .$$

Multiplying both sides by $2s + 1$ we see that

$$(s^2 + \sqrt{2}\,s + 1)Z_\delta(s) = 2s + 1 \; ,$$

where $Z_\delta(s) = (2s + 1)Y_\delta(s)$. Thus $Z_\delta(s)$ is the transform of the impulse response for case (3). It therefore follows that

$$z_\delta(t) = 2\frac{\mathrm{d}y_\delta}{\mathrm{d}t}(t) + y_\delta(t) \; ,$$

where $y_\delta(t) = \left[\sqrt{2}e^{-\frac{t}{\sqrt{2}}}\sin\left(\frac{t}{\sqrt{2}}\right)\right]\zeta(t)$ is the impulse response of system (2). Thus

$$z_\delta(t) = 2\sqrt{2}\left[-\frac{1}{\sqrt{2}}e^{-\frac{t}{\sqrt{2}}}\sin\left(\frac{t}{\sqrt{2}}\right)\zeta(t)\right.$$

$$+ \frac{1}{\sqrt{2}}e^{-\frac{t}{\sqrt{2}}}\cos\left(\frac{t}{\sqrt{2}}\right)\zeta(t) + e^{-\frac{t}{\sqrt{2}}}\sin\left(\frac{t}{\sqrt{2}}\right)\delta(t)\right]$$

$$+ \sqrt{2}e^{-\frac{t}{\sqrt{2}}}\sin\left(\frac{t}{\sqrt{2}}\right)\zeta(t) ,$$

where we have used the result that $d\zeta(t)/dt = \delta(t)$, (see Example 1.6). If we now use Equation (1.28), we obtain $f(t)\delta(t) = f(0)\delta(t)$, and the above simplifies to

$$z_\delta(t) = e^{-\frac{t}{\sqrt{2}}}\left[2\cos\left(\frac{t}{\sqrt{2}}\right) - (2-\sqrt{2})\sin\left(\frac{t}{\sqrt{2}}\right)\right]\zeta(t) ,$$

which agrees with our previous result. We again see that the impulse response decays to zero and so the system is stable.

(4) This time the transfer function is given by

$$H(s) = \frac{1}{s^4 + 3s^3 + 3s^2 + 3s + 2} ,$$

and so

$$Y_\delta(s) = \frac{1}{s^4 + 3s^3 + 3s^2 + 3s + 2}$$

$$= \frac{1}{(s^2+1)(s^2+3s+2)}$$

$$= \frac{1}{(s^2+1)(s+1)(s+2)}$$

$$= \frac{\frac{1}{2}}{s+1} - \frac{\frac{1}{5}}{s+2} - \frac{1}{10}\frac{3s-1}{s^2+1} .$$

Inverting the Laplace transform then gives

$$y_\delta(t) = \left[\frac{1}{2}e^{-t} - \frac{1}{5}e^{-2t} - \frac{3}{10}\cos(t) + \frac{1}{10}\sin(t)\right]\zeta(t) .$$

We see that the impulse response remains bounded but does not decay to zero. Therefore, this system is marginally stable.

(5) Again by inspection, we obtain the transfer function

$$H(s) = \frac{s+2}{s^2(s+1)} = \frac{2}{s^2} - \frac{1}{s} + \frac{1}{s+1} .$$

It follows that the impulse response is given by

$$y_\delta(t) = (2t - 1 + e^{-t})\zeta(t) .$$

We see that $y_\delta(t) \to \infty$ as $t \to \infty$ and so this system is unstable.

□

2.4   THE STEP RESPONSE

In the last section, we investigated the concept of the impulse response, finding it a useful device for 'characterizing' the system because it isolated the system dynamics. A second 'standard' response is the **step response**, that is, the system response from zero initial conditions, (at time $t = 0^-$), to the input

$$u(t) = \zeta(t) = \begin{cases} 0, & \text{for } t < 0 \\ 1, & \text{for } t \geq 0, \end{cases}$$

where $\zeta(t)$ is the Heaviside unit step function (see Figure 2.3). Conceptually, the step function is easier than the impulse 'function' and approximations to step inputs can be made in the laboratory.

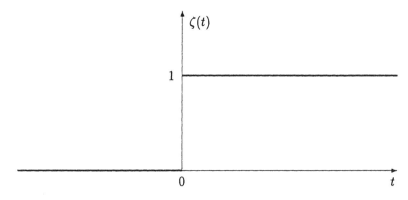

Figure 2.3: The step function $\zeta(t)$.

Example 2.3
Calculate the step response (see Figure 2.4) for the system

$$\frac{d^2 y}{dt^2}(t) + \frac{3}{4}\frac{dy}{dt}(t) + \frac{1}{8}y(t) = u(t) .$$

If we denote the Laplace transform of the step response by $Y_\zeta(s)$, then we see that

$$\left(s^2 + \frac{3}{4}s + \frac{1}{8}\right) Y_\zeta(s) = \mathcal{L}\{\zeta(t)\} = \frac{1}{s} ,$$

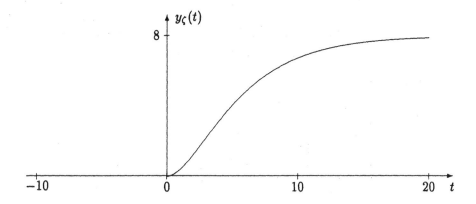

Figure 2.4: Step response for system of Example 2.3.

that is

$$Y_\zeta(s) = \frac{1}{s(s^2 + \frac{3}{4}s + \frac{1}{8})} = \frac{1}{s(s + \frac{1}{2})(s + \frac{1}{4})}$$

$$= \frac{8}{s} + \frac{8}{s + \frac{1}{2}} - \frac{16}{s + \frac{1}{4}} .$$

Thus, taking the step response as being causal, we obtain
$y_\zeta(t) = 8(1 + e^{-t/2} - 2e^{-t/4})\zeta(t)$.

□

Let us note that for a general system with transfer function, $H(s)$, the transform of the response, $Y_\zeta(s)$, to a unit step input $u(t) = \zeta(t)$, assuming the system is initially quiescent, is given by

$$Y_\zeta(s) = H(s)U(s) = \frac{H(s)}{s} .$$

It therefore follows that the step response is given by

$$y_\zeta(t) = \mathcal{L}^{-1}\left\{\frac{H(s)}{s}\right\} . \qquad (2.12)$$

We recall that the impulse response, $y_\delta(t)$, is given by $y_\delta(t) = \mathcal{L}^{-1}\{H(s)\}$ and so it follows that

$$Y_\delta(s) = H(s) = sY_\zeta(s) .$$

This means that *the impulse response is equal to the derivative of the step response.* This result also follows using the fact that the impulse 'function' is the derivative of the step function together with the linearity of differentiation. Thus, for Example 2.3, we can calculate the impulse response as

$$y_\delta(t) = \frac{d}{dt} y_\zeta(t)$$

$$= 8(-\tfrac{1}{2}e^{-t/2} + \tfrac{1}{2}e^{-t/4})\zeta(t) + 8(1 + e^{-t/2} - 2e^{-t/4})\delta(t)$$
$$= 4(e^{-t/4} - e^{-t/2})\zeta(t) ,$$

which can be checked by direct calculation.

Example 2.4
Determine the step and impulse responses for the system

$$\frac{dy}{dt}(t) + y(t) = au(t) + \frac{du}{dt}(t) .$$

where a is a constant.

For the step response, we have

$$Y_\zeta(s) = \frac{H(s)}{s} = \frac{s+a}{s(s+1)} = \frac{a}{s} + \frac{1-a}{s+1} .$$

Thus $y_\zeta(t) = (a + (1 - a)e^{-t})\zeta(t)$. Let us note that this formula gives $y_\zeta(0) = 1$. This illustrates the point that zero initial conditions refer to time $t = 0^-$ and not $t = 0$. For this particular system, the impulse on the right hand side of the differential equation, obtained by differentiating the input $u(t) = \zeta(t)$, has caused the response, $y_\zeta(t)$, to jump from 0 to 1 at time $t = 0$.

The impulse response may then be found, by differentiation, as

$$y_\delta(t) = (a - 1)e^{-t}\zeta(t) + (a + (1 - a)e^{-t})\delta(t)$$
$$= (a - 1)e^{-t}\zeta(t) + \delta(t) .$$

As a check, we note that, taking the Laplace transform of $y_\delta(t)$, we obtain

$$Y_\delta(s) = \frac{a-1}{s+1} + 1 = \frac{a - 1 + s + 1}{s+1}$$
$$= \frac{s+a}{s+1} = H(s)$$

as desired. We note that the impulse response contains an impulse at $t = 0$. Figure 2.5 shows a simulation diagram for this system. From this diagram, we see that the impulse applied at $t = 0$ has a direct path to the output $y(t)$ which explains the presence of an impulse in $y_\delta(t)$.

□

2.5　SIGNAL DECOMPOSITION AND CONVOLUTION

This section focuses on a concept which, in conjunction with the ideas on impulse response developed earlier, leads to a succinct picture of signal processing for a linear time-invariant system.

Let us first recall that the 'sifting property' of the unit impulse, $\delta(t)$, states that for any continuous signal, $u(t)$, we have

$$u(t) = \int_{-\infty}^{\infty} u(\tau)\delta(t - \tau)\,d\tau . \tag{2.13}$$

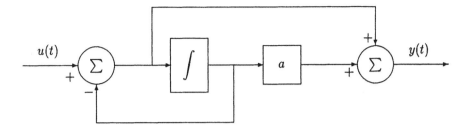

Figure 2.5: Simulation diagram for system of Example 2.4.

Since the product $u(\tau)\delta(t-\tau)$ represents an impulse at time $t = \tau$ weighted by a factor of $u(\tau)$, we interpret (2.13) as the superposition of a continuum of appropriately weighted impulse functions. In other words, $u(t)$ can be *decomposed* as a superposition of impulses. We shall see that this is a key idea in the understanding of system responses.

Let us begin our discussion by approaching (2.13) via approximating $u(t)$ as a piecewise constant ('staircase') signal as depicted in Figure 2.6.

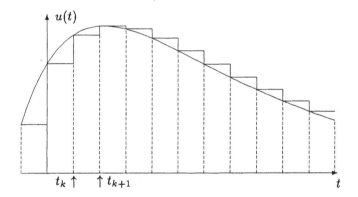

Figure 2.6: Rectangular pulse approximation to the signal $u(t)$.

Now, since

$$\zeta(t - t_k) = \begin{cases} 0, & \text{for } t < t_k \\ 1, & \text{for } t \geq t_k \end{cases}$$

and

$$\zeta(t - t_{k+1}) = \begin{cases} 0, & \text{for } t < t_{k+1} \\ 1, & \text{for } t \geq t_{k+1} \end{cases}$$

we may write for $t_k \leq t < t_{k+1}$

$$u(t) \approx u(t_k) \left[\zeta(t - t_k) - \zeta(t - t_{k+1}) \right] . \qquad (2.14)$$

Thus our approximation to $u(t)$ can be written as the sum

$$u(t) \approx \sum_k u(t_k) \left[\zeta(t - t_k) - \zeta(t - t_{k+1}) \right] . \qquad (2.15)$$

Let us now examine the result of supplying the scaled pulse (2.14) as the input to a linear, time-invariant system. The resultant output will be

$$y(t) = u(t_k) \left[y_\zeta(t - t_k) - y_\zeta(t - t_{k+1}) \right] , \qquad (2.16)$$

where $y_\zeta(t)$ is the step response of the system.

Suppose that the step response, $y_\zeta(t)$, has a continuous derivative, $\dot{y}_\zeta(t)$, then the mean value theorem of differential calculus[1] gives

$$y(t) = u(t_k)\dot{y}_\zeta(p_k)\Delta\tau_k \qquad (2.17)$$

where $\Delta\tau_k = t_{k+1} - t_k$ and $t - t_{k+1} < p_k < t - t_k$.

In §2.4, we saw that $\dot{y}_\zeta(t)$, the derivative of the step response, is just the impulse response, $y_\delta(t)$, discussed earlier in §2.3. Thus we see that (2.17) becomes

$$y(t) = u(t_k)y_\delta(p_k)\Delta\tau_k . \qquad (2.18)$$

The response in (2.18) arises from the single pulse (2.14) as input to our system. If we now supply the approximating pulse train (2.15) to the system, linearity and time invariance give the output as

$$y(t) = \sum_k u(t_k)y_\delta(p_k)\Delta\tau_k .$$

Strictly speaking, this result only follows directly if the above sum is finite, as would be the case if $u(t_k)$ were non-zero only for a finite number of sample times, t_k. Given suitable asymptotic behaviour for $u(t)$, the result can often be extended to the case of an infinite sum.

Now taking the limit as $\Delta\tau_k \to 0$, implying that the approximating pulses become smaller and smaller in duration, and with $t_k \to \tau$, a continuous variable, we see that

$$y(t) = \lim_{\Delta\tau_k \to 0} \sum_k u(t_k)y_\delta(p_k)\Delta\tau_k = \int_{-\infty}^{\infty} u(\tau)y_\delta(t - \tau)\,\mathrm{d}\tau . \qquad (2.19)$$

In obtaining (2.19), we have used the assumed continuity of $y_\delta(t)$. This cannot be considered as being entirely rigorous since $y_\delta(t)$ will usually have a discontinuity at $t = 0$. The finer details are omitted.

In our initial introduction to the impulse function, we saw that $\delta(t)$ could, in some sense, be considered as a limit of sequences of unit pulses as the pulse-width decreased to zero. Taking the impulse as occuring at time $t = \tau$ gives

$$\delta(t - \tau) = \lim_{\Delta\tau_k \to 0} \frac{1}{\Delta\tau_k} \left[\zeta(t - t_k) - \zeta(t - t_{k+1}) \right] .$$

In this sense we see that, for $\Delta\tau_k$ sufficiently small, we can approximate the staircase approximation (2.15) to $u(t)$ by the impulse train approximation

$$u(t) \approx \sum_k \Delta\tau_k u(t_k)\delta(t - t_k) . \qquad (2.20)$$

---

[1] If f is continuous for $a \le t \le b$ and differentiable for $a < t < b$ then there exists a value p with $a < p < b$ such that $f(b) - f(a) = f'(p)(b - a)$.

The point to bear in mind here is that, although $u(t)$ does not look like a train of impulses, the response to such a train does look like the response to $u(t)$ as is evident from the above discussion.

So, to summarize, for $\Delta \tau_k$ sufficiently small we can make sense of approximating $u(t)$ by the impulse train (2.20). This gives rise to an approximation of the resultant system output as given by

$$y(t) \approx \sum_k \Delta \tau_k u(t_k) y_\delta (t - t_k) . \tag{2.21}$$

If we then let $\Delta \tau_k \to 0$ in (2.20) and (2.21), we obtain the equalities

$$u(t) = \int_{-\infty}^{\infty} u(\tau) \delta(t - \tau) \, d\tau \qquad \text{(Sifting Property)} \tag{2.22}$$

$$y(t) = \int_{-\infty}^{\infty} u(\tau) y_\delta (t - \tau) \, d\tau . \tag{2.23}$$

In general, given two time signals, $f(t)$ and $g(t)$ say, we define the **convolution** of $f(t)$ with $g(t)$, written $(f * g)(t)$, to be the time signal given by

$$(f * g)(t) = \int_{-\infty}^{\infty} f(t - \tau) g(\tau) \, d\tau. \tag{2.24}$$

We leave it as a simple exercise to show that $f * g = g * f$. As a consequence of (2.24), it follows that (2.22) and (2.23) can be rewritten in the form,

$$u(t) = (\delta * u)(t) \tag{2.25}$$

$$y(t) = (y_\delta * u)(t) . \tag{2.26}$$

We are often interested in convolving two causal functions, $f(t)$ and $g(t)$ say. In this case, we note that

$$
\begin{aligned}
(f * g)(t) &= \int_{-\infty}^{\infty} f(t - \tau) g(\tau) \, d\tau \\
&= \int_{0}^{\infty} f(t - \tau) g(\tau) \, d\tau \qquad \text{(since } g \text{ is causal)} \\
&= \int_{0}^{t} f(t - \tau) g(\tau) \, d\tau \qquad \text{(for } t \geq 0, \text{ since } f \text{ is causal).}
\end{aligned}
$$

In addition, we see that $(f * g)(t) = 0$ for $t < 0$. In conclusion, if both f and g are causal, then $f * g$ is also causal and we have

$$(f * g)(t) = \int_{0}^{t} f(t - \tau) g(\tau) \, d\tau \qquad (t \geq 0). \tag{2.27}$$

At this point, let us recall from §1.8 the important convolution property of the Laplace transform, i.e.

$$\mathcal{L}[(f * g)(t)] = F(s)G(s).$$

In particular, it is important to note that this means that

$$\mathcal{L}^{-1}[F(s)G(s)] = (f * g)(t) , \quad \text{when } t \geq 0, \tag{2.28}$$

and not $f(t)g(t)$. Equation (2.28) provides the main method for calculating convolutions of causal signals by hand. The integral definition (2.27) is, however, useful both for analytical and conceptual purposes.

Example 2.5
For the causal, time-invariant system

$$\frac{d^2 y}{dt^2}(t) + 5\frac{dy}{dt}(t) + 6y(t) = u(t) \tag{2.29}$$

calculate the impulse response and use the convolution result (2.26) to find the zero initial state response to the input

$$u(t) = e^{-t}\zeta(t).$$

The transform of the impulse response is given by

$$Y_\delta(s) = H(s)$$

where

$$H(s) = \frac{1}{s^2 + 5s + 6} = \frac{1}{(s+3)(s+2)}$$
$$= \frac{1}{s+2} - \frac{1}{s+3} .$$

Hence $y_\delta(t) = h(t) = (e^{-2t} - e^{-3t})\zeta(t)$.
Using (2.26), we see that the response to the input $u(t) = e^{-t}\zeta(t)$, for $t \geq 0$, is given by

$$y(t) = \int_0^t y_\delta(t-\tau)u(\tau)\,d\tau = \int_0^t u(t-\tau)y_\delta(\tau)\,d\tau$$
$$= \int_0^t e^{-(t-\tau)}\{e^{-2\tau} - e^{-3\tau}\}\,d\tau$$
$$= e^{-t}\int_0^t (e^{-\tau} - e^{-2\tau})\,d\tau$$
$$= e^{-t}\left[-e^{-\tau} + \frac{e^{-2\tau}}{2}\right]_0^t$$
$$= e^{-t}\left[-e^{-t} + \frac{e^{-2t}}{2} + 1 - \frac{1}{2}\right] = \frac{1}{2}(e^{-t} + e^{-3t}) - e^{-2t}.$$

Since $y(t)$ is causal, we have $y(t) = 0$ for $t < 0$. Thus, the response is

$$y(t) = (\tfrac{1}{2}e^{-t} + \tfrac{1}{2}e^{-3t} - e^{-2t})\zeta(t)$$

and this result can be checked by direct calculation. Taking Laplace transforms in (2.29) with zero initial conditions, we obtain

$$(s^2 + 5s + 6)Y(s) = U(s) = \frac{1}{s+1} \ .$$

That is,

$$Y(s) = \frac{1}{(s+1)(s+2)(s+3)}$$

$$= \frac{\frac{1}{2}}{s+1} - \frac{1}{s+2} + \frac{\frac{1}{2}}{s+3} \ ,$$

and so $y(t) = (\frac{1}{2}e^{-t} - e^{-2t} + \frac{1}{2}e^{-3t})\zeta(t)$, as before.

Clearly, *the direct calculation of the response is more efficient* and thus reinforces our point that *convolution integration is an aid to understanding and manipulation rather than a tool for solution.*

□

Example 2.6
Calculate the impulse response, and hence the step response, for the system

$$\frac{d^2y}{dt^2}(t) + 4\frac{dy}{dt}(t) + 3y(t) = 2u(t) + 3\frac{du}{dt}(t) \ .$$

The impulse response has transform

$$Y_\delta(s) = \frac{2+3s}{s^2+4s+3} = \frac{2+3s}{(s+1)(s+3)}$$

$$= \frac{\frac{7}{2}}{s+3} - \frac{\frac{1}{2}}{s+1} \ .$$

Thus, the impulse response is

$$y_\delta(t) = (\tfrac{7}{2}e^{-3t} - \tfrac{1}{2}e^{-t})\zeta(t).$$

For the step response, $y_\zeta(t)$, we set $u(t) = \zeta(t)$ and, by convolution, we see that for $t \geq 0$

$$y_\zeta(t) = \int_0^t y_\delta(t-\tau).1\,d\tau$$

$$= \int_0^t y_\delta(\tau)\,d\tau \ ,$$

by the commutativity property of convolution. That is, the step response is just the integral of the impulse response. Performing this integration, we see that

$$y_\zeta(t) = \int_0^t (\tfrac{7}{2}e^{-3\tau} - \tfrac{1}{2}e^{-\tau})\,d\tau$$

$$= [\tfrac{1}{2}e^{-\tau} - \tfrac{7}{6}e^{-3\tau}]_0^t = \tfrac{1}{2}e^{-t} - \tfrac{7}{6}e^{-3t} + \tfrac{2}{3} \qquad t \geq 0.$$

Again the response is zero for $t < 0$, and so we write

$$y_\zeta(t) = (\tfrac{1}{2}e^{-t} - \tfrac{7}{6}e^{-3t} + \tfrac{2}{3})\zeta(t).$$

This result can also be checked by direct calculation. As usual, we interpret zero initial conditions as referring to time $t = 0^-$ and interpret the derivative term, du/dt, in the distributional sense, i.e. $du/dt = d\zeta/dt = \delta(t)$. With this in mind, taking Laplace transforms in the original differential equation leads to

$$(s^2 + 4s + 3)Y_\zeta(s) = (2 + 3s)\frac{1}{s}.$$

That is

$$Y_\zeta(s) = \frac{2 + 3s}{s(s^2 + 4s + 3)}$$

$$= \frac{1}{s}\left[\frac{2 + 3s}{s^2 + 4s + 3}\right] = \frac{1}{s}Y_\delta(s).$$

We may infer at this stage that our result will be consistent with that obtained above in view of the $1/s$ factor. This is the transform domain integration operator, and confirms the general result that the step response is the time integral of the impulse response. Since

$$\frac{1}{s}\frac{2 + 3s}{s^2 + 4s + 3} = \frac{\tfrac{2}{3}}{s} - \frac{\tfrac{7}{6}}{s + 3} + \frac{\tfrac{1}{2}}{s + 1}$$

we see that $y_\zeta(t) = (\tfrac{1}{2}e^{-t} - \tfrac{7}{6}e^{-3t} + \tfrac{2}{3})\zeta(t)$, as before. Before leaving this example, let us note that differentiating the above step response gives

$$\dot{y}_\zeta(t) = (-\tfrac{1}{2}e^{-t} + \tfrac{7}{2}e^{-3t})\zeta(t) + (\tfrac{1}{2}e^{-t} - \tfrac{7}{6}e^{-3t} + \tfrac{2}{3})\delta(t)$$

$$= (\tfrac{7}{2}e^{-3t} - \tfrac{1}{2}e^{-t})\zeta(t) = y_\delta(t)$$

as obtained previously. Notice in particular that $\dot{y}_\zeta(0) = y_\delta(0) = 3$. This does not contradict the assumption of zero initial conditions, which forms part of the definition of the impulse response, since these initial conditions are associated with $t = 0^-$, rather than $t = 0$.

□

Example 2.7
Calculate the impulse response for the system

$$\frac{d^2y}{dt^2}(t) + 3\frac{dy}{dt}(t) + 2y(t) = u(t). \tag{2.30}$$

Hence, find the response to the input $u(t) = 2e^{-3t}\zeta(t)$ if $y(0^-) = 1$ and $\dot{y}(0^-) = 0$. What are $y(0)$ and $\dot{y}(0)$?

The impulse response is defined as a response from the quiescent state, with its Laplace transform $Y_\delta(s)$ given by

$$Y_\delta(s) = H(s) = \frac{1}{s^2 + 3s + 2} = \frac{1}{(s + 1)(s + 2)} = \frac{1}{s + 1} - \frac{1}{s + 2}$$

where $H(s)$ is the transfer function. Inverting, we obtain the impulse response as

$$y_\delta(t) = \mathcal{L}^{-1}\{H(s)\} = (e^{-t} - e^{-2t})\zeta(t).$$

We now wish to calculate the response to the input $u(t) = 2e^{-3t}\zeta(t)$ with the initial conditions $y(0^-) = 1$, $\dot{y}(0^-) = 0$. Taking Laplace transforms in (2.30), as usual with 0^- as the lower limit of integration, we see that

$$s^2 Y(s) - s + 3(sY(s) - 1) + 2Y(s) = U(s)$$

or

$$(s^2 + 3s + 2)Y(s) = 3 + s + U(s).$$

That is

$$\begin{aligned}
Y(s) &= (3 + s + U(s))\,\frac{1}{s^2 + 3s + 2} \\
&= (3 + s + U(s))H(s) \\
&= (3 + s + U(s))Y_\delta(s)
\end{aligned}$$

where $H(s)$ is the transfer function and $Y_\delta(s)$ is the Laplace transform of the impulse response, which are, of course, equal. Thus $Y(s) = (3 + s)Y_\delta(s) + U(s)Y_\delta(s)$, and so

$$y(t) = \left(\mathcal{L}^{-1}\{3 + s\} * y_\delta\right)(t) + (u * y_\delta)(t). \tag{2.31}$$

In (2.31), we see that the system response can be expressed as the sum of two convolutions, the first of which involves a time signal whose Laplace transform is $F(s) = s$. We argue that, since the operation of differentiation in the time domain corresponds to multiplication by s in the Laplace transform domain, then

$$\begin{aligned}
\mathcal{L}^{-1}\{s\} &= \frac{d}{dt}\mathcal{L}^{-1}\{1\} = \frac{d}{dt}\delta(t) \\
&= \dot{\delta}(t).
\end{aligned}$$

This conjecture can be rigorized using the concept of distributional derivatives. Formally, we have

$$\begin{aligned}
\mathcal{L}\{\dot{\delta}(t)\} &= \int_{0^-}^{\infty} \dot{\delta}(t)e^{-st}\,dt \\
&= e^{-st}\delta(t)\Big|_{0^-}^{\infty} + \int_{0^-}^{\infty} se^{-st}\delta(t)\,dt \\
&= s,
\end{aligned}$$

as required.

As before, let us first calculate $y(t)$ using the convolution integrals directly and then, more succinctly, using the Laplace transform. We have, for $t \geq 0$,

$$\begin{aligned}
y(t) &= ([3\delta + \dot{\delta}] * y_\delta)(t) + (u * y_\delta)(t) \\
&= \int_{0^-}^{t} 3\delta(\tau)y_\delta(t - \tau)\,d\tau + \int_{0^-}^{t} \dot{\delta}(\tau)y_\delta(t - \tau)\,d\tau + \int_{0^-}^{t} 2e^{-3\tau}y_\delta(t - \tau)\,d\tau \\
&= 3y_\delta(t) + \int_{0^-}^{t} \dot{\delta}(\tau)y_\delta(t - \tau)\,d\tau + \int_{0^-}^{t} 2e^{-3\tau}\left(e^{-(t-\tau)} - e^{-2(t-\tau)}\right)\,d\tau.
\end{aligned}$$

The second term can be (formally) integrated by parts to obtain

$$\int_{0-}^{t} \dot{\delta}(\tau) y_\delta(t-\tau) \, d\tau = y_\delta(t-\tau)\delta(\tau)|_{0-}^{t} + \int_{0-}^{t} \dot{y}_\delta(t-\tau)\delta(\tau) \, d\tau$$

$$= \dot{y}_\delta(t)$$

$$= \frac{d}{dt}\left[(e^{-t} - e^{-2t})\zeta(t)\right]$$

$$= (e^{-t} - e^{-2t})\delta(t) + (2e^{-2t} - e^{-t})\zeta(t)$$

$$= (2e^{-2t} - e^{-t})\zeta(t),$$

using the equivalence $f(t)\delta(t) = f(0)\delta(t)$.

Finally, we obtain for $t \geq 0$

$$y(t) = 3(e^{-t} - e^{-2t}) + (2e^{-2t} - e^{-t}) + \int_{0-}^{t} 2e^{-t}e^{-2\tau} - 2e^{-2t}e^{-\tau} \, d\tau$$

which simplifies to

$$y(t) = 3e^{-t} - 3e^{-2t} + e^{-3t}.$$

It should, perhaps, be pointed out that we have used the causal form for convolution in the above analysis. However, since the system is not initially quiescent, we cannot infer that $y(t)$ is causal. This does not negate the above analysis which, in effect, calculates the causal signal $y(t)\zeta(t)$, but we have no means of finding $y(t)$ for $t < 0$ without further knowledge as to how the given initial conditions arose. (Since the stated input was causal, we cannot assume the differential equation as given was valid for $t < 0$. For example, it could represent an electrical circuit, with $y(t)$ representing charge say, with the circuit being switched on at time $t = 0$.)

From the formula for $y(t)$, $t \geq 0$, we see that $y(0) = 1$. We can obtain the right hand derivative, $\dot{y}(0^+)$, as

$$\dot{y}(0^+) = \lim_{t \to 0+} -3e^{-t} + 6e^{-2t} - 3e^{-3t} = 0$$

and so, as would be expected here, both $y(t)$ and $\dot{y}(t)$ are continuous at $t = 0$.

It is obvious that a direct calculation using Laplace transforms is a more efficient solution technique. Taking Laplace transforms in (2.30), with $u(t) = e^{-3t}\zeta(t)$, we obtain

$$Y(s) = \frac{s+3}{(s+1)(s+2)} + \frac{2}{(s+1)(s+2)(s+3)}$$

$$= \frac{3}{s+1} - \frac{3}{s+2} + \frac{1}{s+3}.$$

Inverting gives $y(t) = 3e^{-t} - 3e^{-2t} + e^{-3t}$, $t \geq 0$ as before.

$\square$

2.6 FREQUENCY RESPONSE

In this section, we lay the foundations for the characterization of systems in the so-called frequency domain. In order to achieve this purpose, we examine the response of stable, linear, time-invariant systems to inputs of the form

$$u(t) = Ae^{j\,\omega t}\zeta(t) \tag{2.32}$$

where A is a (possibly) complex constant and ω is a real constant which may be positive or negative (or zero). By writing A in the complex exponential form,

$$A = |A|e^{j\,\arg A}\,,\quad\text{(see Appendix A)},$$

we see that $u(t)$ can be rewritten as

$$u(t) = |A|e^{j\,(\omega t + \arg A)}\zeta(t) \tag{2.33a}$$
$$= |A|[\cos(\omega t + \arg A) + j\,\sin(\omega t + \arg A)]\zeta(t)\,. \tag{2.33b}$$

The resulting response, $y(t)$, is then complex-valued in general and we can write

$$y(t) = y_c(t) + j\,y_s(t)$$

where $y_c(t)$ and $y_s(t)$ are real-valued and provide the system responses to the inputs, $u(t) = |A|\cos(\omega t + \arg A)$ and $u(t) = |A|\sin(\omega t + \arg A)$ respectively. Such trigonometric signals, (**sinusoids**), are fundamental to frequency-domain characterization but *the use of complex exponentials eases the mathematical analysis*. After such an analysis has been performed, the relevant sinusoidal results can always be obtained by taking the real and/or imaginary part. We say that the signal $u(t) = Ae^{j\,\omega t}$ has **amplitude** $|A|$, **frequency** ω in radians per second and **phase** $\arg A$ in radians. The frequency may be converted into Hz. (Hertz) through division by 2π. Similarly the phase may be converted into degrees through multiplication by $180/\pi$. Also the amplitude can be converted into decibels (dB) by finding $20\log_{10}|A|$.

Now suppose a stable system with transfer function

$$H(s) = \frac{b(s)}{a(s)}\,,\qquad \deg b \leq \deg a,$$

is subjected to the input, $u(t) = Ae^{j\,\omega t}\zeta(t)$ and, for the sake of simplicity, let us assume for the moment that the system is initially quiescent. It follows that the output, $y(t)$, has a Laplace transform given by

$$Y(s) = H(s)U(s) = \frac{b(s)}{a(s)}\,\frac{A}{s - j\,\omega}\,.$$

Using partial fraction techniques, we can write $Y(s)$ in the form

$$Y(s) = \frac{\beta}{s - j\,\omega} + \frac{\tilde{b}(s)}{a(s)}$$

for some appropriately chosen constant β and polynomial $\tilde{b}(s)$ with degree strictly less than the degree of $a(s)$. Since we are assuming that $H(s)$ is stable, it follows that the inverse Laplace transform of the term $\tilde{b}(s)/a(s)$ gives rise to a time signal that decays to zero as $t \to \infty$. In fact, if we had allowed non-zero initial conditions, the extra terms that would appear in $Y(s)$ would also give rise to decaying time signals. The net result of this is that the output $y(t)$ settles down towards the **steady-state signal** $y_{ss}(t)$ defined by

$$y_{ss}(t) = \mathcal{L}^{-1}\left\{\frac{\beta}{s - j\,\omega}\right\}$$
$$= \beta e^{j\,\omega t} \ .$$

Using the cover-up rule, or otherwise, it can be seen that $\beta = H(j\,\omega)A$ and so

$$y_{ss}(t) = H(j\,\omega)Ae^{j\,\omega t} = H(j\,\omega)u(t) \ . \tag{2.34}$$

The function, $H(j\,\omega)$, of frequency ω, is sometimes referred to as the system **frequency response** due to the property given in (2.34). It also goes by other names and, in particular, it is sometimes referred to as the **(Fourier) transfer function** of the system.

Using the terminology introduced above, let us note that the steady-state response, $y_{ss}(t) = H(j\,\omega)Ae^{j\,\omega t}$, has amplitude given by

$$|H(j\,\omega)A| = |H(j\,\omega)|\,|A| \ .$$

That is to say, *the amplitude of the steady-state output equals that of the input scaled by a factor,* $|H(j\,\omega)|$. The variation of this gain factor with frequency is a very important system property. It is commonly referred to as the system's **amplitude response**. Alternative names include **magnitude response** and **gain spectrum**. The word 'spectrum' essentially refers to a function of frequency, (in much the same way as we have been using the word 'signal' to refer to a function of time), which explains the latter terminology.

Similarly, we note that $y_{ss}(t)$ has phase given by

$$\arg H(j\,\omega)A = \arg H(j\,\omega) + \arg A \ .$$

This means that *the phase of the steady-state output equals that of the input with an additive phase-shift of* $\arg H(j\,\omega)$. The variation of this phase-shift with frequency is also an important system property and is commonly referred to as the system's **phase response**. An alternative name is **phase-shift spectrum** for obvious reasons.

We should also note that *the steady state output has the same frequency as the input.*

As mentioned previously, we can always obtain sinusoidal results by taking real and imaginary parts in the above. For example, by taking real parts we deduce that the steady-state response of a stable system, with transfer function $H(s)$, to an input $u(t) = K\cos(\omega t + \phi)$ is given by $y_{ss}(t) = |H(j\,\omega)|K\cos(\omega t + \phi + \arg H(j\,\omega))$. Taking imaginary parts provides the same result with cos replaced by sin. (This does not give any extra information since $\sin(\omega t + \phi) = \cos(\omega t + \theta)$ where $\theta = \phi - \pi/2$.)

At this stage, we have only analysed the steady-state response of a stable system to a pure sinusoid, i.e. a signal composed of a single frequency, ω. The full value of the frequency response as a system characteristic may only be appreciated once it is shown how a large class of practical signals may be decomposed into pure sinusoids of various frequencies. This is the topic of **Fourier analysis** which is the subject of the next chapter. As with the frequency response analysis presented above, complex exponentials will be used in place of sinusoids in order to simplify the analysis.

We conclude this section with some examples of the determination of frequency domain responses.

Example 2.8

Calculate the frequency response of the linear time-invariant system defined by

$$\frac{d^2 y}{dt^2}(t) + 2\frac{dy}{dt}(t) + 5y(t) = 5u(t) .$$

Find also the amplitude and phase responses.

We first check the location of the system poles. On taking Laplace transforms, with the usual notation, we see that

$$Y(s) = H(s)U(s)$$

where $H(s) = 5/(s^2 + 2s + 5)$.

The system poles are located at the zeros of the characteristic polynomial, $a(s)$, where

$$a(s) = s^2 + 2s + 5 .$$

Now $a(s) = 0$ implies $s = -1 \pm \sqrt{-4} = -1 \pm j2$, so that $\lambda_1 = -1 + j2$ and $\lambda_2 = -1 - j2$. Thus, both poles lie in the open left half of the complex plane and the system is stable. The frequency response is then

$$H(j\omega) = \frac{5}{(j\omega)^2 + 2(j\omega) + 5} = \frac{5}{(5 - \omega)^2 + j\,2\omega} .$$

Before proceeding, let us note that if z is a complex quantity that can be written in the form

$$z = \frac{z_1 z_2 \cdots z_m}{p_1 p_2 \cdots p_n} ,$$

where z_i and p_i are complex quantities, then we have

$$|z| = \left| \frac{z_1 z_2 \cdots z_m}{p_1 p_2 \cdots p_n} \right| = \frac{|z_1| \cdot |z_2| \cdot \cdots \cdot |z_m|}{|p_1| \cdot |p_2| \cdot \cdots \cdot |p_n|} \tag{2.35}$$

and

$$\arg z = \arg \frac{z_1 z_2 \cdots z_m}{p_1 p_2 \cdots p_n}$$
$$= \arg z_1 + \arg z_2 + \cdots + \arg z_m$$
$$- \arg p_1 - \arg p_2 - \cdots - \arg p_n . \tag{2.36}$$

It follows from (2.35) that the amplitude response is given by

$$|H(j\omega)| = \left| \frac{5}{(5 - \omega^2) + j\,2\omega} \right| = \frac{|5|}{|(5 - \omega^2) + j\,2\omega|}$$
$$= \frac{5}{\sqrt{(5 - \omega^2)^2 + 4\omega^2}} = \frac{5}{\sqrt{\omega^4 - 6\omega^2 + 25}} .$$

Similarly, it follows from (2.36) that the phase response is given by

$$\arg H(j\,\omega) = \arg\left(\frac{5}{(5-\omega^2)+j\,2\omega}\right) = \arg(5) - \arg\left((5-\omega^2)+j\,2\omega\right)$$
$$= 0 - \arg\left((5-\omega^2)+j\,2\omega\right) .$$

We can note that $-\arg\left((5-\omega^2)+j\,2\omega\right) = \arg\left((5-\omega^2)-j\,2\omega\right)$ which can be found using a 4-quadrant arctangent function. In terms of the usual arctangent, we have

$$\arg H(j\,\omega) = \begin{cases} -\tan^{-1}\left(\dfrac{2\omega}{5-\omega^2}\right) & \text{for } \omega^2 < 5 \\[2mm] -\dfrac{\pi}{2} & \text{for } \omega = \sqrt{5} \\[2mm] \dfrac{\pi}{2} & \text{for } \omega = -\sqrt{5} \\[2mm] \tan^{-1}\left(\dfrac{2\omega}{\omega^2-5}\right) - \pi & \text{for } \omega > \sqrt{5} \\[2mm] \pi - \tan^{-1}\left(\dfrac{2\omega}{5-\omega^2}\right) & \text{for } \omega < -\sqrt{5}. \end{cases}$$

It should be stressed that this rather complicated looking formula is only due to the fact that we have not used a 4-quadrant arctangent. This latter function is often available in software packages (and languages) and can be obtained on calculators that have the facility to convert rectangular coordinates to polar coordinates.

Figure 2.7 shows plots of the amplitude and phase responses. In particular, we can see from the amplitude response that at one extreme, signals of low frequency, (small values of ω), are passed with little change in amplitude whereas at the other extreme, high-frequency (large values of ω) signals are virtually eliminated.

□

Example 2.9
Find the steady-state response of the system

$$\frac{d^2y}{dt^2}(t) + 2\frac{dy}{dt}(t) + 5y(t) = 5\left(u(t) + \frac{du}{dt}(t)\right)$$

to the inputs $u_1(t) = 10\cos(t)\zeta(t)$ and $u_2(t) = 10\cos(10t)\zeta(t)$.

The system transfer function is

$$H(s) = \frac{5(s+1)}{s^2 + 2s + 5}$$

and, since the pole locations are as for Example 2.8, the system is stable. Thus, there is a frequency response given by

$$H(j\,\omega) = \frac{5(1+j\,\omega)}{(5-\omega^2)+j\,2\omega} .$$

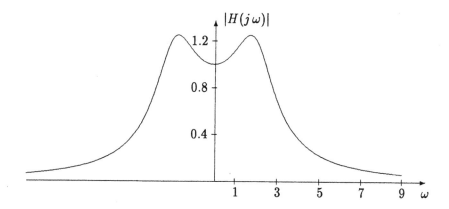

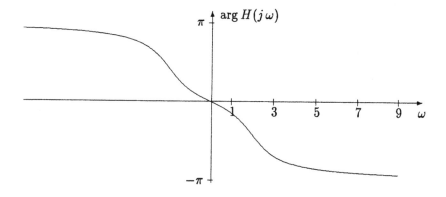

Figure 2.7: Amplitude and phase responses for system of Example 2.8.

The signal $u_1(t)$ has a frequency of $\omega = 1$ radian/sec. It follows that the steady-state response is given by

$$y_{ss}(t) = 10|H(j)| \cos(t + \arg H(j)) .$$

Now, the amplitude response is given by

$$|H(j\omega)| = \frac{5|1 + j\omega|}{|(5 - \omega^2) + j\, 2\omega|} = \frac{5\sqrt{1 + \omega^2}}{\sqrt{\omega^4 - 6\omega^2 + 25}}$$

and the phase response is given by

$$\arg H(j\omega) = \arg(1 + j\omega) + \arg(5 - \omega^2 - j\, 2\omega)$$
$$= \tan^{-1}(\omega) + \theta ,$$

where θ is the same as the phase response of the system given in Example 2.8. We therefore deduce that the system gain at the frequency $\omega = 1$ is given by $|H(j)| = 5\sqrt{2}/\sqrt{20} = \sqrt{(5/2)}$ and the corresponding phase-shift is $\arg H(j) = \tan^{-1}(1) - \tan^{-1}(1/2) \approx 0.3218$. Thus, with input $u_1(t)$,

$$y_{ss}(t) \approx 10\sqrt{\frac{5}{2}} \cos(t + 0.3218) \approx 15.8114 \cos(t + 0.3218) .$$

On the other hand, with input $u_2(t)$, we obtain

$$y_{ss}(t) = 10|H(j\,10)| \cos(10t + \arg H(j\,10)),$$

where

$$|H(j\,10)| = \frac{5\sqrt{101}}{\sqrt{9425}} \approx 0.5176$$

and

$$\arg H(j\,10) = \tan^{-1}(10) + \tan^{-1}\left(\frac{20}{95}\right) - \pi$$
$$\approx -1.4630 .$$

Thus, with input $u_2(t)$,

$$y_{ss}(t) \approx 10 \times 0.5176 \times \cos(10t - 1.4630)$$
$$\approx 5.176 \cos(10t - 1.463) .$$

It is now possible to compare how the two input signals are processed by the system. Both inputs are subject to substantial modification, both in amplitude and phase. In particular, although both $u_1(t)$ and $u_2(t)$ have the same amplitudes, the amplitude of the steady-state response in the second case is only about 33% of that of the first case.

Since the system under consideration is linear, the steady-state response that would result from the input $u(t) = u_1(t) + u_2(t)$ may be found by summing the two individual responses.

□

In this last section, we developed a characterization of a linear system in terms of its frequency domain behaviour. At the end of the above example, we saw how this theory can by applied to an input given by the sum of two sinusoids. This theory can be generalized still further by decomposing arbitrary signals, (subject to some mild restrictions), in terms of sinusoids, or equivalently, in terms of complex exponentials. This frequency domain decomposition of signals is the subject of the next chapter.

2.7   EXERCISES

1. Find the Laplace transfer function for the following systems, and hence the impulse response in each case. Draw a simulation diagram for each system.

   (a) $\dfrac{d^2y}{dt^2}(t) + 5\dfrac{dy}{dt}(t) + 4y(t) = u(t)$.

   (b) $\dfrac{d^2y}{dt^2}(t) + 2\dfrac{dy}{dt}(t) + 2y(t) = u(t)$.

   (c) $8\dfrac{d^2y}{dt^2}(t) + 6\dfrac{dy}{dt}(t) + y(t) = u(t)$.

   (d) $3\dfrac{d^2y}{dt^2}(t) + 2\dfrac{dy}{dt}(t) + 2y(t) = 2u(t) + \dfrac{du}{dt}(t)$.

2. Find the response, $y(t)$, $t \geq 0$, of the system

$$2\frac{d^2y}{dt^2}(t) + 3\frac{dy}{dt}(t) + y(t) = u(t)$$

   to an input, $u(t) = e^{-at}$, $(a > 0)$, assuming zero initial conditions, i.e. $y(0) = \dfrac{dy}{dt}(0) = 0$. Use both the direct, (Laplace transform), method and then use the convolution integral after first calculating the impulse response.

3. Characterize the stability of the systems whose characteristic polynomials are given by:
   (a) $s^2 + 4s + 5$;
   (b) $s^2 + 2s + 3$;
   (c) $4s^2 - s + 1$;
   (d) $s^3 + s^2 + s + 1$;
   (e) $s^4 + 2s^2 + 1$.

4. Calculate the step response of the system

$$4\frac{d^2y}{dt^2}(t) + 4\frac{dy}{dt}(t) + 5y(t) = u(t) \ .$$

What is the impulse response?

5. Determine the amplitude and phase responses of the system

$$\frac{d^3y}{dt^3}(t) + 2\frac{d^2y}{dt^2}(t) + 2\frac{dy}{dt}(t) + y(t) = \frac{du}{dt}(t) \ .$$

Illustrate with a diagram, perhaps obtained using a computer.

6. Find the response of the system

$$\frac{d^2y}{dt^2}(t) + 2\frac{dy}{dt}(t) + 5y(t) = u(t) \qquad (t \geq 0),$$

if $u(t) = e^{-t}\zeta(t)$ and $y(0^-) = 0$, $\dot{y}(0^-) = 1$.

7. Where applicable, find the steady-state response to the input $u(t) = \cos(2t)$, for the systems with the following Laplace transfer functions:

(a) $H(s) = \dfrac{2}{s+2}$;

(b) $H(s) = \dfrac{s-2}{s^2 + 3s + 2}$;

(c) $H(s) = \dfrac{5}{s^3 + s + 10}$;

(d) $H(s) = \dfrac{(s-2)^2}{(s+2)^3}$.

3

Fourier methods

3.1 INTRODUCTION.

In the last chapter, we examined various time-domain system responses. We saw that considerable insight could be gained using the concept of the system impulse response, because signals can be decomposed into a superposition of impulses. Later, we turned our attention to the frequency domain and we found that a so-called frequency response function could be associated with stable, causal, linear, time-invariant systems. This function enabled us to predict the steady-state output of the system when the input signal was a sine or cosine wave at frequency ω. This was achieved using the complex exponential representation form rather than the trigonometric functions, sine and cosine.

The principle of superposition discussed in Chapter 1 allows us to combine together such signals of different frequencies and to again predict the steady-state response. In this chapter we examine a frequency-domain decomposition of signals which permits us to make full use of the frequency response concept in understanding the frequency-domain view of signal processing.

3.2 FOURIER SERIES

We begin our treatment with a consideration of **periodic functions**. Our aim in this section is to achieve the decomposition of periodic signals into a form suitable for use with a frequency response function in determining the steady-state response of a system to a periodic signal as input.

Suppose that $f(t)$ is a periodic function, or signal, of period $P = 2A$. Such a signal is illustrated in Figure 3.1.

Clearly, we can define this signal by the two relations

$$f(t) = \begin{cases} 1, & -A < t < 0 \\ 0, & 0 \leq t \leq A \end{cases}$$

and

$$f(t + P) = f(t),$$

where $P = 2A$ is the period.

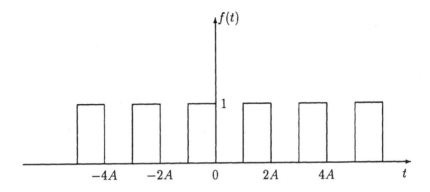

Figure 3.1: A periodic pulse of period $2A$.

In some applications, notably the solution of partial differential equations, it is necessary to seek *a representation of such a function in terms of a combination of sine and cosine functions of different frequencies ω.* Our requirements are slightly different, and we obtain an expansion or representation using the complex exponential function $e^{j\omega t}$ instead.

The detailed question of when such a representation exists is beyond the scope of this text. However, when the **Dirichlet conditions** are satisfied, we can be certain that the periodic function, or signal, $f(t)$ of period P, does possess a **Fourier expansion**. The Dirichlet conditions can be stated as follows.

1. *Within each period, $f(t)$ is bounded, so that $f(t)$ is absolutely integrable on $[0, P]$.*

2. *Within each period, there are at most a finite number of maxima and minima.*

3. *Within each period, there are at most a finite number of finite discontinuities.*

If these conditions, which are 'sufficient' conditions, are satisfied, then $f(t)$ has a Fourier expansion which can be expressed in the form,

$$f(t) \sim \sum_{n=-\infty}^{\infty} F_n e^{j\,n\omega_0 t} , \qquad (3.1)$$

where $\omega_0 = 2\pi/P$ and is referred to as the **fundamental frequency**. Here ω_0 is in radians/sec. Equivalently, we have $\omega_0 = 2\pi f_0$, where $f_0 = 1/P$ is the fundamental frequency in Hertz (Hz).

We use the '$\sim$' notation to denote 'is represented by', and (3.1) means that the signal $f(t)$ is represented by its Fourier series expansion. This is an important idea, since it is not true that at every value of t, the left-hand side of the expression always takes the same value as the right-hand side. In particular, at a point of discontinuity, the Fourier expansion converges to the mean of the left and right hand limits of $f(t)$, irrespective of any definition of the value of $f(t)$ actually at the point.

The process of determining the Fourier expansion is now reduced to the determination of the coefficients F_n for all integer values of n. In fact this task is not particularly onerous, and is possible in view of the **orthogonality** relationship:

$$\frac{1}{P}\int_{t_0}^{t_0+P} e^{j\,n\omega_0 t}e^{-j\,m\omega_0 t}\,dt = \delta_{m,n} \stackrel{\text{def}}{=} \left\{ \begin{array}{ll} 0, & m \neq n \\ 1, & m = n. \end{array} \right. \qquad (3.2)$$

The reader is invited to verify this result in the exercises at the end of this chapter. The technique for determining the expansion now involves multiplying (3.1) by

$$\frac{1}{P}e^{-j\,m\omega_0 t},$$

where m is an integer, and integrating over one full period. If we require that the same result be obtained for $f(t)$ or its Fourier representation, we must have:-

$$\frac{1}{P}\int_{t_0}^{t_0+P} f(t)e^{-j\,m\omega_0 t}dt = \frac{1}{P}\int_{t_0}^{t_0+P}\left[\sum_{n=-\infty}^{\infty} F_n e^{j\,n\omega_0 t}\right]e^{-j\,m\omega_0 t}dt$$

$$= \frac{1}{P}\sum_{n=-\infty}^{\infty} F_n \int_{t_0}^{t_0+P} e^{j\,n\omega_0 t}e^{-j\,m\omega_0 t}dt$$

$$= \sum_{n=-\infty}^{\infty} F_n \delta_{m,n} \qquad \text{(using (3.2))}$$

$$= F_m.$$

We have assumed that interchanging the order of integration and summation is permissable in obtaining our result. We have thus established a formula for the generation of the coefficients in the Fourier expansion of the periodic function $f(t)$ as

$$F_n = \frac{1}{P}\int_{t_0}^{t_0+P} f(t)e^{-j\,n\omega_0 t}dt\;, \qquad (3.3)$$

where $\omega_0 = 2\pi/P$. Note that the value of F_n is independent of the choice for t_0. The values for t_0 most commonly used in examples are $t_0 = -P/2$ and $t_0 = 0$, whichever is the most convenient for performing the integration. The method is demonstrated in the following example.

Example 3.1
Calculate the coefficients of the Fourier series which represents the periodic function of Figure 3.1.

The expansion coefficients are given by (3.3) as

$$F_n = \frac{1}{P}\int_{t_0}^{t_0+P} f(t)e^{-j\,n\omega_0 t}dt,$$

where $\omega_0 = \dfrac{2\pi}{P} = \dfrac{\pi}{A}$. Thus, choosing $t_0 = -P/2 = -A$, we obtain

$$F_n = \frac{1}{2A} \int_{-A}^{0} e^{-j\,n\omega_0 t}\, dt$$

$$= \frac{1}{-2j\,n\pi} \left[1 - e^{j\,n\pi}\right]$$

for $n \neq 0$, where we have used $\omega_0 A = \pi$. Noting that $e^{j\,n\pi} = (e^{j\,\pi})^n = (-1)^n$, we have

$$F_n = \frac{-1}{2j\,n\pi} \left(1 - (-1)^n\right) \qquad \text{for } n \neq 0$$

$$= \begin{cases} 0, & \text{for } n \text{ even, } n \neq 0 \\[2mm] \dfrac{j}{n\pi}, & \text{for } n \text{ odd.} \end{cases} \tag{3.4}$$

When $n = 0$, we find that

$$F_0 = \frac{1}{2A} \int_{-A}^{0} dt = \frac{1}{2}. \tag{3.5}$$

Thus, the required coefficients are given by (3.4) and (3.5), and it is now possible to write down the Fourier series which represents $f(t)$ as

$$f(t) \sim \frac{1}{2} + \frac{j}{\pi} \sum_{\substack{n=-\infty \\ n \text{ odd}}}^{\infty} \frac{e^{j\,n\pi t/A}}{n}\ .$$

The original signal $f(t)$ is a real signal and this means that the Fourier series is in fact real, despite appearances to the contrary. The interested reader may wish to establish that this is the case.

□

At this point, it may be worth a short digression in the cause of better understanding. The process we have just gone through in Example 3.1 is exactly analogous to the process of computing the components of a vector in a vector space. Suppose that we have a vector $\mathbf{r}$ in a three dimensional Euclidean space. We are familiar with the task of computing the components of such a vector relative to the orthogonal base vectors $\mathbf{i}$, $\mathbf{j}$, and $\mathbf{k}$. These components are actually obtained by taking the 'dot', or scalar, product with each base vector in turn, that is we form

$$r_x = \mathbf{r}.\mathbf{i}, \ r_y = \mathbf{r}.\mathbf{j}, \text{ and } r_z = \mathbf{r}.\mathbf{k}.$$

The process of decomposition of a signal using the techniques of Fourier analysis can be thought of in the same way. In effect, we are treating a signal as *a vector in an infinite dimensional vector space*, and computing the components of the signal in the 'harmonic directions' defined by the basis vectors $e^{j\,n\omega_0 t}$. The scalar, or inner, product in this vector space is now

$$\frac{1}{P} \int_0^P f(t)\,\overline{g(t)}\,\mathrm{d}t,$$

where $f(t)$ and $g(t)$ are signals or vectors in the signal space, and $1/P$ is a normalizing factor, chosen so that the inner product of a basis vector with itself is unity. (The 'bar' over the signal $g(t)$ in the above integral refers to complex conjugation.) This structure is called a scalar product because the result of the calculation is a scalar, or number, exactly as we obtain from the 'dot' product of two vectors in Euclidean space.

Our previous work on frequency response can now be seen as an investigation of how a linear system processes the basis vectors of our signal space. Intuitively, we can appreciate that if this is known, then it is reasonable to suppose that we can predict the response to any signal vector in the space.

Having found the Fourier coefficients of a periodic signal $f(t)$, we now interpret their meaning. Since $e^{j\,n\omega_0 t} = \cos(n\omega_0 t) + j\,\sin(n\omega_0 t)$, where n is an integer, our decomposition has yielded a prescription for constructing a periodic signal $f(t)$ from the elementary harmonic functions $\cos(n\omega_0 t)$ and $\sin(n\omega_0 t)$. We have chosen to express this prescription using the complex exponential representation, and this choice has two implications. First, the Fourier coefficients are, in general, complex numbers, and secondly, we have to use both positive and negative values of the index n. It is important to appreciate that both these implications are direct consequences of our choice of basis functions used to span the signal space. From another point of view, had we chosen to use as basis functions, the real set $\{\cos(n\omega_0 t)\,(n \geq 0),\ \sin(n\omega_0 t)\,(n > 0)\}$, then real signals $f(t)$ could have been 'resolved' into real components relative to these real 'harmonic directions'. *Our choice of a set of complex basis vectors is made in order to simplify later manipulation.*

Let us now consider in more detail the Fourier coefficients relative to our chosen base set $\{e^{j\,n\omega_0 t}\}$, n an integer. First, notice that if $f(t)$ is periodic of period P, then it has frequency components at the frequencies $n\omega_0$, $n = 0, \pm 1, \pm 2, \pm 3 \ldots$, and $\omega_0 = 2\pi/P$. This is a discrete set of components located at multiples of the 'fundamental' frequency ω_0. Each component F_n is, in general, a complex number, so that $F_n = r_n e^{j\,\phi_n}$, where $r_n = |F_n|$ and $\phi_n = \arg F_n$. Thus, each term in the Fourier series takes the form

$$
\begin{aligned}
F_n e^{j\,n\omega_0 t} &= \left(r_n e^{j\,\phi_n}\right) e^{j\,n\omega_0 t} \\
&= r_n e^{j\,(n\omega_0 t + \phi_n)} \\
&= |F_n| e^{j\,(n\omega_0 t + \arg F_n)} .
\end{aligned}
$$

We see that each coefficient of the Fourier series provides the required information on the amplitude, $|F_n|$, and the phase, $\arg F_n$, of the corresponding harmonic. This information is displayed graphically using two diagrams. The first is called the signal **amplitude spectrum**, and is a specification of $|F_n|$ as a function of n, or $n\omega_0$. The second diagram is called the **phase spectrum**, and specifies $\arg F_n$ again as a function of n or $n\omega_0$.

We now consider these spectra for the signal $f(t)$ of Example 3.1. From (3.4) and (3.5), we have that

$$|F_n| = \begin{cases} 0, & \text{for } n \text{ even, } n \neq 0, \\ \dfrac{1}{|n|\pi}, & \text{for } n \text{ odd,} \\ \dfrac{1}{2}, & \text{for } n = 0, \end{cases}$$

and

$$\arg(F_n) = \begin{cases} \text{undefined}, & \text{for } n \text{ even, } n \neq 0, \\ \dfrac{\pi}{2}, & \text{for } n \text{ odd, } n > 0, \\ -\dfrac{\pi}{2}, & \text{for } n \text{ odd, } n < 0, \\ 0, & \text{for } n = 0. \end{cases}$$

These spectra are plotted in Figure 3.2.

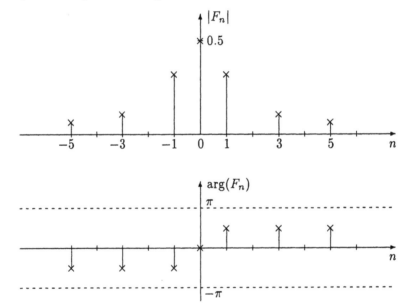

Figure 3.2: Amplitude and phase spectra for the signal of Example 3.1.

It is interesting to observe the effect of time shifting on the spectrum of periodic signals. Suppose that $f(t)$ is periodic of period P and has Fourier coefficients F_n, say, where $n = 0, \pm1, \pm2, \ldots$. If we calculate the coefficients G_n of the time shifted version, $f(t + \tau)$, we find that

$$G_n = \frac{1}{P} \int_0^P f(t + \tau) e^{-j\,n\omega_o t}\, dt$$

$$= \frac{1}{P} \int_\tau^{\tau+P} f(u) e^{-j\, n\omega_0 (u-\tau)} \, du$$

$$= \frac{e^{j\, n\omega_0 \tau}}{P} \int_\tau^{\tau+P} f(u) e^{-j\, n\omega_0 u} \, du$$

$$= e^{j\, n\omega_0 \tau} F_n,$$

using (3.3) with $t_0 = \tau$. We see that, since $|e^{j\, n\omega_0 \tau}| = 1$ we have $|G_n| = |F_n|$ for all n. This means that *the time shift has no effect at all on the amplitude spectrum.* However, $\arg G_n = \arg F_n + n\omega_0 \tau$, showing a phase shift proportional to the frequency $n\omega_0$ for a given time shift τ. (Note: this relation needs an obvious modification if we require that arguments lie within some specified range, such as $(-\pi,\, \pi]$.)

3.3 THE FOURIER TRANSFORM

Having established a technique for the representation of periodic signals in the frequency domain, we must admit that not many signals of interest are likely to be periodic! It must now be our task to attempt to *develop a method which achieves a similar representation or decomposition, for non-periodic signals*, defined on $-\infty < t < \infty$. Figure 3.3 illustrates such a signal, $f(t)$, and shows a portion of the signal as visible through a **time window** of duration T', placed symmetrically about the origin.

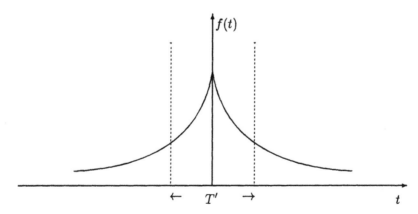

Figure 3.3: A non-periodic signal and the view through the window of duration T'.

Concentrating only on that portion of signal 'visible' through the window, we could pretend that outside the window our signal consisted of periodic repeats of the portion we observe through the window. This would make no difference at all to the portion of signal that we can 'see', that is, the segment in the window alone. We could then go on to perform a Fourier analysis of the periodic signal in Figure 3.4 using the methods of section 3.2. The clever part of the operation is then to investigate the behaviour of the Fourier expansion as the duration of the window,

T', increases without bound. Loosely speaking, we are adopting the picture that non-periodic signals are actually periodic signals, but that their period is infinite in duration!

To explore these ideas, set

$$g(t) = \begin{cases} f(t), & |t| < T'/2 \\ g(t - nT'), & |t| \geq T'/2, \end{cases}$$

where n is an integer. This defines $g(t)$ as the periodic extension of that portion of $f(t)$, visible through our window, and is illustrated in Figure 3.4.

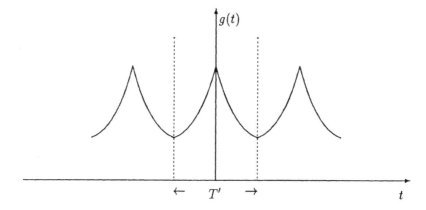

Figure 3.4: The periodic function $g(t)$, the periodic extension of the view through the window.

Clearly, when $|t| < T'/2$, the graphs of $f(t)$ and $g(t)$ are identical, but, since $g(t)$ is periodic of period T', we can use the methods of §3.2 to write

$$g(t) \sim \sum_{n=-\infty}^{\infty} G_n e^{j\,n\omega_0 t} \tag{3.6}$$

with $\omega_0 = 2\pi/T'$ and

$$G_n = \frac{1}{T'} \int_{-T'/2}^{T'/2} g(\tau) e^{-j\,n\omega_0\tau}\, d\tau \;. \tag{3.7}$$

We proceed by substituting (3.7) into (3.6) to obtain

$$g(t) \sim \sum_{n=-\infty}^{\infty} \left[\frac{1}{T'} \int_{-T'/2}^{T'/2} g(\tau) e^{-j\,n\omega_0\tau}\, d\tau \right] e^{j\,n\omega_0 t}. \tag{3.8}$$

We now ask what happens as $T' \to \infty$. Obviously our window widens, so that eventualy $g(t)$ agrees with $f(t)$ for all values of t. Also, the summation in (3.8) becomes an integral, a point which becomes apparent if we write (3.8) in the form

$$g(t) \sim \sum_{n=-\infty}^{\infty} \frac{\omega_0}{2\pi} e^{j\,n\omega_0 t} \int_{-T'/2}^{T'/2} g(\tau) e^{-j\,n\omega_0\tau}\, d\tau. \tag{3.9}$$

Each time the summation index n changes by 1, $n\omega_0$ changes by an amount ω_0. However, since $\omega_0 = 2\pi/T' \to 0$ as $T' \to \infty$, we are lead to write $n\omega_0 = \omega$, a continuous variable, with $\omega_0 = d\omega$. Note that for non-periodic signals, the separation between harmonic components in the frequency domain goes to zero as $T' \to \infty$. Thus, summation over n, corresponding to the process of summation over distinct frequency components at separation ω_0, is replaced by integration over the continuous frequency variable ω. Recall that if $T' \to \infty$, then $g(t) = f(t)$ everywhere and we can write (3.9) in the form

$$f(t) \sim \int_{-\infty}^{\infty} \frac{d\omega}{2\pi} e^{j\,\omega t} \int_{-\infty}^{\infty} f(\tau) e^{-j\,\omega\tau}\,d\tau \ . \tag{3.10}$$

This is known as the Fourier Integral representation of $f(t)$, and if we define

$$F(j\,\omega) = \int_{-\infty}^{\infty} f(t)\, e^{-j\,\omega t}\,dt \tag{3.11}$$

then (3.10) can be written as

$$f(t) \sim \frac{1}{2\pi} \int_{-\infty}^{\infty} F(j\,\omega)\, e^{j\,\omega t}\,d\omega. \tag{3.12}$$

Here $F(j\,\omega)$ is called the **Fourier Transform** of the time signal $f(t)$. Whenever the defining integral exists, $F(j\,\omega)$ plays the rôle of the Fourier coefficients of §3.2 and is our desired frequency domain representation of $f(t)$. *We note that non-periodic signals have frequency components at all values of the continuous frequency variable* ω. This is in contrast to the situation we observed for signals of finite period P, for which the frequency components occurred only at distinct values $n\omega_0$ of the frequency variable.

We note the connection between $F(j\,\omega)$, as given by (3.11), and the bilateral Laplace transform $F(s)$ evaluated at $s = j\,\omega$. We will see that we can sometimes still make sense of the Fourier transform of a signal $f(t)$ even if the defining integral does not exist. In such situations, $F(j\,\omega)$ is not given by setting $s = j\,\omega$ in the bilateral Laplace transform $F(s)$. For example, if $f(t) = \zeta(t)$, the Heaviside unit step function, then $F(s) = 1/s$ but $F(j\,\omega) \neq 1/j\,\omega$. In fact, as seen in the exercises at the end of this chapter, we have $F(j\,\omega) = \dfrac{1}{j\,\omega} + \pi\delta(\omega)$.

Equation (3.12) shows how to construct a *representation* of the time signal $f(t)$ from a knowledge of its Fourier Transform $F(j\,\omega)$. By retaining the '$\sim$' notation, we stress that this is a representation of $f(t)$, noting in particular the behaviour of this representation at any points of discontinuity. We recall from §3.2 that the Fourier representation converges to the mean of left and right hand limits of $f(t)$ at a point of discontinuity, irrespective of any definition of $f(t)$ actually at the point.

Based on the results (3.11) and (3.12), we can construct a bi-directional path between time and frequency domains. In this connection, it is useful to have a notation for the Fourier Transform operation, and we use the symbol $\mathcal{F}$ for this purpose. Thus, we may write (3.11) as

$$\mathcal{F}\{f(t)\} = F(j\,\omega) = \int_{-\infty}^{\infty} f(t)\, e^{-j\,\omega t}\,dt \tag{3.13}$$

and (3.13) defines the Fourier Transform of $f(t)$ whenever the integral exists. The path from the frequency domain to the time domain makes use of the inverse Fourier Transform, as defined by the integral on the right side of (3.12). Thus we write,

$$\mathcal{F}^{-1}\{G(j\omega)\} = g(t) = \frac{1}{2\pi} \int_{-\infty}^{\infty} G(j\omega)\, e^{j\,\omega t}\, d\omega. \tag{3.14}$$

In (3.14), we use an '=' sign, because the time domain signal $g(t)$ is defined as the result of performing the integration on the right-hand side.

Example 3.2
Find the Fourier Transform of the causal function

$$f(t) = e^{-at}\zeta(t), \quad a > 0.$$

Using (3.13), we have,

$$\mathcal{F}\{e^{-at}\zeta(t)\} = \int_{-\infty}^{\infty} e^{-at}\zeta(t)\, e^{-j\,\omega t}\, dt$$

$$= \int_{0}^{\infty} e^{-(a+j\omega)t}\, dt$$

$$= \frac{-1}{(a+j\omega)} \left[e^{-(a+j\omega)t} \right]_{0}^{\infty}$$

$$= \frac{1}{(a+j\omega)} = \frac{a - j\omega}{a^2 + \omega^2}$$

$$= F(j\omega)$$

(Note: in this example, the region of convergence for the Laplace transform includes the imaginary axis and so it is permissible to set $s = j\omega$ in the Laplace transform $F(s) = 1/(s+a)$.)

□

The Fourier Transform of Example 3.2 was calculated without difficulty. However, we note that attempts to calculate the transforms of e^{at} and e^{-at}, $a > 0$, would both fail because the defining integral would not exist in either case. In fact, we find that we are unable to obtain the transforms of many 'elementary' functions by this means. Dirichlet gave a set of conditions which are sufficient for the existence of the integral in the definition of the Fourier Transform of the signal $f(t)$. These are:-

1. $f(t)$ *must be absolutely integrable on* $(-\infty, \infty)$, *that is*

$$\int_{-\infty}^{\infty} |f(t)|\, dt \quad \text{is finite, and}$$

2. $f(t)$ *has, at most, a finite number of finite discontinuities, and a finite number of maxima and minima in any finite interval.*

These conditions are seen to be a natural extension to the conditions given earlier for the existence of a Fourier Series representation of a periodic function.

In Example 3.3 we examine a function or signal which clearly satisfies these conditions.

Example 3.3
Calculate the Fourier Transform of the rectangular pulse defined by

$$f(t) = \begin{cases} A, & |t| \le T/2 \\ 0, & |t| > T/2 \end{cases}$$

and depicted, for the case $A > 0$, in Figure 3.5.

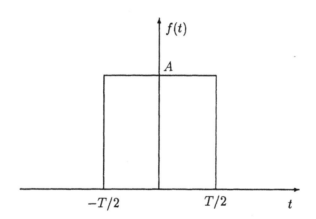

Figure 3.5: The rectangular pulse of Example 3.3.

Again, using the defining integral (3.13), we have

$$\mathcal{F}\{f(t)\} = \int_{-\infty}^{\infty} f(t)\, e^{-j\omega t}\, dt$$

$$= \int_{-T/2}^{T/2} A e^{-j\omega t}\, dt$$

$$= \frac{-A}{j\omega}\left[e^{-j\omega t}\right]_{-T/2}^{T/2} \quad \omega \ne 0$$

$$= \frac{A}{j\omega}\left[e^{j\omega T/2} - e^{-j\omega T/2}\right] \quad \omega \ne 0$$

$$= AT\frac{\sin(\omega T/2)}{\omega T/2} \quad \omega \ne 0.$$

$f(t)$	Fourier Transform $\mathcal{F}\{f(t)\}$
$e^{-at}\zeta(t),\ a > 0$	$\dfrac{1}{a + j\,\omega}$
$te^{-at}\zeta(t),\ a > 0$	$\dfrac{1}{(a + j\,\omega)^2}$
$e^{-a\lvert t\rvert},\ a > 0$	$\dfrac{2a}{a^2 + \omega^2}$
$\begin{cases} A, & \lvert t\rvert \le T/2 \\ 0, & \lvert t\rvert > T/2 \end{cases}$	$AT\mathrm{sinc}(\omega T/2)$

Table 3.1: Some elementary functions and their Fourier Transforms.

The quantity $\sin(x)/x$ occurs frequently in the analysis of signals and it is useful to define the function $\mathrm{sinc}(x)$[1] as

$$\mathrm{sinc}(x) \overset{\text{def}}{=} \begin{cases} \dfrac{\sin x}{x}, & \text{for } x \ne 0, \\[2mm] 1, & \text{for } x = 0. \end{cases}$$

The motivation for the value at $x = 0$ comes from the value of the limit of $\sin(x)/x$ as $x \to 0$. Since, when $\omega = 0$, we have

$$\mathcal{F}\{f(t)\} = \int_{-\infty}^{\infty} f(t)\,\mathrm{d}t = AT,$$

we can express the result as

$$\mathcal{F}\{f(t)\} = F(j\,\omega) = AT\mathrm{sinc}(\omega T/2)$$

for all values of ω.

$\square$

The results of Example 3.2 and Example 3.3 are contained in Table 3.1, and the reader is invited to confirm the other transforms in the exercises.

3.4   THE FOURIER SPECTRUM

In §3.2, we discussed the two spectra associated with the Fourier decomposition of periodic signals. There we saw that both the amplitude and phase spectra consisted of a discrete set of components, located at multiples of the fundamental frequency

---

[1] Many texts define $\mathrm{sinc}(x)$ by $\mathrm{sinc}(x) = \sin(\pi x)/(\pi x)$ for $x \ne 0$. Since the definition does vary, readers should beware when reading other texts, or using packages such as Matlab.

ω_0. In the case of non-periodic signals, the Fourier Transform takes the rôle of the Fourier coefficients and if $\mathcal{F}\{f(t)\} = F(j\omega)$ is the transform of a non-periodic signal $f(t)$, then $F(j\omega)$ is known as the (complex) **Fourier spectrum** of $f(t)$. Since $F(j\omega)$ is, in general, a complex valued function of the real frequency variable ω, we write

$$F(j\omega) = |F(j\omega)|e^{j\,\theta(j\omega)}.$$

Here, the two real-valued functions of ω, $|F(j\omega)|$ and $\theta(j\omega)$, are called respectively the **amplitude spectrum** and **phase spectrum**.

Example 3.4
Calculate the amplitude and phase spectra for the signal of Example 3.2.

In Example 3.2, we found that if $f(t) = e^{-at}\zeta(t)$ $(a > 0)$, then

$$F(j\omega) = \frac{1}{a + j\omega}.$$

The amplitude spectrum is then given by

$$|F(j\omega)| = \frac{1}{\sqrt{a^2 + \omega^2}}\,,$$

whereas $\theta(j\omega)$, the phase spectrum, is defined by

$$\theta(j\omega) = -\tan^{-1}(\omega/a)\,.$$

These spectra are illustrated in Figure 3.6.

$\square$

The illustrations in Figure 3.6 serve to stress that the amplitude and phase spectra are defined for all values of the continuous variable ω.

Sometimes, as in Example 3.3, we find that the Fourier Transform is a purely real quantity. In Example 3.3, we saw that

$$\mathcal{F}\{f(t)\} = F(j\omega) = AT\mathrm{sinc}(\omega T/2).$$

Here, $|F(j\omega)| = AT|\mathrm{sinc}(\omega T/2)|$, and

$$\theta(j\omega) = \begin{cases} 0, & \mathrm{sinc}(\omega T/2) \geq 0 \\ \pi, & \mathrm{sinc}(\omega T/2) < 0. \end{cases}$$

(Note: strictly speaking $\theta(j\omega)$ is not well-defined when $\mathrm{sinc}(\omega T/2) = 0$. We have arbitrarily set it to zero in the above.) Clearly, if we plotted the graph of the real valued function $F(j\omega) = AT\mathrm{sinc}(\omega T/2)$, then we could convey all the information in a single graph. We do not make use of this representation, although it is in common use elsewhere.

We can make two observations based on our discussion of the Fourier Transform so far. First note that in the examples so far considered, the amplitude spectrum

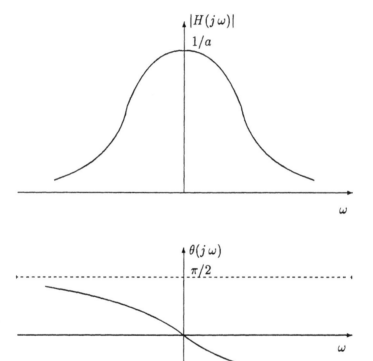

Figure 3.6: The amplitude and phase spectra for the signal $e^{-at}\zeta(t)$ of Example 3.4.

has been an **even function**[2] of the frequency variable ω. This is not a coincidence. Indeed, we see from (3.13) that

$$F(-j\omega) = \overline{F(j\omega)}$$

whenever $f(t)$ is a real signal. It then follows that

$$|F(-j\omega)| = |F(j\omega)|$$

which is precisely what is meant by saying the amplitude spectrum is even. We can also note that the phase spectrum is 'essentially' an **odd function** since

$$\arg F(-j\omega) = -\arg F(j\omega).$$

It is not necessarily odd in the strict mathematical sense, since we usually replace an argument of $-\pi$ with $+\pi$ and also $\arg F(j0)$ is not necessarily zero.

A second point is the behaviour of the amplitude spectra with increasing ω. In each case, the magnitude of $F(j\omega)$ falls off rapidly as ω increases. This means that most of the information on the shape of the pulse or signal f(t) is conveyed in the frequency domain in an interval around the origin $\omega = 0$. For the rectangular pulse of Example 3.3, a graph of the amplitude spectrum shows that a device capable of passing accurately signals of frequencies less than about $5\pi/T$, would pass a reasonably accurate copy of the pulse. *Such information on the ability of systems to pass signals of different frequencies is contained in the frequency response of the system, and it should now be apparent why this is such an important system property.*

3.5   PROPERTIES OF THE FOURIER TRANSFORM

In this section we do not propose to give an exhaustive discussion of the properties of the Fourier Transform. Rather, we restrict ourselves to a consideration of those properties which assist us in our work on the processing of signals using linear systems.

1. **Linearity**: The first property we consider is the linearity property, which is vital to many applications of the theory. Suppose that two signals, $f(t)$ and $g(t)$, have Fourier Transforms $F(j\omega)$ and $G(j\omega)$ respectively. The Fourier Transform of a linear combination of these two signals is then

$$\mathcal{F}\{af(t) + bg(t)\} = \int_{-\infty}^{\infty} \{af(t) + bg(t)\}\, e^{-j\omega t}\, \mathrm{d}t$$

$$= a\int_{-\infty}^{\infty} f(t)\, e^{-j\omega t}\, \mathrm{d}t + b\int_{-\infty}^{\infty} g(t)\, e^{-j\omega t}\, \mathrm{d}t$$

$$= aF(j\omega) + bG(j\omega). \tag{3.15}$$

Clearly, the linearity result also applies to the inverse transform.

---

[2]A function, $f(x)$ say, of a real variable x, is said to be even if $f(-x) = f(x)$ for all x. It is said to be odd if $f(-x) = -f(x)$ for all x.

2. **Time Differentiation**: The second important property to be discussed is the time-differentiation result. To establish this result, assume that $f(t)$ is suitably smooth and has a Fourier Transform $F(j\omega)$, then from (3.14) we have

$$f(t) = \frac{1}{2\pi} \int_{-\infty}^{\infty} F(j\omega) e^{j\omega t} \, d\omega.$$

Differentiating this result with respect to t gives

$$\frac{df}{dt} = \frac{1}{2\pi} \int_{-\infty}^{\infty} (j\omega) F(j\omega) e^{j\omega t} \, d\omega.$$

Comparison with (3.14) shows us that this means that df/dt is the inverse Fourier transform of $(j\omega)F(j\omega)$, and thus that $\mathcal{F}\{df/dt\} = (j\omega)F(j\omega)$.

If we apply the argument n times, we see that

$$\mathcal{F}\left\{\frac{d^n f}{dt^n}\right\} = (j\omega)^n F(j\omega).$$

3. **Time Shifting**: We have already discussed the effects of a time shift on the Fourier spectrum of a periodic signal. Here, we establish the general property. Suppose that a signal $f(t)$ has a Fourier Transform $F(j\omega)$, the transform of the time-shifted version, $f(t-\tau)$, is

$$\mathcal{F}\{f(t-\tau)\} = \int_{-\infty}^{\infty} f(t-\tau) e^{-j\omega t} \, dt.$$

Writing $x = t - \tau$, we obtain

$$\mathcal{F}\{f(t-\tau)\} = e^{-j\omega\tau} \int_{-\infty}^{\infty} f(x) e^{-j\omega x} \, dx$$

$$= e^{-j\omega\tau} F(j\omega).$$

Note that $|e^{-j\omega\tau} F(j\omega)| = |F(j\omega)|$, showing that the amplitude spectrum of $f(t-\tau)$ is identical with that of $f(t)$. On the other hand, $\arg[e^{-j\omega\tau} F(j\omega)] = \arg[F(j\omega)] - \omega\tau \, (\pm 2n\pi)$, and we see that there is a phase-shift by an amount proportional to frequency ω.

4. **Frequency Shifting**: This result is the basis of the process of **amplitude modulation (AM)**, by which information is transmitted using **carrier signals** at selected frequencies. Again, suppose that a signal $f(t)$ has Fourier Transform $F(j\omega)$ and consider the transform of the signal $g(t) = e^{j\omega_0 t} f(t)$. From the definition of the Fourier Transform we have

$$\mathcal{F}\{g(t)\} = \int_{-\infty}^{\infty} e^{j\omega_0 t} f(t) e^{-j\omega t} \, dt$$

$$= \int_{-\infty}^{\infty} f(t) e^{-j(\omega-\omega_0)t} \, dt. \tag{3.16}$$

Now, by definition,

$$F(j\,\omega) = \int_{-\infty}^{\infty} f(t)\,e^{-j\,\omega t}\,dt \qquad (3.17)$$

and, since substituting $\omega - \omega_0$ for ω in (3.17) yields (3.16), we have shown that

$$\mathcal{F}\left\{f(t)e^{j\,\omega_0 t}\right\} = F\left(j\left[\omega - \omega_0\right]\right).$$

We see that *the effect of multiplication of $f(t)$ by $e^{j\,\omega_0 t}$ is to shift the spectrum of $f(t)$, so that it is centered on $\omega = \omega_0$.*

5. **The symmetry property**: This is the last property that we consider in detail. The symmetry property reflects the symmetry which is apparent in the paths between the time and frequency domains. We shall make significant use of this result when we seek to enlarge our library of Fourier Transforms. Suppose as usual that $F(j\,\omega)$ is the Fourier Transform of some time signal $f(t)$. Then by (3.14) we have

$$f(t) = \frac{1}{2\pi} \int_{-\infty}^{\infty} F(j\,\omega)\,e^{j\,\omega t}\,d\omega$$

or

$$2\pi f(t) = \int_{-\infty}^{\infty} F(j\,x)e^{j\,xt}\,dx.$$

Thus

$$2\pi f(-t) = \int_{-\infty}^{\infty} F(j\,x)e^{-j\,xt}\,dx,$$

and writing ω in place of t, we obtain

$$2\pi f(-\omega) = \int_{-\infty}^{\infty} F(j\,x)e^{-j\,\omega x}\,dx$$

$$= \int_{-\infty}^{\infty} F(j\,t)e^{-j\,\omega t}\,dt = \mathcal{F}\left\{F(j\,t)\right\},$$

the Fourier Transform of the time signal $F(j\,t)$.

There is a convenient notation for conveying the implication of the symmetry result. If $F(j\,\omega)$ is the Fourier Transform of $f(t)$, then we write

$$f(t) \leftrightarrow F(j\,\omega),$$

and we say that $f(t)$ and $F(j\,\omega)$ are a **Fourier Transform pair**. Thus, we may express the symmetry property in the form

$$f(t) \leftrightarrow F(j\,\omega) \quad \text{implies} \quad F(j\,t) \leftrightarrow 2\pi f(-\omega).$$

We now demonstrate these properties in the following examples.

Example 3.5
The input $u(t)$ and output $y(t)$ of a causal, linear, time-invariant system are related
by

$$\frac{d^2y}{dt^2}(t) + 3\frac{dy}{dt}(t) + 2y(t) = \frac{du}{dt}(t) + 4u(t).\qquad(3.18)$$

If $u(t)$ and $y(t)$ have Fourier transforms denoted by

$$\mathcal{F}\{u(t)\} = U(j\omega), \quad \text{and} \quad \mathcal{F}\{y(t)\} = Y(j\omega),$$

find $H(j\omega)$ such that $Y(j\omega) = H(j\omega)U(j\omega)$.

At once, using the time differentiation property 2 above, we have that

$$\mathcal{F}\left\{\frac{du}{dt}(t)\right\} = j\omega\mathcal{F}\{u(t)\} = j\omega U(j\omega),$$

$$\mathcal{F}\left\{\frac{dy}{dt}(t)\right\} = j\omega Y(j\omega), \text{ and}$$

$$\mathcal{F}\left\{\frac{d^2y}{dt^2}(t)\right\} = (j\omega)^2 Y(j\omega).$$

Now taking Fourier Transforms in (3.18), we have

$$\mathcal{F}\left\{\frac{d^2y}{dt^2}(t) + 3\frac{dy}{dt}(t) + 2y(t)\right\} = \mathcal{F}\left\{\frac{du}{dt}(t) + 4u(t)\right\}$$

and using the linearity property 1 above, we can write

$$\mathcal{F}\left\{\frac{d^2y}{dt^2}(t)\right\} + 3\mathcal{F}\left\{\frac{dy}{dt}(t)\right\} + 2\mathcal{F}\{y(t)\} = \mathcal{F}\left\{\frac{du}{dt}(t)\right\} + 4\mathcal{F}\{u(t)\}.$$

That is,

$$(j\omega)^2 Y(j\omega) + 3j\omega Y(j\omega) + 2Y(j\omega) = (j\omega)U(j\omega) + 4U(j\omega),$$

so that

$$Y(j\omega) = \frac{4 + j\omega}{2 - \omega^2 + 3j\omega}U(j\omega).$$

Thus,

$$H(j\omega) = \frac{4 + j\omega}{2 - \omega^2 + 3j\omega}.$$

$\square$

In Example 3.5 we have, in effect, obtained a **Fourier transfer function** for
the linear system (3.18). In §2.6, we identified the same function, $H(j\omega)$, with
the Laplace transfer function of the system $H(s)$, evaluated on the imaginary axis
when this was possible. Such an evaluation was shown to be possible for causal
linear systems which were stable. Such systems possess impulse response functions
$h(t) = \mathcal{L}^{-1}\{H(s)\} = \mathcal{F}^{-1}\{H(j\omega)\}$, which are causal and decay to zero as $t \to \infty$.
By examining the defining integrals for the Laplace and Fourier transforms for such

functions $h(t)$, we see that the Fourier transform (3.13) is correctly obtained in these cases, by writing $s = j\omega$ in the Laplace transform integral (1.21).

Example 3.6
Calculate the Fourier transform of the pulse of Figure 3.7.

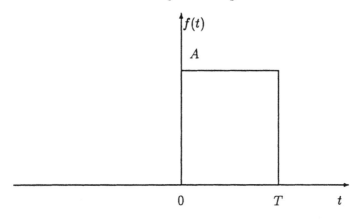

Figure 3.7: The shifted pulse of duration T.

This pulse is simply the pulse of Example 3.3, delayed by $T/2$ seconds. Thus, using the result of Example 3.3 and the time shifting result, property 3 above, we obtain for the Fourier transform

$$F(j\omega) = e^{-j\omega T/2} AT \text{sinc}(\omega T/2).$$

Comparing this result with that of Example 3.3, we observe that the amplitude spectrum of each pulse is identical. The phase spectra, however, exhibit a difference in line with that discussed earlier for the periodic pulse train.

□

Example 3.7
The signal $f(t)$ has Fourier transform $F(j\omega)$, what are the Fourier transforms of the signals

(i) $g_1(t) = f(t)\cos(\omega_c t)$ and

(ii) $g_2(t) = f(t)\sin(\omega_c t)$?

(i) Now $g_1(t) = f(t)\cos(\omega_c t) = f(t)(e^{j\omega_c t} + e^{-j\omega_c t})/2$.

Thus we can write, using first the linearity property, and then the frequency shifting property 4 above,

$$\mathcal{F}\{g_1(t)\} = \tfrac{1}{2}\mathcal{F}\left\{e^{j\omega_c t} f(t)\right\} + \tfrac{1}{2}\mathcal{F}\left\{e^{-j\omega_c t} f(t)\right\}$$
$$= \tfrac{1}{2}F\left(j\left(\omega - \omega_c\right)\right) + \tfrac{1}{2}F\left(j\left(\omega + \omega_c\right)\right).$$

(ii) Similarly,

$$\mathcal{F}\left\{g_2(t)\right\} = \tfrac{1}{2}\frac{1}{j}\mathcal{F}\left\{e^{j\,\omega_c t}f(t)\right\} - \tfrac{1}{2}\frac{1}{j}\mathcal{F}\left\{e^{-j\,\omega_c t}f(t)\right\}$$

$$= \tfrac{1}{2}\frac{1}{j}F\left(j\left(\omega - \omega_c\right)\right) - \tfrac{1}{2}\frac{1}{j}F\left(j\left(\omega + \omega_c\right)\right).$$

□

The result in the first part of Example 3.7 provides the explanation of the process of amplitude modulation. The signal $\cos(\omega_c t)$ is a carrier signal and $f(t)$ is a real, information carrying signal, with amplitude spectrum $|F(j\,\omega)|$ centered on $\omega = 0$. The carrier signal $\cos(\omega_c t)$ is modulated by the process of multiplication at each instant by the signal $f(t)$ to produce the signal $g_1(t)$. We see that $|G(j\,\omega)|$, the amplitude spectrum of $g(t)$, contains two copies of $|F(j\,\omega)|$, one centered on $\omega = \omega_c$, the other on $\omega = -\omega_c$, the frequency of the carrier signal. In our discussion so far, the signals we have considered have had the significant portion of their amplitude spectra restricted to a fairly small band of frequencies, centered on $\omega = 0$. If such signals are used to modulate carrier signals of different, suitably separated frequencies, then the simultaneous transmission of several such signals should be possible in such a way that the information contained in each modulating signal can be recovered. In Chapter 4, we examine devices, called analogue filters, which enable the necessary recovery process.

Example 3.8
Use the symmetry property and the third entry of Table 3.1 to calculate the Fourier transform of the signal

$$g(t) = \frac{1}{a^2 + t^2}.$$

Let $f(t) = e^{-a|t|}$, then $\mathcal{F}\left\{f(t)\right\} = F(j\,\omega) = 2a/(a^2 + \omega^2)$ from Table 3.1. That is,

$$f(t) \leftrightarrow F(j\,\omega)$$

is a Fourier transform pair. Using the symmetry property 5 above, we have that $F(j\,t) \leftrightarrow 2\pi f(-\omega)$ is then another transform pair. Now,

$$F(j\,t) = \frac{2a}{a^2 + t^2},$$

and $f(-\omega) = e^{-a|\omega|}$, and thus we deduce that

$$\frac{2a}{a^2 + t^2} \leftrightarrow 2\pi e^{-a|\omega|}$$

is a transform pair.
Writing $g(t) = 1/(a^2 + t^2)$, we then have $g(t) \leftrightarrow G(j\,\omega)$ where

$$G(j\,\omega) = (\pi/a)\,e^{-a|\omega|}.$$

□

From Example 3.8, we see that the symmetry property provides a method of extending our library of transforms. Actually, the property also serves as an alternative expression of the path between the frequency and time domain, in place of the integral form (3.14). (See, for example Lighthill [15].) We make considerable use of this property in the next section.

We conclude this section with an example which motivates our study of analogue filters in Chapter 4.

Example 3.9
A signal $u(t) = g(t) + h(t)$, consists of two elements. The first element, $g(t)$, is a symmetric unit pulse of duration 2π secs. The other element, $h(t)$, is a copy of this pulse modulating a carrier signal with carrier frequency $\omega_c = 5$ radians/sec. Discuss the transmission of the signal $u(t)$ through the stable linear system with Laplace transfer function

$$H(s) = \frac{1}{s^3 + 2s^2 + 2s + 1}.$$

We first form the Fourier transform $\mathcal{F}\{u(t)\}$ and examine the amplitude spectrum of the input signal.

Using the result of Example 3.3, and applying the linearity and frequency shifting properties, we obtain

$$\mathcal{F}\{u(t)\} = 2\pi\text{sinc}(\omega\pi) + \pi\left[\text{sinc}(\omega - 5)\pi + \text{sinc}(\omega + 5)\pi\right] = U(j\omega).$$

The amplitude spectrum of this signal, $|U(j\omega)|$, is depicted in Figure 3.8.

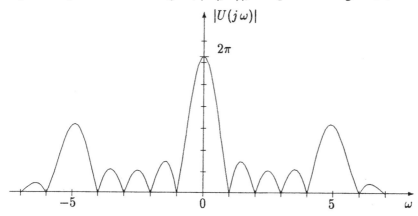

Figure 3.8: The amplitude spectrum of the composite signal $u(t)$ of Example 3.9.

Now we turn our attention to the linear system itself. From $H(s)$, we form $H(j\omega)$, which we can do, since we are considering a stable system.

At once,

$$H(j\omega) = \frac{1}{1 - 2\omega^2 + j\left(2\omega - \omega^3\right)}$$

and the amplitude response, $|H(j\omega)|$, is thus

$$|H(j\omega)| = \frac{1}{\sqrt{\left\{(1 - 2\omega^2)^2 + (2\omega - \omega^3)^2\right\}}}.$$

This response is illustrated in Figure 3.9.

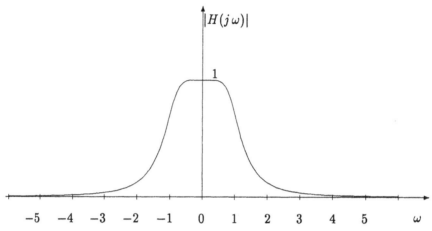

Figure 3.9: The amplitude response of the system of Example 3.9.

If the output signal is $y(t)$, with Fourier transform $Y(j\omega)$, then $Y(j\omega) = H(j\omega)U(j\omega)$, and the amplitude spectrum of the output signal is

$$|Y(j\omega)| = |H(j\omega)\,U(j\omega)| = |H(j\omega)|\,|U(j\omega)|.$$

This amplitude spectrum is shown in Figure 3.10.

From Figure 3.10, we see that the spectrum of the output signal contains a reasonably good copy of the spectrum of $g(t)$, the first element of the input signal. The second element, $h(t)$, has been 'filtered out' by the linear system. In fact, the system used to process $f(t)$ is a third order Butterworth low-pass filter, the design and properties of which we study in detail in Chapter 4.

□

3.6   SIGNAL ENERGY AND POWER

In our discussion of non-periodic signals thus far, we have restricted attention to a fairly small group of signals, all of which satisfy the Dirichlet conditions for the existence of a Fourier transform. It is necessary to extend our work to include a wider class of signals. Before attempting to do this, we introduce two quantities associated with signals, which may be used to provide a convenient classification of signals. We define the **total energy** associated with a signal $f(t)$, defined on $(-\infty, \infty)$ as

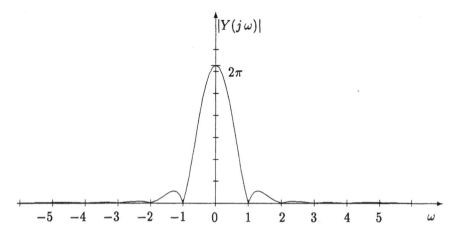

Figure 3.10: The amplitude spectrum of the output signal in Example 3.9.

$$E = \int_{-\infty}^{\infty} |f(t)|^2 \, \mathrm{d}t.$$

The **average power** associated with a signal $f(t)$ is defined as

$$P_{av} = \lim_{T \to \infty} \frac{1}{2T} \int_{-T}^{T} |f(t)|^2 \, \mathrm{d}t. \qquad (3.19)$$

We can make some observations, based on the definition of signal energy and power, for the types of signal that we have already encountered.

Non-periodic signals, which satisfy the Dirichlet conditions, are absolutely integrable and bounded on $(-\infty, \infty)$. Thus, the total energy associated with such a signal is E_0 say, where

$$E_0 = \int_{-\infty}^{\infty} |f(t)|^2 \, \mathrm{d}t$$

and E_0 is a finite number. Clearly, the average power associated with the same signal $f(t)$ is zero, since

$$P_{av} = \lim_{T \to \infty} \frac{1}{2T} E_0 = 0.$$

If we now turn our attention to periodic signals of period P, we notice that if we set $T = rP + \tau$, where r is a positive integer and τ a constant, we can write (3.19) as

$$P_{av} = \lim_{r \to \infty} \frac{1}{2\,(rP + \tau)} \int_{-(rP+\tau)}^{(rP+\tau)} |f(t)|^2 \, \mathrm{d}t$$

$$= \lim_{r \to \infty} \left[\frac{2r}{2\,(rP + \tau)} \int_{\tau}^{P+\tau} |f(t)|^2 \, \mathrm{d}t + \frac{1}{2\,(rP + \tau)} \int_{-\tau}^{\tau} |f(t)|^2 \, \mathrm{d}t \right]$$

$$= \frac{1}{P} \int_\tau^{P+\tau} |f(t)|^2 \, \mathrm{d}t. \tag{3.20}$$

This means that the average power associated with a periodic function can be obtained by integration over a single period.

Example 3.10
Find the average power associated with the periodic signal

$$f(t) = A \sin(\omega_0 t).$$

What is the total energy?

At once,

$$\boldsymbol{P}_{\mathrm{av}} = \frac{\omega_0}{2\pi} \int_0^{2\pi/\omega_0} A^2 \sin^2(\omega_0 t) \, \mathrm{d}t = \frac{A^2}{2}.$$

It is clear that the total energy is unbounded, since the defining integral does not tend to a finite limit.

□

We now have the means to establish our classification of signals, referred to above. This classification is into **energy signals**, which have finite energy, and zero power, and into **power signals**, which have unbounded energy, but finite power associated with them. It follows that those non-periodic signals which satisfy the Dirichlet conditions for the existence of a Fourier transform fall into the first category, while those periodic signals which satisfy the Dirichlet conditions for the existence of a Fourier series are members of the second. In the next section, we try to devise a generalization of the Fourier Transform which allows us to include signals of the second type into a single representation in the frequency domain. Before this, we draw attention to another consequence of the observed symmetry between time and frequency domains. Let us write (3.20) as

$$\boldsymbol{P}_{\mathrm{av}} = \frac{1}{P} \int_0^P |f(t)|^2 \, \mathrm{d}t = \frac{1}{P} \int_0^P f(t) \, \overline{f(t)} \, \mathrm{d}t$$

where $f(t)$ is periodic of period P and $\overline{f(t)}$ denotes the complex conjugate signal. Replacing $\overline{f(t)}$ with its Fourier expansion, we obtain

$$\boldsymbol{P}_{\mathrm{av}} = \frac{1}{P} \int_0^P f(t) \sum_{n=-\infty}^{\infty} \overline{F_n} e^{-j \, n\omega_0 t} \, \mathrm{d}t .$$

Assuming it is permissible to interchange the order of integration and summation, we have

$$\boldsymbol{P}_{\mathrm{av}} = \sum_{n=-\infty}^{\infty} \overline{F_n} \left[\frac{1}{P} \int_0^P f(t) e^{-j \, n\omega_0 t} \, \mathrm{d}t \right]$$

$$= \sum_{n=-\infty}^{\infty} \overline{F_n} \, F_n$$

$$= \sum_{n=-\infty}^{\infty} |F_n|^2 . \qquad (3.21)$$

We have shown that for periodic signals, which possess a Fourier expansion, then it is possible to obtain an expression for the power content of the signal in terms of the frequency domain representation. The reader may wish to obtain a similar result for the energy associated with an energy signal.

Example 3.11
Demonstrate the use of (3.21) above for the periodic signal

$$f(t) = A \sin(\omega_0 t).$$

Now $f(t) = \dfrac{A}{2j}(e^{j\omega_0 t} - e^{-j\omega_0 t})$, and so

$$F_{-1} = -A/2j, \quad F_1 = A/2j \quad \text{and} \quad F_n = 0 \text{ for } n \neq \pm 1.$$

The average power associated with $f(t)$ is then, by (3.21),

$$\boldsymbol{P}_{av} = \{A^2/4 + A^2/4\} = A^2/2,$$

in agreement with our previous calculation.

□

3.7   A GENERALIZATION OF THE FOURIER TRANSFORM

This section is devoted to an attempt to draw together our previous results into a unified treatment. The key step in our development is to examine the Fourier Transform of $\delta(t)$, the Dirac delta function. By definition,

$$\mathcal{F}\{\delta(t)\} = \int_{-\infty}^{\infty} \delta(t) e^{-j\omega t} \, dt = 1$$

using the sifting property of Equation (1.23). This is equivalent to saying that

$$\delta(t) \leftrightarrow 1 \qquad (3.22)$$

is a Fourier Transform pair. We recall that the Laplace transform of $\delta(t)$ is also 1, provided that due attention is paid to the lower limit of integration. It also follows from (1.23), that

$$\mathcal{F}\{\delta(t - t_0)\} = \int_{-\infty}^{\infty} \delta(t - t_0) e^{-j\omega t} \, dt = e^{-j\omega t_0}$$

or,

$$\delta(t - t_0) \leftrightarrow e^{-j\omega t_0} \qquad (3.23)$$

is a Fourier Transform pair. These two transform pairs are represented pictorially in Figure 3.11.

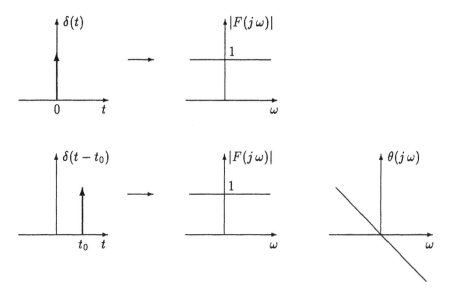

Figure 3.11: The Fourier Transform pairs (3.22) and (3.23).

The next step is to use the symmetry property to deduce further transform pairs from (3.22) and (3.23). From (3.22), we obtain, using the symmetry property,

$$1 \leftrightarrow 2\pi\delta(-\omega) = 2\pi\delta(\omega) \tag{3.24}$$

and (3.23) leads in a similar way to

$$e^{-jtt_0} \leftrightarrow 2\pi\delta(-\omega - t_0) = 2\pi\delta(\omega + t_0). \tag{3.25}$$

Writing $t_0 = -\omega_0$ in (3.25) establishes the transform pair

$$e^{j\omega_0 t} \leftrightarrow 2\pi\delta(\omega - \omega_0). \tag{3.26}$$

Notice that the energy associated with $f(t) = 1$, and $g(t) = e^{j\omega_0 t}$ is unbounded, however, both signals possess finite power. (The calculations are left as an exercise.) In each case, we see that the transforms of the power signals $f(t)$ and $g(t)$ contain δ-functions, and are thus not transforms in the ordinary sense. For this reason, such transforms are called **generalized Fourier Transforms**. Although we have had to follow an indirect path to obtain such generalized transforms, the reader is invited in the exercises to verify that use of the inversion result (3.14) is a straightforward operation. We give a pictorial representation of the Fourier Transforms in (3.24) and (3.26) in Figure 3.12.

It is now possible to contemplate an attempt to calculate a generalized Fourier Transform for the power signal $f(t) = A\cos(\omega_0 t)$.

Since $f(t) = A\cos(\omega_0 t) = A(e^{j\omega_0 t} + e^{-j\omega_0 t})/2$, we have, using linearity and (3.26)

$$\mathcal{F}\{A\cos(\omega_0 t)\} = A\mathcal{F}\{e^{j\omega_0 t} + e^{-j\omega_0 t}\}/2$$

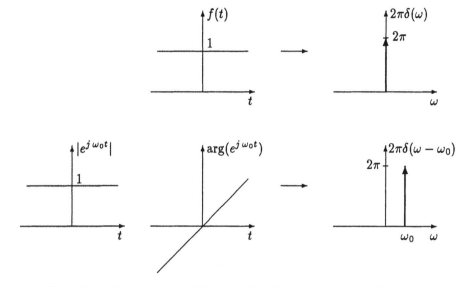

Figure 3.12: Representation of the generalised Fourier transforms of (3.24) and (3.26).

$$= A \left\{ 2\pi\delta(\omega - \omega_0) + 2\pi\delta(\omega + \omega_0) \right\} / 2$$
$$= A\pi \left\{ \delta(\omega - \omega_0) + \delta(\omega + \omega_0) \right\}. \tag{3.27}$$

From (3.27), we see that the spectrum of $A\cos(\omega_0 t)$ consists of two impulses, located at $\omega = \omega_0$ and $\omega = -\omega_0$. The amplitude spectrum is thus zero everywhere except at $\omega = \pm\omega_0$, where it is undefined. We give a pictorial representation (not a graph!) in Figure 3.13.

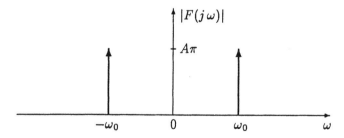

Figure 3.13: A pictorial representation of the amplitude spectrum of the signal $A\cos(\omega_0 t)$.

The result (3.26) provides the means of producing the Fourier transform of a periodic signal $f(t)$, of period P, which has a Fourier series representation. Suppose that the Fourier series has been calculated, so that

$$f(t) \sim \sum_{n=-\infty}^{\infty} F_n e^{j\,n\omega_0 t}.$$

We then find the generalized Fourier transform of $f(t)$ as

$$\mathcal{F}\{f(t)\} = \mathcal{F}\left\{ \sum_{n=-\infty}^{\infty} F_n e^{j\,n\omega_0 t} \right\}$$

$$= \sum_{n=-\infty}^{\infty} \mathcal{F}\left\{ F_n e^{j\,n\omega_0 t} \right\}$$

$$= 2\pi \sum_{n=-\infty}^{\infty} F_n \delta(\omega - n\omega_0), \qquad (3.28)$$

assuming that we may interchange the order of summation and transformation. The generalized Fourier Transform of $f(t)$ thus consists of impulses located at multiples of the fundamental frequency $\omega_0 = 2\pi/P$. Associated with each impulse is the appropriate coefficient from the Fourier Series expansion.

Example 3.12
Obtain the generalized Fourier Transform for the pulse train $f(t)$ of Figure 3.1.

In Example 3.1, we established that

$$f(t) \sim \sum_{n=-\infty}^{\infty} F_n e^{j\,n\omega_0 t}$$

where $\omega_0 = \pi/A$, and

$$F_n = \begin{cases} \dfrac{j}{n\pi}, & \text{for } n \text{ odd} \\[2mm] 0, & \text{for } n \text{ even and } n \neq 0 \\[2mm] \dfrac{1}{2}, & \text{for } n = 0. \end{cases}$$

The generalized Fourier Transform is then

$$\pi\delta(\omega) + 2j \sum_{\substack{n=-\infty \\ n \text{ odd}}}^{\infty} \frac{1}{n}\delta(\omega - n\pi/A).$$

□

Clearly, all the information on the spectrum of the periodic signal of Example 3.12 is available after the Fourier Series calculation. The Fourier Transform also contains this same information, with each discrete spectral component now associated with an impulse located at an appropriate point on the frequency axis. In our discussion of sampling of continuous-time signals, we need to make use of the so-called **unit impulse train** or **comb signal**

$$c(t) = \sum_{n=-\infty}^{\infty} \delta(t - nT) \tag{3.29}$$

a representation of which is shown in Figure 3.14. Here $c(t)$ is periodic, of period

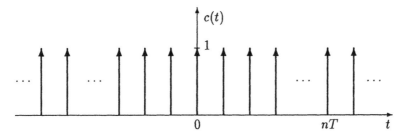

Figure 3.14: A pictorial representation of the unit impulse train.

T, and thus we represent $c(t)$ by a Fourier Series of the form

$$c(t) \sim \sum_{n=-\infty}^{\infty} C_n e^{j\,n\omega_0 t},$$

with $\omega_0 = 2\pi/T$. The coefficients C_n are then given by

$$C_n = \frac{1}{T} \int_{-T/2}^{T/2} c(t) e^{-j\,n\omega_0 t}\,\mathrm{d}t.$$

Within the single period $[-T/2, T/2]$, $c(t) = \delta(t)$, and so

$$C_n = \frac{1}{T} \int_{-T/2}^{T/2} \delta(t) e^{-j\,n\omega_0 t}\,\mathrm{d}t. = \frac{1}{T}.$$

It then follows from (3.28), that the generalized Fourier Transform of $c(t)$ is

$$2\pi \sum_{n=-\infty}^{\infty} C_n \delta(\omega - n\omega_0) = \frac{2\pi}{T} \sum_{n=-\infty}^{\infty} \delta(\omega - n\omega_0).$$

Writing $\omega_0 = 2\pi/T$, we have established the transform pair

$$\sum_{n=-\infty}^{\infty} \delta(t - nT) \leftrightarrow \omega_0 \sum_{n=-\infty}^{\infty} \delta(\omega - n\omega_0) \tag{3.30}$$

and we see that the time domain impulse train, with period T, has as its image, an impulse train in the frequency domain with period $\omega_0 = 2\pi/T$.

3.8   THE CONVOLUTION THEOREMS

In our discussion of the properties of the Fourier Transform, we did not discuss convolution. This was a deliberate choice, because it is felt that the power of these results is easier to appreciate when their use can be demonstrated. An important use is demonstrated in the concluding section of this chapter, when the sampling of time signals is discussed. In the meantime, we give here two convolution results.

Using essentially the same argument as that used in a Laplace transform setting in §1.8, we can show

$$\mathcal{F}\left\{(f*g)(t)\right\} = F(j\,\omega)G(j\,\omega). \tag{3.31}$$

Similarly, we can define a **frequency domain convolution** so that it corresponds to a time domain product. We have

$$
\begin{aligned}
\mathcal{F}\left\{f(t)g(t)\right\} &= \int_{-\infty}^{\infty} f(t)g(t)e^{-j\,\omega t}\,\mathrm{d}t \\
&= \int_{-\infty}^{\infty} f(t)\left[\frac{1}{2\pi}\int_{-\infty}^{\infty} G(j\,\sigma)e^{j\,\sigma t}\,\mathrm{d}\sigma\right]e^{-j\,\omega t}\,\mathrm{d}t \\
&= \frac{1}{2\pi}\int_{-\infty}^{\infty}\left[\int_{-\infty}^{\infty} f(t)e^{-j\,(\omega-\sigma)t}\,\mathrm{d}t\right]G(j\,\sigma)\,\mathrm{d}\sigma \\
&= \frac{1}{2\pi}\int_{-\infty}^{\infty} F(j\,(\omega-\sigma)\,)G(j\,\sigma)\,\mathrm{d}\sigma.
\end{aligned}
$$

We therefore define the frequency domain convolution by

$$(F*G)(j\,\omega) = \frac{1}{2\pi}\int_{-\infty}^{\infty} F(j\,(\omega-\sigma)\,)G(j\,\sigma)\,\mathrm{d}\sigma \tag{3.32}$$

and we then have the transform pair

$$f(t)g(t) \leftrightarrow (F*G)(j\,\omega). \tag{3.33}$$

We note that the factor $1/2\pi$ appearing in (3.32) is an artefact of our choice of using radians/sec. as our units of frequency. This factor disappears if we convert to Hertz.

3.9   SAMPLING OF TIME SIGNALS AND ITS IMPLICATIONS

The later chapters of this book are concerned with discrete-time systems which process signals obtained by **sampling** continuous-time signals. The sampling process produces signal sequences, representing signals defined only at discrete time intervals. In this section, we examine a representation of the sampling process and determine the Fourier spectrum of the signal sequences so obtained.

Suppose that a continuous-time signal $u(t)$ is sampled at equal intervals, T, to produce the signal sequence

$$\{u(kT)\} = \{u(0), u(T), u(2T), \ldots, u(kT), \ldots\}. \tag{3.34}$$

(The reader who is unfamiliar with sequence notation should note that this topic is discussed in Chapter 5.)

We now consider this sequence as representing a train of impulses and therefore replace the sequence with the signal

$$u_s(t) = \sum_{k=0}^{\infty} u(kT)\delta(t - kT).$$
(3.35)

Using the formula (1.28), with T replaced by kT, we have $f(t)\delta(t - kT) = f(kT)\delta(t - kT)$. Using this formula in reverse, we can write $u_s(t)$ in the product form

$$u_s(t) = u(t)\sum_{k=0}^{\infty} \delta(t - kT).$$
(3.36)

We can generalize (3.36), to include the possibility of non-causal signals, by writing

$$u_s(t) = u(t)\sum_{k=-\infty}^{\infty} \delta(t - kT).$$
(3.37)

The relation (3.37) leads us to picture $u_s(t)$ as in Figure 3.15, where the solid line refers to the continuous-time signal, $u(t)$, prior to sampling.

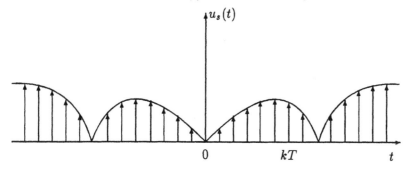

Figure 3.15: A pictorial representation of the sampled signal $u_s(t)$.

We now seek to determine the spectrum of such a representation of sampled time signals and, in particular, to investigate how this spectrum relates to that of the unsampled signal, $u(t)$. We find that the investment of time in the study of generalized functions and convolution is now fully rewarded!

Let $U_s(j\omega)$ be the Fourier Transform of $u_s(t)$, then

$$U_s(j\omega) = \mathcal{F}\{u_s(t)\} = \mathcal{F}\left\{u(t)\sum_{k=-\infty}^{\infty} \delta(t - kT)\right\}$$

$$= \mathcal{F}\{u(t)\} * \mathcal{F}\left\{\sum_{k=-\infty}^{\infty} \delta(t - kT)\right\}$$

using (3.33). Now,

$$\mathcal{F}\left\{\sum_{k=-\infty}^{\infty} \delta(t - kT)\right\} = \omega_0 \sum_{k=-\infty}^{\infty} \delta(\omega - n\omega_0),$$

from (3.30), where $\omega_0 = 2\pi/T$.

Thus, if $U(j\,\omega)$ is the Fourier Transform of $u(t)$,

$$U_s(j\,\omega) = U(j\,\omega) * \frac{2\pi}{T} \sum_{n=-\infty}^{\infty} \delta(\omega - n\omega_0) \qquad (3.38)$$

$$= \frac{1}{T} \int_{-\infty}^{\infty} U(j\,\omega - j\,\sigma) \sum_{n=-\infty}^{\infty} \delta(\sigma - n\omega_0)\, d\sigma.$$

Assuming we can interchange summation and integration, we obtain

$$U_s(j\,\omega) = \frac{1}{T} \sum_{n=-\infty}^{\infty} \int_{-\infty}^{\infty} U(j\,\omega - j\,\sigma)\,\delta(\sigma - n\omega_0)\, d\sigma$$

$$= \frac{1}{T} \sum_{n=-\infty}^{\infty} U(j\,\omega - j\,n\omega_0)$$

$$= \frac{1}{T} \sum_{n=-\infty}^{\infty} U(j\,(\omega - 2\pi n/T)). \qquad (3.39)$$

From (3.39) above, we see that *the spectrum of the sampled signal $u_s(t)$ consists of scaled periodic repeats of the spectrum of the unsampled version $u(t)$ added together.* These repeats are centered on the frequencies $n\omega_0 = 2\pi n/T, n = 0, \pm1, \pm2, \ldots$ and an example illustrating the resulting amplitude spectrum is shown in Figure 3.16.

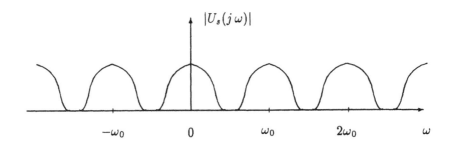

Figure 3.16: The amplitude spectrum of the sampled signal $u_s(t)$.

The separation in frequency between the centre of successive repeats of the spectrum of the unsampled signal is $\Delta\omega = \omega_0 = 2\pi/T$, which depends on the sampling interval, T. Suppose that the spectrum of an unsampled continuous-time signal $u(t)$ is bounded in the sense that the amplitude spectrum $|U(j\,\omega)|$ satisfies

$$|U(j\,\omega)| = 0, \quad |\omega| > \omega_m \text{ say.}$$

Such a signal is said to be **band-limited**, and it follows that the successive scaled repeats of $|U(j\,\omega)|$ in the amplitude spectrum $|U_s(j\,\omega)|$ of the sampled version $u_s(t)$

do not overlap if

$$\Delta\omega > 2\omega_m,$$

that is, $2\pi/T > 2\omega_m$.

This means that we must sample $u(t)$ at such a rate that

$$T < \pi/\omega_m. \tag{3.40}$$

If we define $f_s = 1/T$, the **sampling frequency** measured in Hertz, and $f_m = \omega_m/2\pi$ as the **band limit** of $u(t)$, also now measured in Hertz, we can write (3.40) as

$$f_s > 2f_m.$$

This condition on the sampling frequency, known as the **Nyquist condition** or **Nyquist-Shannon condition**, implies that *we must take samples from $u(t)$ at a frequency greater than twice the band-limiting frequency, if 'overlaps' are to be avoided.* We shall see that, if this condition is satisfied, then it will be possible to reconstruct $u(t)$ from its samples. The (theoretical) minimum sampling frequency $f_N = 2f_m$ for which this recovery is possible is sometimes called the **Nyquist frequency** or **Nyquist rate**. Figure 3.16 has been drawn on the assumption that the sampling frequency exceeded the Nyquist rate. If the sampling rate had been less than the Nyquist rate, then the amplitude spectrum $|U_s(j\omega)|$ would be as illustrated in Figure 3.17, where the regions of 'overlap' represent regions where the periodic repeats interfere through the summation in (3.39). It should be noted that

$$|U_s(j\omega)| = \frac{1}{T}\left| \sum_{n=-\infty}^{\infty} U(j(\omega - 2n\pi/T)) \right|,$$

which does *not* in general equal

$$\frac{1}{T} \sum_{n=-\infty}^{\infty} |U(j(\omega - 2n\pi/T))|.$$

Therefore, without phase information, Figure 3.17 does not give the whole picture.

This 'interference' due to overlapping is known as **aliasing**. Such a situation precludes the exact reconstruction of the unsampled signal from its samples. The recovery process is achieved by the use of a low-pass filter, of bandwidth ω_b, say. This is a device which passes signals, or components of signals, of frequency less than or equal to ω_b, and rejects all components of frequency above ω_b. In the frequency domain, the filtering process is achieved by the multiplication of the spectrum of the signal to be filtered by such a device, with frequency response

$$H(j\omega) = \begin{cases} 1, & |\omega| \le \omega_b \\ 0, & |\omega| > \omega_b. \end{cases}$$

(In Chapter 4, we shall see that such a filter cannot be realized in practice and, hence, we refer to this frequency response as being **ideal**.) For the signal with amplitude spectrum as in Figure 3.16, we would choose ω_b to be approximately $\omega_0/2$.

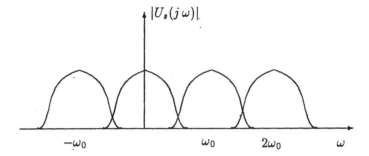

Figure 3.17: Amplitude spectrum exhibiting aliasing effects.

To illustrate the process, suppose that a signal $g(t)$ has been sampled at a frequency f above the Nyquist frequency f_N, so that $f = A f_N = 2A f_m$, $A > 1$. We multiply the spectrum of the sampled sequence by $H(j\omega)$ and invert the result. The resulting time domain signal is $(1/T)\tilde{g}(t)$ where $\tilde{g}(t)$ is the reconstructed version of $g(t)$. We have,

$$\frac{1}{T}\tilde{g}(t) = \mathcal{F}^{-1}\{G_s(j\omega)\,H(j\omega)\} = \mathcal{F}^{-1}[G_s(j\omega)] * \mathcal{F}^{-1}\{H(j\omega)\}$$

$$= g_s(t) * \frac{\omega_b}{\pi}\text{sinc}(\omega_b t).$$

Here we have used the result that $\mathcal{F}^{-1}\{H(j\omega)\} = (\omega_b/\pi)\text{sinc}(\omega_b t)$, which follows easily from the symmetry property and which the reader is invited to establish in the exercises. Note that the ideal filter's impulse response, $\mathcal{F}^{-1}\{H(j\omega)\}$, is not causal. We have more to say on this matter in the next chapter.

Let us select the band-width of the filter to be $\omega_b = \omega_0/2$, so that the filtered signal contains exactly one copy of the spectrum of $u(t)$, to obtain

$$\frac{1}{T}\tilde{g}(t) = \int_{-\infty}^{\infty} g_s(\tau)\frac{\omega_0}{2\pi}\text{sinc}\left(\frac{\omega_0}{2}(t-\tau)\right)\,\mathrm{d}\tau.$$

Since $\dfrac{1}{T} = \dfrac{\omega_0}{2\pi}$ and $g_s(t) = \displaystyle\sum_{n=-\infty}^{\infty} g(nT)\delta(t-nT)$, we obtain

$$\tilde{g}(t) = \int_{-\infty}^{\infty} \sum_{n=-\infty}^{\infty} g(nT)\delta(\tau-nT)\text{sinc}\left(\frac{\omega_0}{2}(t-\tau)\right)\,\mathrm{d}\tau.$$

Thus, assuming we can interchange summation and integration, we have

$$\tilde{g}(t) = \sum_{n=-\infty}^{\infty} g(nT)\,\text{sinc}\left(\frac{\omega_0}{2}(t-nT)\right)$$

and we see that the signal is reconstructed using a superposition of sinc functions, each one centered on a sample point. Since $T = 2\pi/\omega_0$, it follows that

$$\text{sinc}\left(\frac{\omega_0}{2}(t - nT)\right) = \text{sinc}\left(\frac{\omega_0 t}{2} - n\pi\right).$$

In particular, if $t = kT$, then we have $\text{sinc}((k - n)\pi)$, which in turn equals $\delta_{k,n}$ (defined in (3.2)) and so

$$\tilde{g}(kT) = \sum_{n=-\infty}^{\infty} g(nT)\delta_{k,n} = g(kT).$$

This means that the reconstruction gives the exact value at each sample point, with the accuracy of the interpolation between sample points depending on the sampling rate. It is an illustrative exercise to construct a computer programme to investigate the reconstruction process with different sampling rates. The exercise becomes even more relevant when signals which are not strictly band-limited are considered. In fact, all finite duration signals will have amplitude spectra which do not vanish outside a finite interval defined by $|\omega| < \omega_m$. Usually, such signals have amplitude spectra which decay rapidly with increasing ω, and the value of ω at which a particular amplitude spectrum has become negligible is a matter of judgement. This judgement is used to set a Nyquist-like rate, implying that above a certain value of ω, the signal contains no further significant components. The sampling rate is then chosen, based on this Nyquist-like rate, and even if sampling is sufficiently fast and good reconstructions are obtained, the process can never be completely exact.

3.10   EXERCISES

1. Verify the orthogonality result

$$\frac{1}{T}\int_{t_0}^{t_0+T} e^{j\,n\omega_0 t}\, e^{-j\,m\omega_0 t}\, \mathrm{d}t = \delta_{n,m} = \begin{cases} 0, & m \neq n, \\ 1, & m = n. \end{cases}$$

2. Determine the Fourier coefficients in the expansion of the periodic pulse shown in Figure 3.18. Sketch the amplitude spectrum for this signal and compare with Figure 3.2. Apply the time-shifting result in §3.2 to deduce the result in Example 3.1.

3. Verify the results for the Fourier transforms of

$$f(t) = te^{-at}\zeta(t),\ a > 0,\ \text{and}$$
$$g(t) = e^{-a|t|},\ a > 0,$$

given in Table 3.1.

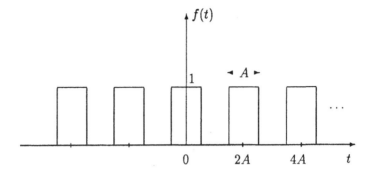

Figure 3.18: Figure for Exercise 2.

4. Use the result that the Fourier transform of

$$h(t) = \frac{1}{(a^2 + t^2)}, \quad a > 0,$$

is $(\pi/a)e^{-a|\omega|}$, and the frequency shifting property to deduce the transform of $\cos(bt)/(1+t^2)$. Hence, find the transform of $\sin(bt)/(1+t^2)$.

5. Show that the Fourier transform of the symmetric unit pulse of duration T, given by

$$f(t) = \begin{cases} 1, & |t| \leq T/2, \\ 0, & |t| > T/2, \end{cases}$$

is $T\text{sinc}(\omega T/2)$. Use the symmetry property to find the transform of $h(t) = (\omega_b/\pi)\,\text{sinc}(\omega_b t)$, where $\omega_b = T/2$. Sketch the amplitude spectrum of this signal. If this signal is the impulse response of a linear system, how would you describe the frequency response, and what could be its use?

6. From Example 1.6, we know that $\delta(t) = \dfrac{d\zeta}{dt}(t)$. It follows that $j\omega\mathcal{F}\{\zeta(t)\} = 1$. However, it does *not* follow that $\mathcal{F}\{\zeta(t)\} = 1/j\omega$. In fact, letting $F(j\omega) = \mathcal{F}\{\zeta(t)\}$, we have

$$F(j\omega) = \frac{1}{j\omega} + \pi\delta(\omega).$$

(a) Show $j\omega F(j\omega) = 1$.

(b) Let $\Theta(j\omega)$ be an arbitrary testing function of the frequency variable ω and let $\theta(t) = \mathcal{F}^{-1}\{\Theta(j\omega)\}$.

    i. Show that

$$\int_{-\infty}^{\infty} \pi\delta(\omega)\Theta(j\,\omega)\,\mathrm{d}\omega = \pi \int_{-\infty}^{\infty} \theta(t)\,\mathrm{d}t.$$

   ii. Define

$$\int_{-\infty}^{\infty} \frac{\Theta(j\,\omega)}{j\,\omega}\,\mathrm{d}\omega \stackrel{\mathrm{def}}{=} \lim_{\varepsilon\downarrow 0}\left[\int_{-\infty}^{-\varepsilon} \frac{\Theta(j\,\omega)}{j\,\omega}\,\mathrm{d}\omega + \int_{\varepsilon}^{\infty} \frac{\Theta(j\,\omega)}{j\,\omega}\,\mathrm{d}\omega\right].$$

Show that this gives

$$\int_{-\infty}^{\infty} \frac{\Theta(j\,\omega)}{j\,\omega}\,\mathrm{d}\omega = -\int_{-\infty}^{\infty} 2t\theta(t)\int_{0}^{\infty} \mathrm{sinc}(\omega t)\,\mathrm{d}\omega\,\mathrm{d}t.$$

(You may assume that interchanging the order of integration is permissible.)

  iii. It can be shown that $\int_{0}^{\infty} \mathrm{sinc}(x)\,\mathrm{d}x = \pi/2$, (see Exercise 9). Use this to show that

$$\int_{-\infty}^{\infty} \frac{\Theta(j\,\omega)}{j\,\omega}\,\mathrm{d}\omega = -\pi\int_{-\infty}^{\infty} \theta(t)\mathrm{sgn}(t)\,\mathrm{d}t,$$

where the 'signum' function $\mathrm{sgn}(t)$ is defined by

$$\mathrm{sgn}(t) = \begin{cases} -1 & \text{for } t < 0, \\ 1 & \text{for } t \geq 0. \end{cases}$$

(c) The Fourier transform of $\zeta(t)$, i.e. $F(j\,\omega)$, is defined to be that generalized function of ω which satisfies

$$\int_{-\infty}^{\infty}\int_{-\infty}^{\infty} \zeta(t)\Theta(j\,\omega)e^{-j\,\omega t}\,\mathrm{d}\omega\,\mathrm{d}t = \int_{-\infty}^{\infty} F(j\,\omega)\Theta(j\,\omega)\,\mathrm{d}\omega$$

for all testing functions, $\Theta(j\,\omega)$. Show that the left hand side of this equation simplifies to $\pi \int_{-\infty}^{0} \theta(t)\,\mathrm{d}t$. Hence show $F(j\,\omega) = \dfrac{1}{j\,\omega} + \pi\delta(\omega)$.

Finally, use the frequency convolution theorem to deduce the transforms of the causal functions:

$$f(t) = A\cos(\omega_0 t)\,\zeta(t) \quad \text{and} \quad g(t) = A\sin(\omega_0 t)\,\zeta(t).$$

7. Obtain the amplitude response of the time-invariant linear system with Laplace transfer function

$$H(s) = \frac{1}{s^2 + \sqrt{2}s + 1},$$

and illustrate with a sketch. Follow the method of Example 3.9 to discuss the transmission of the signal consisting of a symmetric unit pulse of duration 1 sec., together with a copy of the pulse modulating a carrier signal at frequency $\omega_0 = 10$ rad/sec.

8. Establish Parseval's formula

$$\int_{-\infty}^{\infty} f(t)G(j\,t)\,\mathrm{d}t = \int_{-\infty}^{\infty} g(t)F(j\,t)\,\mathrm{d}t$$

where $f(t) \leftrightarrow F(j\omega)$ and $g(t) \leftrightarrow G(j\omega)$ are Fourier transform pairs. Also show that

$$\int_{-\infty}^{\infty} |f(t)|^2\,\mathrm{d}t = \frac{1}{2\pi}\int_{-\infty}^{\infty} |F(j\omega)|^2\,\mathrm{d}\omega.$$

9. Show that

(a)

$$\int_{-\infty}^{\infty} \operatorname{sinc}^2(x)\,\mathrm{d}x = \int_{-\infty}^{\infty} \operatorname{sinc}(x)\,\mathrm{d}x.$$

(b)

$$\int_{0}^{\infty}\int_{0}^{\infty} e^{-ax}\sin(x)\,\mathrm{d}x\,\mathrm{d}a = \pi/2,$$

and, by reversing the order of integration, show that

$$\int_{0}^{\infty}\int_{0}^{\infty} e^{-ax}\sin(x)\,\mathrm{d}a\,\mathrm{d}x = \int_{0}^{\infty} \operatorname{sinc}(x)\,\mathrm{d}x.$$

Hence, deduce that

$$\int_{-\infty}^{\infty} \operatorname{sinc}(x)\,\mathrm{d}x = \pi.$$

10. Show that

$$\int_{-\infty}^{\infty} \operatorname{sinc}(\omega_b(t - nT))\operatorname{sinc}(\omega_b(t - mT))\,\mathrm{d}t = 0,$$

where m, n are integers and $m \neq n$, and where $\omega_b = \pi/T$. Use this result with that of Exercise 9 to show that the set $\{f_n(t)\} = \{(1/\sqrt{T})\operatorname{sinc}(\omega_b(t - nT))\}$ is orthonormal on $-\infty < t < \infty$.

4

Analogue filters

4.1 INTRODUCTION

In this chapter we make use of our understanding of the representation of signals in the frequency domain and the concept of system frequency response. Bringing these ideas together allows us to form a model of how systems operate on signals. This concept was discussed in Chapters 2 and 3. At this point, we turn our attention to the **design** or **synthesis** of systems which process signals in a predetermined manner. This is in contrast to our discussion so far, where we have concentrated on the analysis of systems, assumed 'given' in the traditional mathematical sense. We will, however, draw on this experience!

We now consider the problem of how to design **filters**, that is, devices which are capable of processing input signals in such a way that the output signal depends, in a pre-determined manner, on the frequency of the input signal. In this chapter we examine the design of **analogue filters**, which operate on continuous-time signals to produce continuous-time signals as output. In Chapter 8, we discuss the design of **digital filters** which operate on discrete-time signals.

4.2 ANALOGUE FILTER TYPES.

Analogue filters are classified according to their capability in either passing or rejecting signals of different frequencies. An **ideal low-pass filter** passes all signals of frequencies less than a **cut-off**, or **critical frequency** ω_c, and rejects all signals of frequency greater than ω_c. From our work on the frequency domain decomposition of signals, using the methods of Fourier analysis, we can see that a low-pass filter would have the effect of filtering out the higher frequency components in a composite signal. The amplitude response of an ideal low-pass filter is illustrated in Figure 4.1.

A filter with the opposite effect to that of an ideal low-pass filter, is an **ideal high-pass filter**. Such a filter will reject all signals (or components of signals) of frequency less than the cut-off, or critical frequency ω_c, and pass signals of frequency above ω_c. The amplitude response of such an ideal high-pass filter is shown in Figure 4.2.

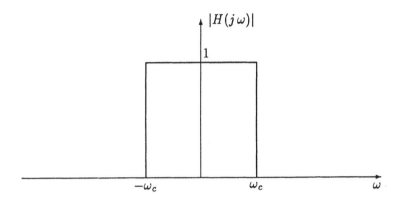

Figure 4.1: Ideal low-pass filter amplitude response.

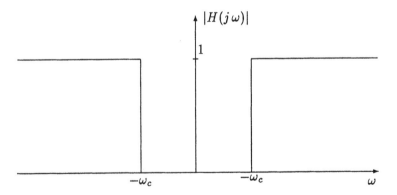

Figure 4.2: Ideal high-pass filter amplitude response.

An **ideal band-reject filter** rejects signals of frequency ω, where

$$\omega_l < \omega < \omega_u.$$

The ideal amplitude response of this filter is illustrated in Figure 4.3.

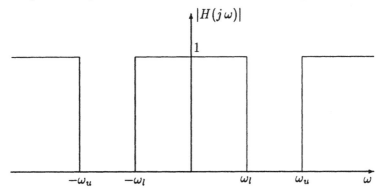

Figure 4.3: Amplitude response of the ideal band-reject filter, with lower cut-off frequency ω_l and upper cut-off frequency ω_u.

The final type we consider is the **ideal band-pass filter**. Such a filter rejects all signals except those with frequencies between the lower critical frequency ω_l and the upper critical frequency ω_u. The amplitude response of such a filter is shown in Figure 4.4.

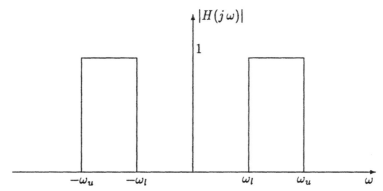

Figure 4.4: The amplitude response of the ideal band-pass filter, with critical frequencies ω_l and ω_u.

It is not necessary to study each of these filter types separately. We shall see in a later section that we can find **filter transformations**, that is transformations which can be used to map low-pass filter designs into any of the above types. For this reason, we restrict our attention to methods of designing satisfactory low-pass filters.

4.3   A CLASS OF LOW-PASS FILTERS

In this section, we consider the design of an important class of filters known as **Butterworth filters**. Viewed from the frequency domain, the design problem is to determine a system whose amplitude response approximates the ideal amplitude response of Figure 4.1 above. There is a second stage to the process, the so-called **realization problem**, the solution of which involves the specification of a circuit which implements the filter design. First we examine the approximation stage, and we see that the ideal low-pass response is given by

$$|H(j\omega)| = \begin{cases} 1, & |\omega| < \omega_c \\ 0, & |\omega| \geq \omega_c. \end{cases}$$

The Butterworth approximation to this ideal response is to set

$$|H(j\omega)| = \frac{1}{\sqrt{1 + (\omega/\omega_c)^{2n}}} \ . \qquad (4.1)$$

It is easy to see that if $|\omega| < \omega_c$ then as $n \to \infty$, $|H(j\omega)| \to 1$, however if $|\omega| > \omega_c$, $|H(j\omega)| \to 0$ as $n \to \infty$, thus for large values of n, we obtain a good approximation to the desired response. The amplitude response of some filters designed on this basis is depicted in Figure 4.5.

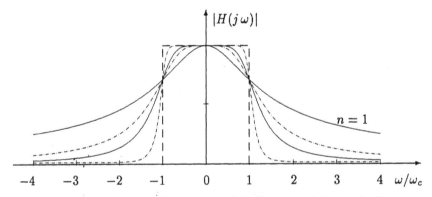

Figure 4.5: The amplitude responses of the Butterworth filters of orders 1, 2, 3 and 9.

Observe that we may write (4.1) in the form

$$|H(j\omega)|^2 = \frac{1}{1 + (\omega/\omega_c)^{2n}} \ . \qquad (4.2)$$

We then argue that this is the square of the amplitude response of a system with Laplace transfer function $H(s)$, obtained by replacing s with $j\omega$. This assumes that the system is a stable system, and the process is only valid if such a stable system can be found.

Note that in general,

$$|H(j\omega)|^2 = H(j\omega)\overline{H}(j\omega)$$

and if $H(s)$ is to have real coefficients then

$$\overline{H}(j\,\omega) = H(-j\,\omega). \tag{4.3}$$

Then, we see that we obtain (4.2), on putting $s = j\,\omega$, if $H(s)$ is such that

$$H(s)H(-s) = \frac{1}{1 + (s/j\,\omega_c)^{2n}}. \tag{4.4}$$

This process involves **analytic continuation**, which forms part of the theory of functions of a complex variable, and is beyond the scope of this text. We now find the poles of this function by setting

$$1 + (s/j\,\omega_c)^{2n} = 0,$$

meaning that

$$(s/j\,\omega_c)^{2n} = \exp(j\,(2k+1)\pi),$$

with k an integer. Thus,

$$s/j\,\omega_c = \exp(j\,(2k+1)\pi/2n),$$

and we obtain,

$$s = \omega_c \exp(j\,((2k+1)\pi/2n + \pi/2)), \tag{4.5}$$

where k is an integer. We examine the locations of these poles for the cases $n = 1, 2, 3$ and 5 in Figure 4.6.

From Figure 4.6, we see that, in each case, there are $2n$ poles equally spaced around the circle of radius ω_c in the complex plane. Also, we note that there are no poles on the imaginary axis and that this will be the case for all values of n. Clearly, if $s = s_p$ is a pole of $H(s)H(-s)$, then so is $s = -s_p$. This means that we may select as the poles corresponding to $H(s)$ those which lie in the left-half plane, that is, in the stable region. The remaining poles, in the right half-plane, are then associated with $H(-s)$. Thus, a design is possible satisfying the criterion that the proposed system should be a stable system.

We next obtain the transfer functions, $H(s)$, in a few cases. When $n = 1$, (4.4) becomes

$$\begin{aligned}
H(s)H(-s) &= \frac{1}{1 + (s/j\,\omega_c)^2} \\
&= \frac{\omega_c^2}{\omega_c^2 - s^2} \\
&= \frac{\omega_c^2}{(\omega_c + s)(\omega_c - s)}.
\end{aligned}$$

We select the pole at $s = -\omega_c$ to be the single pole of our system, thus generating the simple first-order system, defined by the Laplace transfer function

$$H(s) = \frac{\omega_c}{s + \omega_c}. \tag{4.6}$$

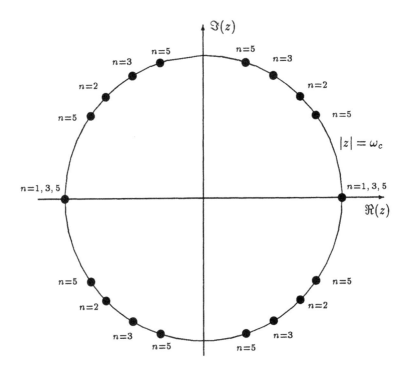

Figure 4.6: Pole locations given by (4.5)

We turn our attention to the case $n = 2$, when (4.4) becomes

$$H(s)H(-s) = \frac{1}{1 + (s/j\,\omega_c)^4}$$

$$= \frac{\omega_c^4}{s^4 + \omega_c^4}.$$

Now factorize $(s^4 + \omega_c^4)$ as

$$s^4 + \omega_c^4 = (s^2 + j\,\omega_c^2)(s^2 - j\,\omega_c^2)$$
$$= (s + (1+j\,)\omega_c/\sqrt{2})(s + (1-j\,)\omega_c/\sqrt{2}) \times$$
$$(s + (-1+j\,)\omega_c/\sqrt{2})(s + (-1-j\,)\omega_c/\sqrt{2}).$$

Thus, selecting the left-half plane poles for the transfer function, $H(s)$, we obtain,

$$H(s) = \frac{\omega_c^2}{(s + (1+j\,)\omega_c/\sqrt{2})(s + (1-j\,)\omega_c/\sqrt{2})}$$

$$= \frac{\omega_c^2}{s^2 + \sqrt{2}\,\omega_c s + \omega_c^2}. \tag{4.7}$$

Proceeding in the same way, it is easy to show that in the case $n = 3$, the transfer function obtained is,

$$H(s) = \frac{\omega_c^3}{(s + \omega_c)(s + (\sqrt{3}+1)\omega_c/2)(s + (\sqrt{3}-1)\omega_c/2)}$$

$$= \frac{\omega_c^3}{s^3 + 2\omega_c s^2 + 2\omega_c^2 s + \omega_c^3}. \tag{4.8}$$

Examining each of the transfer functions obtained, we see that the index n gives the order of the denominator or characteristic polynomial. This, in turn, is the order of the resulting linear system modelled by a differential equation with $H(s)$ as its transfer function. It is natural therefore, to call n the **order of the Butterworth filter**. It would be a tedious operation to continue to calculate the transfer functions of the higher order filters in this manner. Fortunately, we can avoid this chore, as we shall see.

Having obtained the transfer functions of some filters, we turn our attention to the second part of the design process, that is the **realization stage**. As discussed earlier, this involves the specification of a circuit which implements the filter design. In order to see how to do this for the designs we have already obtained, let us construct the time-domain representations of each filter in turn.

For the first-order Butterworth filter, when $n = 1$, we have,

$$H(s) = \frac{\omega_c}{s + \omega_c}. \tag{4.9}$$

If $Y(s) = H(s)U(s)$, we see that this is equivalent to,

$$(s + \omega_c)Y(s) = \omega_c U(s),$$

the Laplace transform of a differential equation. This may now be inverted, with the assumption of a zero initial condition on $y(t)$, to give

$$\frac{\mathrm{d}y}{\mathrm{d}t}(t) + \omega_c y(t) = \omega_c u(t),$$

where

$$y(t) = \mathcal{L}^{-1}\{Y(s)\}, \text{ and } u(t) = \mathcal{L}^{-1}\{U(s)\}.$$

If we set $y(t) = v_c(t)$, $u(t) = e(t)$ and choose the values of R and C so that $1/RC = \omega_c$, then we obtain Equation (1.10). *This means that the first-order Butterworth filter can be realized by the simple C–R circuit of Example 1.2.* In fact, we could have anticipated that this would be the case from the transform domain. This is because Equation (1.31) can be written as

$$Y(s) = \frac{1/RC}{s + 1/RC}U(s) = H(s)U(s)$$

and we see that putting $1/RC = \omega_c$ leads immediately to the transfer function $H(s)$ of (4.9). In a similar, way we can investigate the second-order design, corresponding to the case $n = 2$. Setting $Y(s) = H(s)U(s)$, with the transfer function $H(s)$ given by (4.7) as

$$H(s) = \frac{\omega_c^2}{s^2 + \sqrt{2}\omega_c s + \omega_c^2}, \tag{4.10}$$

we obtain $(s^2 + \sqrt{2}\omega_c s + \omega_c^2)Y(s) = \omega_c^2 U(s)$. Again, inverting this Laplace transform, under the assumption of zero initial conditions on both y and $\mathrm{d}y/\mathrm{d}t$, leads to the time-domain representation of the system. Carrying out the inversion yields

$$\frac{\mathrm{d}^2 y}{\mathrm{d}t^2}(t) + \sqrt{2}\omega_c \frac{\mathrm{d}y}{\mathrm{d}t}(t) + \omega_c^2 y(t) = \omega_c^2 u(t). \tag{4.11}$$

Again our earlier examples are useful. Notice that (1.40) can be re-written as

$$\frac{\mathrm{d}^2 v_c}{\mathrm{d}t^2}(t) + \frac{R}{L}\frac{\mathrm{d}v_c}{\mathrm{d}t}(t) + \frac{1}{LC}v_c(t) = \frac{1}{LC}e(t).$$

Thus if we set $R/L = \sqrt{2}\omega_c$, and $1/LC = \omega_c^2$, we can use the L–C–R circuit of §1.9 as our second-order Butterworth filter. Notice that the input to the system is the voltage $e(t)$, while the output is the voltage drop measured across the capacitor. The value of R may be chosen at will, and the values of L and C are then determined uniquely.

Finally, for the case $n = 3$, the transfer function for the third-order filter is given by (4.8), and it is straight-forward to show that the system generated in the time domain is

$$\frac{\mathrm{d}^3 y}{\mathrm{d}t^3}(t) + 2\omega_c \frac{\mathrm{d}^2 y}{\mathrm{d}t^2}(t) + 2\omega_c^2 \frac{\mathrm{d}y}{\mathrm{d}t}(t) + \omega_c^3 y(t) = \omega_c^3 u(t). \tag{4.12}$$

Observing that this is a third-order differential equation, we return to Chapter 1 and examine the system of Example 1.11. In Equation (1.65), we obtained the time domain representation of the so-called ladder network of Figure 1.19. This circuit is re-drawn here as Figure 4.7.

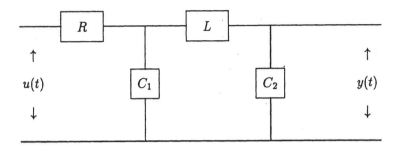

Figure 4.7: The ladder network of Figure 1.19

This circuit can be represented in the time domain by the equation

$$\frac{d^3y}{dt^3}(t) + \frac{1}{C_1R}\frac{d^2y}{dt^2}(t) + \frac{C_1+C_2}{LC_1C_2}\frac{dy}{dt}(t) + \frac{1}{LC_1C_2R}y(t) = \frac{1}{LC_1C_2R}u(t).$$

The only possibility open to us is to see if the component values in this equation can be chosen in such a way as to produce a form equivalent to (4.12). Clearly the input is the voltage $u(t)$, and the output the voltage measured across the capacitor C_2.

Evidently, we must have

$$2\omega_c = \frac{1}{C_1R} \qquad 2\omega_c^2 = \frac{C_1+C_2}{LC_1C_2} \qquad \omega_c^3 = \frac{1}{LC_1C_2R}$$

from which it follows that,

$$C_1 = \frac{1}{2R\omega_c}, \quad C_2 = \frac{3}{2R\omega_c}, \quad \text{and } L = \frac{4R}{3\omega_c}.$$

We again see that R can be chosen at will, with the other circuit parameters then determined uniquely.

It is clear that we cannot continue in this way. Valuable as our previous experience based on the analysis of elementary circuits has been, it is time to seek a more general approach. This is the purpose of the next section.

4.4 BUTTERWORTH FILTERS, THE GENERAL CASE

We approach the generalization problem in two stages. First, we show that the transfer functions themselves can be generated easily for arbitrary filter orders. This is achieved by using a **recurrence relation**, or **generating function**, for the coefficients of the appropriate denominators, called the **Butterworth polynomials**. The approach used is similar to that given in Weinberg [30]. When we have completed this task, we address the realization problem in the general case. At that point we show that *it is always possible to give a circuit construction which will operate as a Butterworth filter of any specified order.*

In order to deal with the first stage, recall from (4.4) that if $H(s)$ is the transfer function of the nth-order filter, then

$$H(s)H(-s) = \frac{1}{1 + (s/j\,\omega_c)^{2n}} \, .$$

Now, writing $x = s/\omega_c$, we have

$$1 + (s/j\,\omega_c)^{2n} = 1 + x^{2n}/(-1)^n = (-1)^n \prod_{k=0}^{2n-1} (\lambda_k - x) \tag{4.13}$$

$$= (-1)^n \prod_{k=0}^{2n-1} (x - \lambda_k) \, ,$$

where $\lambda_k = \exp(j\,(2k+1+n)\pi/2n)$, $k = 0, 1, \ldots 2n-1$, are the zeros of $1 + x^{2n}/(-1)^n$. As we observed, none of these zeros lie on the imaginary axis and this means that $|H(j\omega)|^2$ is always finite for all values of ω.

Noting that

$$\lambda_{k+n} = \exp(j\,(2k + 2n + 1 + n)\pi/2n)$$
$$= \exp(j\,\pi)\,\exp(j\,(2k + 1 + n)\pi/2n)$$
$$= -\lambda_k,$$

we see that we can write (4.13) as

$$1 + x^{2n}/(-1)^n = (-1)^n \prod_{k=0}^{n-1} (\lambda_k - x) \prod_{k=0}^{n-1} (-\lambda_k - x)$$

$$= \prod_{k=0}^{n-1} (\lambda_k - x) \prod_{k=0}^{n-1} (\lambda_k + x) \, . \tag{4.14}$$

Now, define $B_n(x)$ as

$$B_n(x) = \prod_{k=0}^{n-1} (x - \lambda_k) \tag{4.15}$$

$$= (-1)^n \prod_{k=0}^{n-1} (\lambda_k - x) \, .$$

Note that $\arg \lambda_k = (2k + 1 + n)\pi/2n$ for $k = 0, \ldots, n-1$ which, for a given value of n, increases monotonically with k. When $k = 0$ the argument of λ_k is

$$\frac{n+1}{2n}\,\pi = \left(1 + \frac{1}{n}\right)\frac{\pi}{2} > \frac{\pi}{2} \quad \text{for all } n.$$

When $k = n - 1$ the value is

$$\frac{3n-1}{2n}\,\pi = \left(3 - \frac{1}{n}\right)\frac{\pi}{2} < \frac{3\pi}{2} \quad \text{for all } n.$$

Thus, $\pi/2 < \arg(\lambda_k) < 3\pi/2$ for $k = 0, \ldots, n-1$, and these are the zeros in the left half plane. It follows that

$$B_n(-x) = (-1)^n \prod_{k=0}^{n-1} (\lambda_k + x)$$

and so

$$1 + x^{2n}/(-1)^n = B_n(x) B_n(-x).$$

We are now able to write $H(s)H(-s)$ in the alternative form

$$H(s)H(-s) = \frac{1}{B_n(x) B_n(-x)},$$

and we choose

$$H(s) = 1/B_n(x),$$

in order to generate a stable system. The process continues with an examination of $B_n(x)$ itself. Clearly $B_n(x)$ is a polynomial of degree n in x, the normalized variable. Thus, we write

$$B_n(x) = \sum_{k=0}^{n} a_k x^k, \qquad (4.16)$$

and our task is now to determine the values of the coefficients a_k, $k = 0, 1, \ldots, n$. First, we rewrite the product formula for $B_n(x)$ in a more useful form.

Writing $W = \exp(j\,\pi/2n)$, we have

$$B_n(x) = \prod_{k=0}^{n-1} (x - W^{2k+1+n}). \qquad (4.17)$$

Putting $x = 0$ in (4.16) and (4.17) shows that

$$a_0 = \prod_{k=0}^{n-1} (-W^{2k+1+n})$$

and recalling that $W = \exp(j\,\pi/2n)$, we have

$$a_0 = \prod_{k=0}^{n-1} \{-\exp((2k+1+n)\,j\,\pi/2n)\} = \prod_{k=0}^{n-1} \{\exp((2k+1+3n)\,j\,\pi/2n)\}.$$

This is just the product of n complex numbers each of unit modulus (or magnitude), and thus the product also has modulus 1. The argument of the product is given by the sum of the arguments of each term of the product as

$$\arg(a_0) = \sum_{k=0}^{n-1} (2k+1+3n)\,\pi/2n$$

$$= \left(\sum_{k=0}^{n-1} k\pi/n\right) + n(1+3n)\pi/2n$$

$$= n(n-1)\,\pi/2n + (1+3n)\,\pi/2$$

$$= 2n\pi.$$

Thus, we have established that for all filter orders n, the argument of a_0 is an integral multiple of 2π, and so $\boxed{a_0 = 1}$, for all integers n.

Returning to the product formula (4.17) for $B_n(x)$, we write this as

$$B_n(x) = (x - W^{n+1}) \prod_{k=1}^{n-1} (x - W^{2k+1+n})$$

$$= \frac{x - W^{n+1}}{x - W^{3n+1}} \prod_{k=1}^{n} (x - W^{2k+1+n}).$$

Now letting $q = k - 1$, and noting that $W^n = e^{j\pi/2} = j$, we get

$$B_n(x) = \frac{x - jW}{x + jW} W^{2n} \prod_{q=0}^{n-1} (x/W^2 - W^{2q+1+n})$$

$$= -\frac{x - jW}{x + jW} \prod_{k=0}^{n-1} (x/W^2 - W^{2k+1+n})$$

$$= -\frac{x - jW}{x + jW} B_n(x/W^2),$$

that is,

$$(x + jW)B_n(x) + (x - jW)B_n(x/W^2) = 0. \tag{4.18}$$

We now substitute (4.16) into (4.18) and obtain,

$$(x + jW) \sum_{k=0}^{n} a_k x^k + (x - jW) \sum_{k=0}^{n} a_k W^{-2k} x^k = 0.$$

Equating coefficients of like powers of x, we find that

$$a_k + jW a_{k+1} + W^{-2k} a_k - jW^{-2k-1} a_{k+1} = 0, \quad 0 \le k \le n-1,$$

or,

$$a_{k+1} = a_k \frac{1 + W^{-2k}}{j(W^{-2k-1} - W)}$$

$$= a_k \frac{W^k + W^{-k}}{j(W^{-k-1} - W^{k+1})}$$

$$= a_k \frac{\cos(k\alpha)}{\sin\{(k+1)\alpha\}} \tag{4.19}$$

where $\alpha = \pi/2n$.

It is easy to see that *using the reccurence relation* (4.19), *with* $a_0 = 1$, *we can generate the filter coefficients for any order* n. Also, it is easy to see that the coefficients are given explicitly by the formula:

$$a_k = \prod_{r=1}^{k} \frac{\cos\{(r-1)\alpha\}}{\sin(r\alpha)}, \tag{4.20}$$

by repeated use of (4.19). Here, k takes the values $1, 2, \ldots, n$, but if we adopt the usual convention that if the upper limit of the product is less than the lower, then the value of the product is one, the result holds for $k = 0$, also.

It is also possible to establish that $a_k = a_{n-k}$, a result which can be used to avoid a significant amount of calculation. This is achieved using (4.20), as follows. At once, we have for $0 \leq k \leq (n-1)/2$

$$a_k = \prod_{r=1}^{k} \frac{\cos\{(r-1)\alpha\}}{\sin(r\alpha)}$$

$$= \prod_{r=1}^{n-k} \frac{\cos\{(r-1)\alpha\}}{\sin(r\alpha)} \Bigg/ \prod_{r=k+1}^{n-k} \frac{\cos\{(r-1)\alpha\}}{\sin(r\alpha)}$$

$$= a_{n-k} \frac{\displaystyle\prod_{r=k+1}^{n-k} \sin(r\alpha)}{\displaystyle\prod_{r=k+1}^{n-k} \cos\{(r-1)\}\alpha}.$$

But, letting $p = n - r + 1$, we see that

$$\prod_{r=k+1}^{n-k} \cos\{(r-1)\alpha\} = \prod_{p=k+1}^{n-k} \cos\{(n-p)\alpha\}.$$

Since $n\alpha = \pi/2$, it follows that $\cos\{(n-p)\alpha\} = \sin(p\alpha)$ and so

$$\prod_{r=k+1}^{n-k} \cos\{(r-1)\alpha\} = \prod_{r=k+1}^{n-k} \sin(r\alpha)$$

from which it follows that

$$a_k = a_{n-k}, \quad k = 0, 1, \ldots, n. \tag{4.21}$$

Example 4.1
Calculate the coefficients of the fourth-order Butterworth polynomial.

Since $n = 4$, we have that $\alpha = \pi/2n = \pi/8$, meaning that the coefficients are given by (4.20) as

$$a_k = \prod_{r=1}^{k} \cos\{(r-1)\pi/8\}/\sin(r\pi/8), \quad k = 0, \ldots, 4.$$

$a_0 = 1$, as discussed above, and when $k = 1$, we obtain

$$a_1 = \prod_{r=1}^{1} \cos\{(r-1)\pi/8\}/\sin(r\pi/8)$$
$$= 1/\sin(\pi/8)$$
$$= 2.6131.$$

n	a_0	a_1	a_2	a_3	a_4	a_5	a_6	a_7	a_8
1	1.0000	1.0000							
2	1.0000	1.4142	1.0000						
3	1.0000	2.0000	2.0000	1.0000					
4	1.0000	2.6131	3.4142	2.6131	1.0000				
5	1.0000	3.2361	5.2361	5.2361	3.2361	1.0000			
6	1.0000	3.8637	7.4641	9.1416	7.4641	3.8637	1.0000		
7	1.0000	4.4940	10.0978	14.5918	14.5918	10.0978	4.4940	1.0000	
8	1.0000	5.1258	13.1371	21.8462	25.6884	21.8462	13.1371	5.1258	1.0000

Table 4.1: Coefficients of the Butterworth polynomials, for the cases $n = 1$ to $n = 8$.

Also,

$$a_2 = \prod_{r=1}^{2} \cos\{(r-1)\pi/8\}/\sin(r\pi/8)$$

$$= \frac{1}{\sin(\pi/8)} \frac{\cos(\pi/8)}{\sin(\pi/4)}$$

$$= 3.4142.$$

It would be possible to find a_3 and a_4 in this way; however, (4.21) allows us to write

$$a_3 = a_{4-3} = a_1 = 2.6131, \text{ and}$$

$$a_4 = a_{4-4} = a_0 = 1.$$

We may deduce that the transfer function of the fourth-order low-pass Butterworth filter with cut-off frequency ω_c is

$$H(s) = \frac{\omega_c^4}{s^4 + 2.6131\omega_c s^3 + 3.4142\omega_c^2 s^2 + 2.6131\omega_c^3 s + \omega_c^4} \, .$$

$\square$

In order to avoid the tedium of repeating this type of calculation, the coefficients for the polynomials of orders $n = 1, 2, \dots 8$, are given in Table 4.1.

The class of Butterworth filters provide a practical solution to the low-pass filter design problem. We show later that it is possible to give a procedure for the implementation of a Butterworth filter of arbitrary order. One of the desirable properties of such filters is the particularly flat nature of the amplitude response in the pass-band. This effect is evident in Figure 4.5, and can be explained as follows. Since

$$|H_n(j\omega)| = \frac{1}{\sqrt{(1 + (\omega/\omega_c)^{2n})}}$$

$$= [1 + (\omega/\omega_c)^{2n}]^{-1/2}$$

$$= 1 - \tfrac{1}{2}(\omega/\omega_c)^{2n} + \tfrac{3}{8}(\omega/\omega_c)^{4n} - \cdots,$$

when $|\omega| < \omega_c$, we see that the first $2n-1$ derivatives of $|H(j\omega)|$ are zero at $\omega = 0$. This is the reason for the flat shape of the amplitude response near $\omega = 0$ and is known as the **maximally-flat property**.

The total filter design process is beyond the scope of this text, and forms a part of a study of electrical engineering. However, we may consider how simple design criteria are used to determine a suitable order for a Butterworth filter. (The choice of a Butterworth filter is, of course, a design decision in itself!)

Example 4.2
A low-pass Butterworth filter with cut-off frequency ω_c is to be designed to meet the following specifications.

(i) The amplitude response magnitude at $\omega = 3\omega_c/4$ must be not less than 95% of the maximum value (the dc gain at $\omega = 0$).

(ii) At $\omega = 5\omega_c/4$, the amplitude response magnitude must be less than 20% of the maximum value.

For the Butterworth filter of order n, the amplitude response is given by

$$|H(j\omega)| = \frac{1}{\sqrt{1 + (\omega/\omega_c)^{2n}}},$$

with a maximum value of 1 at $\omega = 0$.

Our first requirement is thus that

$$|H(j\,3\omega_c/4)| = \frac{1}{\sqrt{1 + (3/4)^{2n}}} \geq \frac{95}{100}.$$

This means that

$$1 + (3/4)^{2n} \leq \frac{10000}{9025}$$

or

$$(3/4)^{2n} \leq 0.1080$$

and we seek the smallest integer n for which the inequality holds.

Taking natural logs, we have that

$$2n\ln(3/4) \leq \ln(0.1080),$$

giving $n \geq 3.868$. Thus, the smallest integer value for n is $n = 4$.

The second requirement means that we must have

$$|H(j\,5\omega_c/4)| = \frac{1}{\sqrt{1 + (5/4)^{2n}}} < \frac{1}{5}.$$

Rearranging and taking natural logs leads to

$$n > \ln(24)/2\ln(5/4) = 7.121,$$

and thus, the smallest integer value of n is $n = 8$.

In order that both requirements may be simultaneously satisfied, we choose $n \geq \max[4, 8] = 8$, leading us to propose an eighth-order filter design. The coefficients can now be read off from Table 4.1.

$\square$

We are now ready to proceed to the second part of our generalization process. The task is now to show that any Butterworth filter, generated by the above procedure, can be implemented using analogue circuits of the type discussed earlier.

Consider the structure of the transfer function of the nth-order filter. We have

$$H_n(s) = \frac{1}{B_n(x)} = \frac{1}{B_n(s/\omega_c)} \, .$$

Let $\tilde{B}_n(s) = \omega_c^n B_n(x)$ with x replaced by s/ω_c. So

$$H_n(s) = \omega_c^n / \tilde{B}_n(s) = \tilde{B}_n(0)/\tilde{B}_n(s).$$

Also, $\tilde{B}_n(s) = M_n(s) + N_n(s)$, where $M_n(s)$ is an even polynomial, called the even part, and $N_n(s)$ is an odd polynomial, called the odd part. For example, when $n = 3$, we have

$$\begin{aligned}
\tilde{B}_3(s) &= s^3 + 3\omega_c s^2 + 3\omega_c^2 s + \omega_c^3 \\
&= (s^3 + 3\omega_c^2 s) + (3\omega_c s^2 + \omega_c^3)
\end{aligned}$$

so that $M_3(s) = (3\omega_c s^2 + \omega_c^3)$ and $N_3(s) = (s^3 + 3\omega_c^2 s)$.

Clearly $\tilde{B}_n(0) = M_n(0)$, and so we can write,

$$\begin{aligned}
H_n(s) &= M_n(0)/(M_n(s) + N_n(s)) \\
&= \frac{A M_n(0)/N_n(s)}{A + A M_n(s)/N_n(s)},
\end{aligned} \tag{4.22}$$

where A is any constant. Notice that if n, the filter order, is even, then

$$\deg\{M_n(s)\} = \deg\{N_n(s)\} + 1, \tag{4.23}$$

whilst if n is odd, then

$$\deg\{N_n(s)\} = \deg\{M_n(s)\} + 1. \tag{4.24}$$

In Figure 4.8 we show an L–C ladder network, which represents an extension of the type of circuit already seen to be suitable for the implementation of low order filters.

Using the techniques of Chapter 1, we analyse this circuit as follows. For the capacitor, C_1, we have from (1.4),

$$i(t) - i_1(t) = C_1 \frac{\mathrm{d}v_1}{\mathrm{d}t}(t)$$

and, on taking the Laplace transform, we can express this relation as

$$I(s) = C_1 s V_1(s) + I_1(s).$$

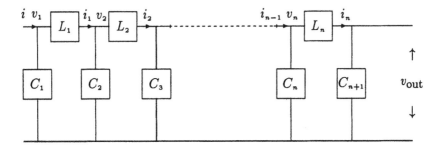

Figure 4.8: An L–C ladder network.

For the first inductance, L_1, we use (1.2) and obtain

$$v_1(t) - v_2(t) = L_1 \frac{\mathrm{d}i_1}{\mathrm{d}t}(t),$$

which, in the transform domain, becomes

$$V_1(s) = L_1 s I_1(s) + V_2(s).$$

Proceeding in this manner for the other components, we eventually find the set of equations,

$$I(s) = C_1 s V_1(s) + I_1(s) \tag{4.25a}$$
$$V_1(s) = L_1 s I_1(s) + V_2(s) \tag{4.25b}$$
$$I_1(s) = C_2 s V_2(s) + I_2(s) \tag{4.25c}$$

$$\vdots$$

$$I_{n-1}(s) = C_n s V_n(s) + I_n(s) \tag{4.25d}$$
$$V_n(s) = L_n s I_n(s) + V_{\text{out}}(s) \tag{4.25e}$$
$$I_n(s) = C_{n+1} s V_{\text{out}}(s). \tag{4.25f}$$

Then, back-substituting, we see that

$$V_n(s) = [L_n C_{n+1} s^2 + 1] V_{\text{out}}(s)$$
$$I_{n-1}(s) = [C_n C_{n+1} L_n s^3 + (C_n + C_{n+1}) s] V_{\text{out}}(s).$$

Continuing this process, we soon find that

$$V_1(s) = F(s) V_{\text{out}}(s) \tag{4.26}$$
$$I(s) = G(s) V_{\text{out}}(s), \tag{4.27}$$

where $F(s)$ is an even polynomial in s, with coefficients determined by the component values, and $F(0) = 1$.

Similarly, $G(s)$ is an odd polynomial and

$$\deg\{G(s)\} = \deg\{F(s)\} + 1. \tag{4.28}$$

We can write (4.26) and (4.27) as

$$V_1(s)/I(s) = F(s)/G(s) \tag{4.29}$$

and

$$V_{out}(s)/I(s) = 1/G(s) = F(0)/G(s), \tag{4.30}$$

which shows that we can determine $V_{out}(s)/I(s)$ from a knowledge of $V_1(s)/I(s)$.

In Figure 4.9 we show such an L–C ladder network with input voltage v_{in} 'driving' the circuit through a resistance R.

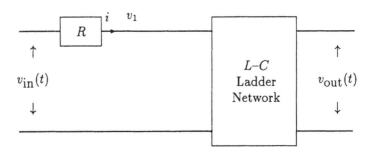

Figure 4.9: A 'type 1' L–C ladder network with input voltage $v_{in}(t)$ and output voltage $v_{out}(t)$

Using (1.1), we can write at once,

$$v_{in}(t) - v_1(t) = Ri(t),$$

which, in the transform domain, becomes

$$V_{in}(s) = V_1(s) + RI(s).$$

Thus,

$$
\begin{aligned}
\frac{V_{out}(s)}{V_{in}(s)} &= \frac{V_{out}(s)}{V_1(s) + RI(s)} \\
&= \frac{V_{out}(s)/I(s)}{R + V_1(s)/I(s)} \\
&= \frac{1/G(s)}{R + F(s)/G(s)}
\end{aligned} \tag{4.31}
$$

from (4.29) and (4.30).

We now compare (4.31) with (4.22), and deduce that, in order to use this type of circuit to implement a Butterworth filter, we should set $A = R$ and choose

$$1/G(s) = RM_n(0)/N_n(s), \tag{4.32}$$

and

$$F(s)/G(s) = RM_n(s)/N_n(s). \tag{4.33}$$

However, from (4.28) we must have

$$\deg\{G(s)\} = \deg\{F(s)\} + 1,$$

that is,

$$\deg\{N_n(s)\} = \deg\{M_n(s)\} + 1,$$

which, from (4.24), means that n, the filter order, must be odd.

Recall that we start from a position of knowing the transfer function of our filter, that is, we know

$$H_n(s) = M_n(0)/(M_n(s) + N_n(s)).$$

Our design method will be to use (4.32) and (4.33) to obtain $G(s)$ and $F(s)$ for the implementation of our desired filter.

Thus, we have established that Butterworth filters of odd order can be implemented using 'type 1' L–C ladder networks. We have not yet shown how to calculate the component values, but we will address this matter after we have considered the case when the filter order, n, is an even integer.

In Figure 4.10 we illustrate a second, or type 2, L–C ladder network. Following

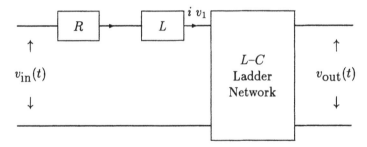

Figure 4.10: A 'type 2' L–C ladder network with input voltage $v_{\text{in}}(t)$ and output voltage $v_{\text{out}}(t)$

the same procedure as for the type 1 network, it is easily shown that

$$\frac{V_{\text{out}}(s)}{V_{\text{in}}(s)} = \frac{V_{\text{out}}(s)}{V_1(s) + RI(s) + LsI(s)}$$

$$= \frac{V_{\text{out}}(s)/I(s)}{R + Ls + V_1(s)/I(s)}$$

$$= \frac{1/G(s)}{R + Ls + F(s)/G(s)}.$$

Comparing with (4.22), shows that we must set $A = R$ and

$$1/G(s) = RM_n(0)/N_n(s)$$

with

$$Ls + F(s)/G(s) = RM_n(s)/N_n(s). \qquad (4.34)$$

It then follows that

$$\deg\{M_n(s)\} = \deg\{N_n(s)\} + 1,$$

meaning that $B_n(s)$ is an even polynomial.

The details of this case are left as an exercise. We have now established that the Butterworth filter can be realized in the general case, by use of an L–C ladder network. The network will be of type 1 or 2, depending upon whether the filter order is odd or even. To complete the design exercise, we must give a method by which the circuit parameters, or component values, may be obtained from a specification of the transfer function. An elegant way of achieving this objective is by use of a technique known as **continued fraction expansion**.

We recall that

$$H_n(s) = \frac{\tilde{B}_n(0)}{\tilde{B}_n(s)} = \frac{M_n(0)}{M_n(s) + N_n(s)}$$

and so we can readily obtain the polynomials $M_n(s)$ and $N_n(s)$ from the desired filter transfer function $H_n(s)$. The approach we adopt involves finding a continued fraction expansion for $RM_n(s)/N_n(s)$, as given by (4.33) when n is odd, and by (4.34) when n is even. Let us begin by considering the case where n is odd. It follows that $RM_n(s)/N_n(s) = F(s)/G(s)$ and, using (4.29), we write

$$F(s)/G(s) = V_1(s)/I(s)$$
$$= \frac{1}{I(s)/V_1(s)}$$
$$= \frac{1}{C_1 s + I_1(s)/V_1(s)} \quad \text{(by (4.25a))}$$
$$= \frac{1}{C_1 s + \dfrac{1}{V_1(s)/I_1(s)}}$$
$$= \frac{1}{C_1 s + \dfrac{1}{L_1 s + V_2(s)/I_1(s)}} \quad \text{(by (4.25b))}$$
$$= \frac{1}{C_1 s + \dfrac{1}{L_1 s + \dfrac{1}{I_1(s)/V_2(s)}}}$$
$$= \frac{1}{C_1 s + \dfrac{1}{L_1 s + \dfrac{1}{C_2 s + \ldots}}} \quad (4.35)$$

and so on. *In this continued fraction form, the parameter values emerge explicitly.* We demonstrate the method by example.

Example 4.3

Determine a realization for the third-order Butterworth filter.

From Table 4.1, we obtain the required transfer function as

$$H_3(s) = \frac{\omega_c^3}{s^3 + 2\omega_c s^2 + 2\omega_c^2 s + \omega_c^3} \ .$$

Thus,

$$M_3(s) = 2\omega_c s^2 + \omega_c^3$$

and

$$N_3(s) = s^3 + 2\omega_c^2 s.$$

Then $F(s)/G(s) = RM_3(s)/N_3(s)$ and, using the substitution $x = s/\omega_c$,

$$\begin{aligned}
F(s)/G(s) &= RM_3(s)/N_3(s) \\
&= R\frac{[2x^2 + 1]}{[x^3 + 2x]} \\
&= \frac{1}{\dfrac{[x^3 + 2x]}{R[2x^2 + 1]}} \\
&= \frac{1}{\dfrac{x[2x^2 + 1 + 3]}{2R[2x^2 + 1]}} \\
&= \frac{1}{\dfrac{x}{2R} + \dfrac{3x}{2R[2x^2 + 1]}} \\
&= \frac{1}{\dfrac{x}{2R} + \dfrac{1}{\dfrac{2R[2x^2 + 1]}{3x}}} \\
&= \frac{1}{\dfrac{x}{2R} + \dfrac{1}{\dfrac{4Rx}{3} + \dfrac{2R}{3x}}} \\
&= \frac{1}{\dfrac{x}{2R} + \dfrac{1}{\dfrac{4Rx}{3} + \dfrac{1}{\dfrac{3x}{2R}}}} \ .
\end{aligned} \tag{4.36}$$

It is now easy to read off corresponding values between (4.35) and (4.36) to obtain

$$C_1 = \frac{1}{2R\omega_c} \qquad L_1 = \frac{4R}{3\omega_c} \qquad C_2 = \frac{3}{2R\omega_c},$$

exactly in accord with the results of §4.3. The circuit design is shown in Figure 4.11.

□

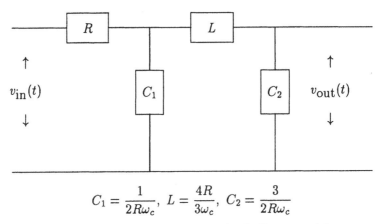

$$C_1 = \frac{1}{2R\omega_c}, \; L = \frac{4R}{3\omega_c}, \; C_2 = \frac{3}{2R\omega_c}$$

Figure 4.11: An implementation of the third-order Butterworth low-pass filter.

To conclude this discussion, we examine the procedure when n is an even integer. From (4.34) and (4.35) we see that in this case,

$$RM_n(s)/N_n(s) = Ls + \cfrac{1}{C_1 s + \cfrac{1}{L_1 s + \cfrac{1}{C_2 s + \cdots}}}. \qquad (4.37)$$

Again, the procedure reduces to a continued fraction development of $RM_n(s)/N_n(s)$.

Example 4.4
Find a realization of the second-order Butterworth filter, given by

$$H_2(s) = \frac{\omega_c^2}{s^2 + \sqrt{2}\,\omega_c s + \omega_c^2}$$

We have $M_2(s) = s^2 + \omega_c^2$, and $N_2(s) = \sqrt{2}\,\omega_c s$. Thus, writing $x = s/\omega_c$, we expand $\dfrac{R[x^2 + 1]}{\sqrt{2}\,x}$ as

$$\frac{R[x^2 + 1]}{\sqrt{2}\,x} = \frac{Rx}{\sqrt{2}} + \frac{R}{\sqrt{2}\,x} = \frac{Rx}{\sqrt{2}} + \frac{1}{\dfrac{\sqrt{2}\,x}{R}}.$$

Then, from (4.37), and remembering that $s = x\omega_c$, we have

$$L = \frac{R}{\sqrt{2}\,\omega_c}, \quad \text{and} \quad C_1 = \frac{\sqrt{2}}{R\omega_c},$$

again in agreement with the result in §4.3. The circuit design is shown in Figure 4.12.

□

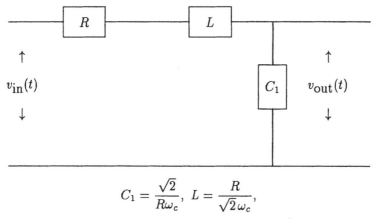

$$C_1 = \frac{\sqrt{2}}{R\omega_c}, \ L = \frac{R}{\sqrt{2}\,\omega_c},$$

Figure 4.12: A realization of the second-order Butterworth low-pass filter.

4.5   FILTER TRANSFORMATIONS

So far, our discussion has concentrated on the design of low-pass filters. For many applications, a filter of another type will be required. For example, we may wish to design a filter to reject a band of frequencies which contains signals interfering with signals we wish to observe. In §4.1, we illustrated the ideal frequency responses of the different filter types. We now wish to design filters of these types, again with frequency responses which are approximations to the ideal response.

It is not necessary to repeat the exercise carried out for the design of low-pass filters because we are able to define filter transformations. These transformations are applied to the transfer functions of low-pass filters to produce filters with high-pass, band-pass, or band-reject characteristics. In fact, it is also unnecessary to re-work the circuit synthesis calculations, since methods are known for the modification of low-pass designs to produce circuits which operate as filters of the other types. This latter topic is beyond the scope of this book; however, the interested reader should consult Kuo [14].

The first transformation we study is that which produces a high-pass filter from a low-pass prototype. It is reasonable to suppose that such a transformation could be achieved by replacing s in the transfer function of the low-pass filter, by $1/s$. We will demonstrate that this is indeed a sensible procedure.

Define the prototype low-pass filter by its transfer function, $H_{\mathrm{LP}}(s)$, and we again introduce the non-dimensional or normalized frequency variable, $x = s/\omega_c$. Now make the transformation $x \to 1/y$, where $y(= s/\omega_\alpha)$ is a second normalized frequency variable. This means that in the transfer function of the low-pass proto-type, $H_{\mathrm{LP}}(s)$, s/ω_c is replaced by ω_α/s. The critical, or cut-off, frequencies of the low-pass prototype filter were located at $\omega = \pm\omega_c$, corresponding to $x = \pm j$. Thus, the normalized critical frequencies of the high-pass design are given by

$$1/y = \pm j, \text{ or } y = \mp j.$$

Since $y = s/\omega_\alpha$, we see that setting $s = j\omega'_c$, where ω'_c is the critical frequency of the high-pass filter, yields

$$\omega'_c = \pm\omega_\alpha.$$

We can examine the effect of this transformation on the Butterworth filters. We have

$$H(s)H(-s) = \frac{1}{1 + (s/j\omega_c)^{2n}},$$

or writing $x = s/\omega_c$,

$$H(s)H(-s) = \frac{1}{1 + (x/j)^{2n}}.$$

Using the transformation $x \to 1/y$, we have, for the transfer function $H'(s)$ of the high-pass filter,

$$H'(s)H'(-s) = \frac{1}{1 + (1/jy)^{2n}}$$

$$= \frac{1}{1 + (\omega_\alpha/js)^{2n}}$$

leading to,

$$|H'(j\omega)| = \frac{1}{\sqrt{1 + (\omega_\alpha/\omega)^{2n}}}.$$

Clearly, if $\omega > \omega_\alpha$, $|H'(j\omega)| \to 1$ as n increases, whilst if $\omega < \omega_\alpha$, $|H'(j\omega)| \to 0$, and thus $H'(s)$ is indeed the transfer function of a high-pass filter.

Example 4.5
Using a fourth-order Butterworth low-pass filter as prototype, design a high-pass filter with cut-off frequency 20Hz.

With a cut-off frequency of 20Hz, we have as cut-off frequency of the high-pass filter $\omega'_c = 2\pi.20 = 125.66$ radians/sec. Selecting the fourth-order Butterworth low-pass filter, with cut-off frequency $\omega_c = 1$, we find from Table 4.1 that the prototype low-pass transfer function is given by

$$H_{LP}(s) = \frac{1}{s^4 + 2.6131s^3 + 3.4142s^2 + 2.6131s + 1}.$$

We now replace s with $40\pi/s$, since for the prototype the cut-off frequency was $\omega_c = 1$, to obtain

$$H(s) = \frac{s^4}{s^4 + 2.6131(40\pi)s^3 + 3.4142(40\pi)^2s^2 + 2.6131(40\pi)^3s + (40\pi)^4}.$$

This is the transfer function of the high-pass filter, and in Figure 4.13, we illustrate the amplitude response.

$\square$

We now consider other filter transformations which produce the band-pass and band-reject filters of §4.3. These transformations can be motivated from a deeper

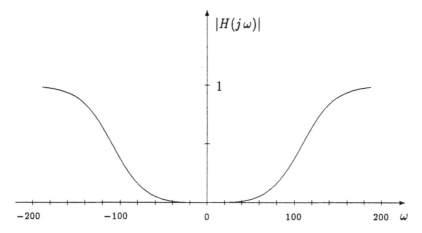

Figure 4.13: The amplitude response of the high-pass design.

knowledge of circuit theory than we have developed in this book. Rather than pursue this here, we consider the transformations alone and content ourselves with a demonstration that each produces the desired filter type.

First we consider a transformation which produces a band-pass filter from a low-pass prototype. Again we work with the two non-dimensional frequency variables x and y, defined as above. The transformation we consider is defined by

$$x \to \frac{y^2 + 1}{y} \frac{\omega_\alpha}{\omega_\beta} \, ,$$

meaning that in the transfer function $H_{\mathrm{LP}}(s)$ of the low-pass filter,

$$s/\omega_c \text{ is replaced by } \frac{s^2 + \omega_\alpha^2}{s\omega_\beta} \, . \tag{4.38}$$

Recall that the critical frequencies of the low-pass filter were defined by $x = \pm j$, so that the critical frequencies of the new design are obtained by solving the equation

$$\frac{y^2 + 1}{y} \frac{\omega_\alpha}{\omega_\beta} = \pm j \, .$$

That is,

$$(y^2 + 1)\omega_\alpha/\omega_\beta = \pm j \, y.$$

We thus have two quadratic equations to solve,

$$\omega_\alpha y^2 - j \, \omega_\beta y + \omega_\alpha = 0 \tag{4.39}$$

and

$$\omega_\alpha y^2 + j \, \omega_\beta y + \omega_\alpha = 0. \tag{4.40}$$

Solving (4.39), we obtain for the critical values of y, y_c say,

$$y_c = \frac{j \omega_\beta \pm j \sqrt{\omega_\beta^2 + 4\omega_\alpha^2}}{2\omega_\alpha}$$

and we note that these values are purely imaginary with one on each side of the origin.

Similarly, from (4.40) we obtain

$$y_c = \frac{-j\,\omega_\beta \pm j\,\sqrt{\omega_\beta^2 + 4\omega_\alpha^2}}{2\omega_\alpha}$$

again obtaining purely imaginary values with one on each side of the origin. Essentially, what has happened here is that each critical frequency of the low-pass prototype has mapped into two critical values for the new design, one on each side of the origin.

Replacing y_c with $j\omega_c'/\omega_\alpha$, we see that the four critical values of the real frequency variable ω are given by

$$\omega_c' = \frac{-\omega_\beta \pm \sqrt{\omega_\beta^2 + 4\omega_\alpha^2}}{2}$$

and

$$\omega_c' = \frac{\omega_\beta \pm \sqrt{\omega_\beta^2 + 4\omega_\alpha^2}}{2}.$$

Now define

$$\omega_u = \frac{\omega_\beta + \sqrt{\omega_\beta^2 + 4\omega_\alpha^2}}{2}$$

and

$$\omega_l = \frac{-\omega_\beta + \sqrt{\omega_\beta^2 + 4\omega_\alpha^2}}{2},$$

so that the critical frequencies ω_c' can be written as

$$\omega_c' = -\omega_u, \ -\omega_l, \ \omega_l, \ \text{and} \ \omega_u.$$

We must now show that we have indeed generated a band-pass filter. First we show that ω_α lies between ω_l and ω_u, and hence that $-\omega_\alpha$ lies between $-\omega_u$ and $-\omega_l$. Since

$$\omega_\alpha^2 = \omega_u \omega_l,$$

it follows that

$$\omega_l^2 \le \omega_\alpha^2 \le \omega_u^2.$$

Taking square roots then gives the desired result. This result enables us to locate the pass-band of the new filter, by investigating the effect of the transformation on the pass-band of the prototype low-pass filter. This was described, using the non-dimensional frequency variable x, by $|x| < 1$. Under the transformation, this becomes

$$|(y^2 + 1)/y|(\omega_\alpha/\omega_\beta) < 1.$$

Since, when $s = j\omega$, $y = j\omega/\omega_\alpha$, we have

$$|(\omega_\alpha^2 - \omega^2)/\omega\omega_\beta| < 1.$$

Restricting ourselves for the moment to the case $\omega > 0$, we have to examine the two cases
(1) $0 < \omega < \omega_\alpha$, and
(2) $\omega > \omega_\alpha$.

In case (1), we have

$$\omega_\alpha^2 - \omega^2 < \omega\omega_\beta,$$

or

$$\omega^2 + \omega\omega_\beta - \omega_\alpha^2 > 0.$$

Since $\omega > 0$, this means that

$$\omega > \frac{-\omega_\beta + \sqrt{\omega_\beta^2 + 4\omega_\alpha^2}}{2} = \omega_l.$$

Thus, if $\omega < \omega_\alpha$, then $\omega > \omega_l$.

On the other hand, if $\omega > \omega_\alpha$, we must have

$$\omega^2 - \omega_\alpha^2 < \omega\omega_\beta,$$

which leads to the requirement that

$$\omega < \omega_u.$$

So we may conclude that the pass-band of the low-pass filter has been mapped to the interval $\omega_l < \omega < \omega_u$.

In a similar way, we can analyse the situation when $\omega < 0$, to establish that the total effect of the transformation is to produce two images of the pass-band of the low-pass filter located in the intervals

$$\omega_l < \omega < \omega_u \text{ and } -\omega_u < \omega < -\omega_l.$$

We can now be sure that the transformation defined by (4.38) does produce a band-pass filter when applied to a low-pass prototype.

Before leaving this discussion, notice that

$$\omega_u - \omega_l = \omega_\beta,$$

and for this reason, ω_β is called the **bandwidth** of the filter. Also,

$$\omega_u\omega_l = \omega_\alpha^2, \text{ or } \omega_\alpha = \sqrt{\omega_u\omega_l},$$

and ω_α is thus called the **geometric centre frequency** of the band-pass filter. This means that the low-pass to band-pass transformation can be written as

$$s/\omega_c \rightarrow \frac{s^2 + \omega_u\omega_l}{s(\omega_u - \omega_l)}.$$

Let us also note that if $\omega_\beta \ll \omega_\alpha$, then

$$\omega_u \approx \omega_\alpha + \omega_\beta/2$$

and
$$\omega_l \approx \omega_\alpha - \omega_\beta/2,$$
so that
$$\omega_\alpha \approx (\omega_u + \omega_l)/2,$$
that is, the geometric centre frequency is, in this case, approximately the same as the algebraic centre frequency.

The use of a design procedure based on this transformation is demonstrated in Example 4.6.

Example 4.6

Using as prototype a second-order Butterworth low-pass filter, design a band-pass filter which has the following properties:
(1) the geometric centre frequency is specified as $f_\alpha = 20\text{Hz}$; and
(2) the bandwidth is to be $f_\beta = 1\text{Hz}$.

Here, $\omega_\alpha = 2\pi f_\alpha = 40\pi$ radians/sec, and $\omega_\beta = 2\pi f_\beta = 2\pi$ radians/sec. The prototype low-pass filter is defined by its transfer function as

$$H_{\text{LP}}(s) = \frac{1}{(s/\omega_c)^2 + 1.4142(s/\omega_c) + 1}.$$

Making the transformation

$$s/\omega_c \rightarrow \frac{s^2 + \omega_\alpha^2}{s\omega_\beta},$$

we obtain the transfer function of the band-pass filter as

$$H(s) = \frac{1}{((s^2 + \omega_\alpha^2)/s\omega_\beta)^2 + 1.4142((s^2 + \omega_\alpha^2)/s\omega_\beta) + 1}$$

$$= \frac{s^2\omega_\beta^2}{s^4 + 1.4142\omega_\beta s^3 + (2\omega_\alpha^2 + \omega_\beta^2)s^2 + 1.4142\omega_\beta\omega_\alpha^2 s + \omega_\alpha^4}.$$

To obtain the transfer function of the design corresponding to our specification, we substitute $\omega_\alpha = 40\pi$, and $\omega_\beta = 2\pi$ to give

$$\frac{(2\pi)^2 s^2}{s^4 + 1.4142(2\pi)s^3 + (2(40\pi)^2 + (2\pi)^2)s^2 + 1.4142(40\pi)^2 2\pi s + (40\pi)^4}.$$

In Figure 4.14, we illustrate the amplitude response of this filter, and, in Figure 4.15, we illustrate the response of a similar design having a band-width of $\omega_\beta = 20\pi$.   □

In view of our discussion of the two previous transformations, it will come as no surprise that the transformation from low-pass to band-reject filter is given by

$$s/\omega_c \rightarrow \frac{s(\omega_u - \omega_l)}{s^2 + \omega_u\omega_l},$$

which is simply the reciprocal of the transformation used to produce band-pass designs. Here ω_l and ω_u denote the lower and upper edges of the stop-band of the filter.

The details of the analysis of this transformation are left as an exercise.

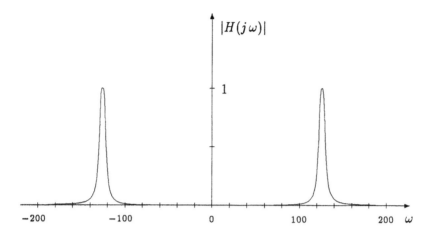

Figure 4.14: The amplitude response of the band-pass filter of Example 4.6.

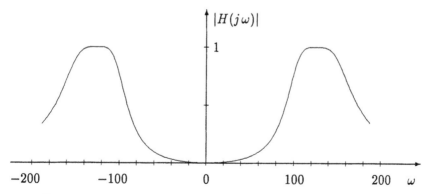

Figure 4.15: Amplitude response of the band-pass filter with band-width
$\omega_\beta = 20\pi$.

4.6　OTHER FILTER DESIGNS

In the discussion so far, we have used the Butterworth filter, both as a low-pass
filter and as the prototype for other designs using the transformations of §4.5.
Whilst the Butterworth filters are widely used, other designs are possible, and
indeed give superior performance in some applications. In all cases, the design of a
low-pass filter reduces to the approximation of the ideal frequency response shape
of Figure 4.1. We have seen that the Butterworth filters have the 'maximally flat'
property; however, other designs have sharper cut-off characteristics. The penalty
for this increased sharpness is 'ripple', perhaps in the pass-band or in the stop-band,
or possibly in both. In this section, we briefly consider another filter design which
is in use, and we illustrate a typical frequency response.

The design we consider is the **Chebyshev design**, which is based on the properties of the **Chebyshev polynomials**. The amplitude response of an nth-order
Chebyshev low-pass filter, with cut-off frequency ω_c, is given by

$$|H_n(j\,\omega)|^2 = \frac{K^2}{1 + \epsilon^2 T_n^2(\omega/\omega_c)}\,, \tag{4.41}$$

where

$$T_n(\omega) = \cos(n\cos^{-1}\omega) \tag{4.42}$$

and K is a constant.

The relation (4.42) defines the Chebyshev polynomials of the first kind, and we
note that when $|\omega| > 1$, this becomes

$$T_n(\omega) = \cosh(n\cosh^{-1}\omega).$$

The motivation for choosing the form specified by (4.41) for the design of a low-pass
filter follows from the well-known properties of the polynomials. These ensure that
as n increases we obtain an increasingly good approximation to the ideal low-pass
frequency response of Figure 4.1. Table 4.2 lists the first few Chebyshev polynomials.

n	$T_n(\omega)$
0	1
1	ω
2	$2\omega^2 - 1$
3	$4\omega^3 - 3\omega$
4	$8\omega^4 - 8\omega^2 + 1$
5	$16\omega^5 - 20\omega^3 + 5\omega$
6	$32\omega^6 - 48\omega^4 + 18\omega^2 - 1$
7	$64\omega^7 - 112\omega^5 + 56\omega^3 - 7\omega$

Table 4.2: The Chebyshev polynomials.

A more useful formula for generating the polynomials can be obtained from
(4.42) as follows.

Write $\alpha = \cos^{-1} w$, so that $T_n(\omega) = \cos(n\alpha)$, and then use the identity,

$$\cos\{(n+1)\alpha\} + \cos\{(n-1)\alpha\} = 2\cos(n\alpha)\cos\alpha. \tag{4.43}$$

But $\cos\{(n+1)\alpha\} = T_{n+1}(\omega)$, and $\cos\{(n-1)\alpha\} = T_{n-1}(\omega)$, and thus we have

$$T_{n+1}(\omega) + T_{n-1}(\omega) = 2\omega T_n(\omega),$$

or

$$T_{n+1}(\omega) = 2\omega T_n(\omega) - T_{n-1}(\omega).$$

From (4.42), we note that $T_n(0) = \cos(n\pi/2)$ and so

$$T_n(0) = \begin{cases} (-1)^{n/2}, & n \text{ even,} \\ 0, & n \text{ odd,} \end{cases}$$

and that $T_n(1) = 1$ for all n.

Furthermore, we can see that

$$T_n(-1) = (-1)^n = \begin{cases} 1, & n \text{ even,} \\ -1, & n \text{ odd.} \end{cases}$$

Also, it is clear from (4.42) that $T_n(\omega)$ oscillates between -1 and 1, when $|\omega| \leq 1$, whilst $T_n(\omega)$ increases monotonically when $|\omega| > 1$. These properties indicate that a generally satisfactory response shape can be obtained; however, there will be an element of ripple in the filter pass-band.

In order to proceed with the design process, we need to determine the filter transfer function. This is achieved by following the general method as derived for the Butterworth filters. As before, set

$$H(s)H(-s) = \frac{K^2}{1 + \varepsilon^2 T_n^2(s/j\,\omega_c)} \,,$$

where $H(s)$ is the transfer function of the desired filter. Now $H(s)H(-s)$ has poles determined by

$$T_n^2(s/j\,\omega_c) = -1/\varepsilon^2.$$

That is,

$$T_n(s/j\,\omega_c) = \pm j\,/\varepsilon,$$

or

$$\cos(n\cos^{-1}(s/j\,\omega_c)) = \pm j\,/\varepsilon.$$

Now, define $\cos^{-1}(s/j\,\omega_c) = a + j\,b$, to give

$$\cos(n(a+j\,b)) = \cos(na)\cosh(nb) - j\,\sin(na)\sinh(nb) = \pm j\,/\varepsilon. \tag{4.44}$$

Equating real and imaginary parts means that we must have

$$\cos(na)\cosh(nb) = 0 \tag{4.45}$$

and

$$\sin(na)\sinh(nb) = \pm 1/\varepsilon. \tag{4.46}$$

Since $\cosh(nb) \neq 0$, (4.45) implies that

$$\cos(na) = 0,$$

or

$$a = (2k+1)\pi/2n, \ k = 0, 1, 2, \ldots, n-1.$$

This means that $\sin(na)$ takes the values ± 1, and so from (4.46) we have,

$$\sinh(nb) = \pm(1/\varepsilon).$$

We can now determine the poles from (4.44) as

$$
\begin{aligned}
s_k &= j\,\omega_c \cos(a + j\,b) \\
&= j\,\omega_c[\cos(a)\cosh(b) - j\,\sin(a)\sinh(b)] \\
&= \omega_c[\sin(a)\sinh(b) + j\,\cos(a)\cosh(b)] \\
&= \omega_c[\sin((2k+1)\pi/2n)\sinh((1/n)\sinh^{-1}(\pm 1/\varepsilon)) \\
&\quad + j\,\cos((2k+1)\pi/2n)\cosh((1/n)\sinh^{-1}(\pm 1/\varepsilon))]
\end{aligned}
\tag{4.47}
$$

with $k = 0, 1, 2, \ldots, n-1$. From (4.47), we see that the poles are placed symmetrically about the imaginary axis, because $\sinh^{-1}(-x) = -\sinh^{-1}(x)$, and lie on the ellipse,

$$\frac{x_k^2}{\{\sinh((1/n)\sinh^{-1}(1/\varepsilon))\}^2} + \frac{y_k^2}{\{\cosh((1/n)\sinh^{-1}(1/\varepsilon))\}^2} = \omega_c^2.$$

We choose as the poles of the transfer function $H(s)$, those poles which lie in the left half plane, that is

$$\lambda_k = \omega_c[-x_k + j\,y_k], \tag{4.48}$$

where

$$x_k = \sin\left((2k+1)\pi/2n\right) . \sinh\left((1/n)\sinh^{-1}(1/\varepsilon)\right),$$

and

$$y_k = \cos\left((2k+1)\pi/2n\right) . \cosh\left((1/n)\sinh^{-1}(1/\varepsilon)\right),$$

with $k = 0, 1, 2, \ldots, n-1$. We can then write the transfer function in the compact form,

$$H(s) = \frac{K'}{\displaystyle\prod_{k=0}^{n-1}(s - \lambda_k)}.$$

Notice that when $s = 0$, corresponding to $\omega = 0$,

$$H(0) = \frac{K'}{\displaystyle\prod_{k=0}^{n-1}(-\lambda_k)}.$$

Also, we have

$$H(0)^2 = \frac{K^2}{1 + \varepsilon^2 T_n^2(0)} = \begin{cases} K^2, & \text{if } n \text{ is odd,} \\ K^2/(1+\varepsilon^2), & \text{if } n \text{ is even} \end{cases}$$

and so

$$K' = K \prod_{k=0}^{n-1} (-\lambda_k) = -K \prod_{k=0}^{n-1} \lambda_k \ , \quad \text{if } n \text{ is odd},$$

and

$$K' = \frac{K}{\sqrt{1+\varepsilon^2}} \prod_{k=0}^{n-1} \lambda_k \ , \quad \text{if } n \text{ is even}.$$

We now look at an example which again highlights some of the features of the design process. When we wish to carry out a design using a Chebyshev filter we must specify the parameter ε, which controls the amplitude of the pass-band ripple. In view of the oscillatory properties of $T_n(\omega)$ when $|\omega| < 1$, we see that the minimum and maximum values of $H_n(j\omega)$ in the pass-band are $K/\sqrt{1+\varepsilon^2}$ and K. A convenient definition of the ripple magnitude, r, is the ratio of these two quantities, that is

$$r = \sqrt{1+\varepsilon^2}.$$

For reasons of greater clarity, particularly in graphical work, it is common to use the decibel (dB) scale in the measurement of such quantities. This, as introduced in Chapter 2, is a logarithmic scale defined by

$$A_{\mathrm{dB}} = 20 \log_{10} A,$$

where A is a scalar quantity, usually related to amplitude response. The ripple magnitude in decibels is then

$$r_{\mathrm{dB}} = 20 \log_{10} \sqrt{1+\varepsilon^2} = 10 \log_{10}(1+\varepsilon^2). \tag{4.49}$$

This scale is also used in other ways to specify the desired performance of filters at the design stage. Specifically, the degree of attenuation to be achieved by the filter at multiples of the cut-off frequency is an important characteristic, often expressed using the decibel scale. We see in the last example for this chapter that this latter requirement effectively specifies the lowest filter order suitable for the task in hand.

Example 4.7
In view of the superior performance near the cut-off frequency $f_c = 5$Hz, a Chebyshev filter is to be designed to meet the following specifications.
(1) The amplitude of the ripple in the pass-band must be no more than 1dB.
(2) At twice the cut-off frequency, the attenuation must be at least 15 dB relative to the gain at the cut-off frequency.

We derive the minimum filter order which meets these specifications, and calculate the transfer function for the resulting design.

From the requirement of a maximum of 1dB ripple in the pass-band, we have from (4.49)

$$1 = 10 \log_{10}(1+\varepsilon^2)$$

so that

$$\varepsilon = \sqrt{10^{0.1} - 1}$$
$$= 0.5088.$$

We chose to use the maximum permitted value of the ripple amplitude, because this parameter also influences the filter order calculation, as we now see.

The ratio of the amplitude response at $2f_c = 10\text{Hz}$ to that at f_c is A, where

$$A = \frac{\sqrt{1 + \varepsilon^2 T_n^2(1)}}{\sqrt{1 + \varepsilon^2 T_n^2(2)}}.$$

Thus,

$$
\begin{aligned}
A_{\text{dB}} &= 20 \log_{10} A \\
&= 10[\log_{10}(1 + \varepsilon^2 T_n^2(1)) - \log_{10}(1 + \varepsilon^2 T_n^2(2))] \\
&\leq -15,
\end{aligned}
$$

or, using the calculated value for ε,

$$\log_{10}[1 + 0.5088^2 T_n^2(2)] - \log_{10}[1 + 0.5088^2 T_n^2(1)] \geq 1.5.$$

Rearranging, and using $T_n(1) = 1$, we obtain

$$T_n(2) \geq 12.244.$$

We now use Table 4.2 to obtain $T_0(2) = 1$, $T_1(2) = 2$, $T_2(2) = 7$ and $T_3(2) = 26$ and so a third-order Chebyshev filter is the design we should implement, and in order to determine the transfer function, we must first find the pole locations. Recall that these were given by (4.48) above, so that, with $\omega_c = 2\pi f_c$,

$$\lambda_k = 2\pi.5[-x_k + j\,y_k]$$

where, since $n = 3$ and $\varepsilon = 0.5088$,

$$
\begin{aligned}
x_k &= \sin((2k+1)\pi/6).\sinh((1/3)\sinh^{-1}(1.9654)) \\
&= 0.4942 \sin((2k+1)\pi/6), \quad k = 0, 1, 2,
\end{aligned}
$$

and

$$
\begin{aligned}
y_k &= \cos((2k+1)\pi/6).\cosh((1/3)\sinh^{-1}(1.9654)) \\
&= 1.1155 \cos((2k+1)\pi/6), \quad k = 0, 1, 2.
\end{aligned}
$$

Thus,

$$
\begin{aligned}
x_0 &= 0.2471, & y_0 &= 0.9661 \\
x_1 &= 0.4942, & y_1 &= 0.0 \\
x_2 &= 0.2471, & y_2 &= -0.9661.
\end{aligned}
$$

Having obtained the pole locations, the transfer function follows at once as

$$
\begin{aligned}
H(s) &= \frac{-K\lambda_0\lambda_1\lambda_2}{(s - \lambda_0)(s - \lambda_1)(s - \lambda_2)} \\
&= \frac{K(15237.6082)}{(s + 15.5258)(s^2 + 15.5258s + 981.4410)}.
\end{aligned}
$$

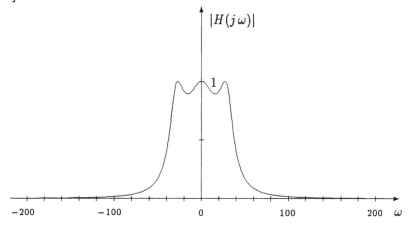

Figure 4.16: Amplitude response of the Chebyshev filter of Example 4.7.
Here $K = 1$.

A plot of the frequency response of this filter is shown in Figure 4.16.

□

This example concludes our discussion of analogue filters and their design. In Chapter 8, we make use of the designs we have obtained here when we construct discrete-time structures which perform similar tasks. The interested reader may wish to consult specialist texts in the area of filter design and synthesis. The material of §4.4 is based on the treatments of Weinberg [30] and Kuo [14], whilst Oppenheim *et al* [22] is a valuable reference work.

4.7   EXERCISES

1. Determine the pole locations for the low-pass Butterworth filter of order 5. Hence find the transfer function.

2. Determine the transfer function of the lowest order Butterworth low-pass filter such that the cut-off frequency is at $f_c = 2\text{kHz}$ and that the attenuation is at least 15dB at twice the cut-off frequency. What order of filter would be required if the 15dB target were to be achieved at one-and-a-half times the cut-off frequency?

3. Using a first-order low-pass Butterworth filter as prototype, design a band-pass filter with geometric centre frequency at $f = 500\text{Hz}$ and bandwidth 25Hz. Sketch the amplitude response.

4. Repeat Exercise 3, this time using a second-order prototype.

5. Using the methods of §4.4, determine a realization of the fourth-order Butterworth filter.

6. Design a second-order high-pass filter with cut-off frequency 100Hz, based on a Butterworth low-pass prototype.

7. Use the low-pass to band-reject transformation to design a fourth-order filter, based on a second-order Butterworth low-pass prototype, such that the geometric centre frequency, $\sqrt{\omega_u \omega_l}$, is located at 100 rad/sec. and the bandwidth, $\omega_u - \omega_l = 10$ rad/sec.

8. Determine a low-pass Chebyshev filter which satisfies the following specifications.

   (a) The pass-band ripple magnitude is 1dB.
   (b) The cut-off frequency is 20kHz.
   (c) The gain at zero frequency is 0dB.
   (d) The attenuation at 100kHz is at least 100dB.

9. Use, as prototype, a second-order Chebyshev filter with 1dB ripple in the pass-band, to design a band-pass filter, with geometric centre frequency 150Hz and bandwidth 20Hz. Adjust your design so that the gain at 150Hz is 1 and plot the amplitude response. Calculate the attenuation in dB at 100, 120 and 140Hz.

5

Discrete-time signals and systems

5.1 INTRODUCTION

In the last four chapters, we have developed a theory of continuous-time linear systems, adequate for the purposes of analysing analogue filter behaviour. The problem of filter design has also been approached, at least in outline. For many years, this was effectively the end of the story so far as relevant mathematical background is concerned. More recently however, fast, robust digital computing equipment has become readily available, and, in line with other areas of engineering, this has changed the technology of signal processing. The implications extend naturally into the field of mathematics, and it is the purpose of the remaining chapters of this book to describe the basic concepts and techniques necessary to understand and apply this developing technology. First, though, we introduce some concepts with which to describe **discrete-time signals and systems**.

5.2 SEQUENCES

We define a **sequence** $\{x\}$ as *an ordered list of real or complex numbers.* The simplest type of sequence is one with only a finite number of terms. As an example, consider the sequence

$$\{x\} = \{1,\ 1,\ \underset{\uparrow}{0},\ 1,\ 0,\ 1\}\ . \tag{5.1}$$

Position in the sequence is important and an **index of position**, usually denoted by k or n (when appropriate), is employed. As we have seen in §3.9, sequences arise in *the sampling of continuous-time signals*, and the position index then relates to an underlying time measurement. This means that we must be able to locate the position 'zero' in the sequence, (provided that such a position exists). Usually this is at the left-hand end, but if this is not the case, the '$\uparrow$' symbol is used as in (5.1).

Thus, we could write, to specify the sequence $\{x\}$ of (5.1),

$$\{x\} = \{x_{-2},\ x_{-1},\ x_0,\ x_1,\ x_2,\ x_3\}$$

where $x_{-2} = 1,\ x_{-1} = 1,\ x_0 = 0,\ x_1 = 1,\ x_2 = 0,\ x_3 = 1$.

Notice that the sequence

$$\{y\} = \{1,\ 1,\ 0,\ 1,\ 0,\ 1\} \tag{5.2}$$

is *not* the same sequence as (5.1). Since no '↑' is present, y_0 is the left-hand member of the sequence and so $y_0 = 1$, $y_1 = 1, y_2 = 0$, etc.

The above method of defining a sequence by listing all its members is not appropriate when the number of members is large. In fact, the majority of sequences that we consider are infinitely large. Another method for specifying a sequence is to *define a general term as a function of the position index*. Thus, if we write

$$\{x\} = \{x_k\}, \qquad k \in \mathbb{Z},\ \text{(the set of all integers)}$$

and

$$x_k = \begin{cases} 0, & \text{for } k < 0 \\ \left(\dfrac{1}{3}\right)^k, & \text{for } k \geq 0, \end{cases} \tag{5.3}$$

we generate the sequence

$$\{x\} = \left\{\ldots,\ 0,\ 0,\ \underset{\uparrow}{1},\ \frac{1}{3},\ \frac{1}{9},\ \frac{1}{27},\ \ldots\right\}.$$

Example 5.1
Let $x(t)$ be the causal sine wave, $x(t) = \sin(\omega t)\zeta(t)$, and suppose this is sampled every $0.5s$ with $t = 0$ being one of these samples. What is the resulting sequence $\{x\}$ of sampled values?

At once $\{x\} = \{x_k\}$, where $x_k = x(0.5k)$. That is,

$$\{x\} = \left\{\ldots,\ 0,\ 0,\ \underset{\uparrow}{0},\ \sin\left(\frac{\omega}{2}\right),\ \sin(\omega),\ \sin\left(\frac{3\omega}{2}\right),\ \ldots\right\}. \tag{5.4}$$

□

By analogy with the continuous-time situation, we define **causal sequences** as *sequences which are zero for $k < 0$*. Thus, the sequences (5.3) and (5.4) are causal sequences. It is often convenient to consider all sequences to have an index that takes on all integer values, with additional zeros used for any 'padding-out'. With this convention, we conclude that (5.1) is not causal but (5.2) is a causal sequence. Also, henceforth, any sequence $\{x\} = \{x_k\}$ is assumed to have an index, k, that takes on all integer values unless specified otherwise.

We define two basic operations on sequences—addition and scaling—and these are summarized in the defining equation

$$\alpha\{x_k\} + \beta\{y_k\} = \{\alpha x_k + \beta y_k\}, \tag{5.5}$$

where α and β are (possibly complex) constants.

Example 5.2
Let the sequences $\{x\}$ and $\{y\}$ be defined by

$$x_k = \left(\frac{1}{2}\right)^k,$$

$$y_k = \begin{cases} 0, & \text{for } k < 0 \\ 1, & \text{for } k \geq 0, \end{cases}$$

and let $\{z\}$ be the finite sequence

$$\{z\} = \{\lambda, \, \mu, \, \nu\} \, .$$

Write down the sequences $\alpha\{x\} + \beta\{y\}$, $\frac{1}{2}\{x\}$, $\frac{1}{4}\{y\}$ and $\{y\} + \{z\}$.

Now $\alpha\{x\} + \beta\{y\} = \{\alpha x_k + \beta y_k\} = \{p_k\}$ say, where

$$p_k = \begin{cases} \alpha\left(\dfrac{1}{2}\right)^k, & \text{for } k < 0 \\[2mm] \alpha\left(\dfrac{1}{2}\right)^k + \beta, & \text{for } k \geq 0. \end{cases}$$

Also

$$\frac{1}{2}\{x\} = \left\{\frac{1}{2}x_k\right\} = \left\{\frac{1}{2} \cdot \frac{1}{2^k}\right\} = \left\{\frac{1}{2^{k+1}}\right\} \, .$$

Similarly, $\frac{1}{4}\{y\} = \left\{\frac{1}{4}y_k\right\} = \{q_k\}$ say, where

$$q_k = \begin{cases} 0, & \text{for } k < 0 \\ \dfrac{1}{4}, & \text{for } k \geq 0. \end{cases}$$

Finally, addition of $\{y\}$ and $\{z\}$ is only possible if we adopt the convention of padding-out $\{z\}$ with zeros. This then gives

$$\{y\} + \{z\} = \{\ldots 0, \, 0, \, 1 + \lambda, \, 1 + \mu, \, 1 + \nu, \, 1, \, 1, \, \ldots\} \, .$$
$$\uparrow$$

$\square$

5.3 LINEAR SYSTEMS

Consider a discrete-time system with input sequence, $\{u\}$, and output sequence, $\{y\}$, governed by the **difference equation**

$$y_{k+2} + 3y_{k+1} - 2y_k = 2u_{k+1} - 3u_k \, , \qquad k \in \mathbb{Z}.$$

In order to help understand this equation, we first rewrite it in the form

$$y_{k+2} = -3y_{k+1} + 2y_k + 2u_{k+1} - 3u_k$$

and then replacing k by $k - 2$ throughout, we obtain

$$y_k = -3y_{k-1} + 2y_{k-2} + 2u_{k-1} - 3u_{k-2} . \tag{5.6}$$

This then gives *a means for calculating the output sequence at index k from previous output and input values.* As a particular example of this, consider the situation where the input sequence is the causal step sequence given by

$$u_k = \zeta_k = \begin{cases} 0, & \text{for } k < 0 \\ 1, & \text{for } k \geq 0, \end{cases}$$

and the system is initially quiescent, which we model by assuming the output sequence, $\{y\}$, is causal. Putting $k = 0$ into (5.6) gives

$$y_0 = -3y_{-1} + 2y_{-2} + 2\zeta_{-1} - 3\zeta_{-2} = 0 .$$

Next, putting $k = 1$ into (5.6) and using $y_0 = 0$ gives

$$y_1 = -3y_0 + 2y_{-1} + 2\zeta_0 - 3\zeta_{-1} = 2\zeta_0 = 2 .$$

Next, putting $k = 2$ into (5.6) and using $y_0 = 0$, $y_1 = 2$ gives

$$y_2 = -3y_1 + 2y_0 + 2\zeta_1 - 3\zeta_0 = -3 \times 2 + 2 \times 0 + 2 \times 1 - 3 \times 1 = -7 .$$

Clearly, we can continue this procedure and calculate further values for the output sequence. *This process does not produce a formula for y_k, but is an essential first step towards understanding the concept of difference equations.* At a later stage in this chapter, we present a method for finding solution formulae for difference equations akin to the Laplace transform method used earlier for solving differential equations.

Let us now introduce the **shift operator, q,** defined by

$$\mathbf{q}(\{x_k\}) = \{x_{k+1}\} . \tag{5.7}$$

So we have

$$\mathbf{q}(\{\ldots, x_{-2}, \ x_{-1}, \ \underset{\uparrow}{x_0}, \ x_1, \ x_2, \ \ldots\}) = \{\ldots, x_{-2}, \ x_{-1}, \ x_0, \ \underset{\uparrow}{x_1}, \ x_2, \ \ldots\}.$$

More generally, we define

$$\mathbf{q}^n(\{x_k\}) = \{x_{k+n}\} , \quad \text{for any integer } n. \tag{5.8}$$

With this in mind, we can write the above difference equation as

$$(\mathbf{q}^2 + 3\mathbf{q} - 2)\{y\} = (2\mathbf{q} - 3)\{u\}.$$

More generally, we consider n^{th} order difference equations given by

$$a(\mathbf{q})\{y\} = b(\mathbf{q})\{u\} , \tag{5.9}$$

where $a(\mathbf{q})$ is an nth order polynomial in $\mathbf{q}$ and $b(\mathbf{q})$ is an mth order polynomial in $\mathbf{q}$ with $m \leq n$.

It should be noted that any polynomial in $\mathbf{q}$ is a linear operator on sequences. Let us illustrate this in the quadratic case. We have

$$(\mathbf{q}^2 + a_1\mathbf{q} + a_0)(\alpha\{x_k\} + \beta\{y_k\})$$
$$= (\mathbf{q}^2 + a_1\mathbf{q} + a_0)(\{\alpha x_k + \beta y_k\}) \quad \text{using (5.5)}$$
$$= \{\alpha x_{k+2} + \beta y_{k+2}\} + a_1\{\alpha x_{k+1} + \beta y_{k+1}\} + a_0\{\alpha x_k + \beta y_k\}$$
$$= \alpha\left(\{x_{k+2}\} + a_1\{x_{k+1}\} + a_0\{x_k\}\right) + \beta\left(\{y_{k+2}\} + a_1\{y_{k+1}\} + a_0\{y_k\}\right)$$
$$= \alpha.(\mathbf{q}^2 + a_1\mathbf{q} + a_0)\{x\} + \beta.(\mathbf{q}^2 + a_1\mathbf{q} + a_0)\{y\}$$

for any constants α and β. For this reason, the difference equation given by (5.9) is said to be **linear**.

Example 5.3
Consider again the second order difference equation

$$y_{k+2} + 3y_{k+1} - 2y_k = 2u_{k+1} - 3u_k , \qquad k \in \mathbb{Z}.$$

Show that this can be rewritten in the form of two coupled equations

$$v_{k+2} + 3v_{k+1} - 2v_k = u_k$$
$$y_k = 2v_{k+1} - 3v_k .$$

If $\{y\}$ is known to satisfy the initial conditions, $y_0 = \lambda$, $y_1 = \mu$, what are the corresponding values for v_0 and v_1?

We can write the difference equation as

$$(\mathbf{q}^2 + 3\mathbf{q} - 2)\{y\} = (2\mathbf{q} - 3)\{u\} .$$

If we now define $\{v\}$ to be any sequence which solves

$$(\mathbf{q}^2 + 3\mathbf{q} - 2)\{v\} = \{u\} ,$$

it then follows that

$$(\mathbf{q}^2 + 3\mathbf{q} - 2)(2\mathbf{q} - 3)\{v\} = (2\mathbf{q} - 3)\{u\}$$

and so $\{y\} = (2\mathbf{q} - 3)\{v\}$ satisfies the original difference equation. It only remains to show that every solution of the original difference equation can be obtained in this manner. Since any solution is uniquely determined from knowledge of y_0 and y_1, all we need to show is that $\{v\}$ can be chosen to match any such given initial values. We have

$$y_0 = 2v_1 - 3v_0 = \lambda$$
$$y_1 = 2v_2 - 3v_1 = \mu.$$

But $v_2 = -3v_1 + 2v_0 + u_0$ and so we obtain

$$2v_1 - 3v_0 = \lambda$$
$$-9v_1 + 4v_0 = \mu - 2u_0.$$

Solving these equations for v_1 and v_0 gives

$$v_1 = \frac{1}{19}(-4\lambda - 3\mu + 6u_0)$$

$$v_0 = \frac{1}{19}(-9\lambda - 2\mu + 4u_0) .$$

□

5.4 SIMULATION DIAGRAMS

Simulation diagrams for discrete-time systems are essentially the same as for the continuous-time case. The main difference lies in the use of **unit delay blocks** to represent the dynamics. For example consider the simulation diagram of Figure 5.1.

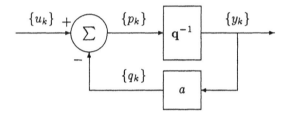

Figure 5.1: A discrete-time system.

It is supposed that the system is regulated by a clock issuing pulses at (equal) intervals. At the instant of the kth pulse, the following events occur.

At the left-hand (or input) side, the kth member of the input sequence, u_k, arrives at the summation block. Instantaneously, it is combined with the kth member of the feedback signal sequence, q_k, to form the kth term of the signal sequence, $p_k = u_k - q_k$. This signal is then held in the unit delay block. Prior to this the previously held value, p_{k-1}, is 'released' to form the kth term of the output sequence, $y_k = p_{k-1}$. Using the notation of (5.8), this is equivalent to $\{y_k\} = \mathbf{q}^{-1}\{p_k\}$, which explains the use of $\mathbf{q}^{-1}$ in the diagram. An exact copy of y_k is fed back and scaled by a factor of a to form the kth term of the feedback sequence, $q_k = ay_k$, which, as previously stated, is combined with u_k at the summation junction. At the next clock pulse, these events are repeated with k replaced by $k + 1$.

It follows from the above description that we must have

$$p_k = y_{k+1} . \tag{5.10}$$

But, as already discussed, we have

$$p_k = u_k - q_k = u_k - ay_k , \tag{5.11}$$

and therefore, it follows that

$$y_{k+1} = u_k - ay_k .$$

In other words, Figure 5.1 represents the difference equation

$$y_{k+1} + ay_k = u_k \ .$$

Example 5.4
Obtain a difference equation for the simulation diagram of Figure 5.2.

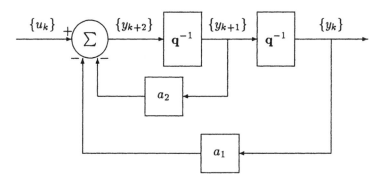

Figure 5.2: Discrete-time system for Example 5.4.

Given that the output sequence on the right-hand side of the diagram is $\{y_k\}$, it immediately follows that the inputs to the two unit delay blocks must be $\{y_{k+1}\}$ and $\{y_{k+2}\}$ as shown in the figure. Therefore, we see from the summation junction that

$$y_{k+2} = u_k - a_1 y_k - a_2 y_{k+1} \ .$$

Re-arranging this gives the desired difference equation

$$y_{k+2} + a_2 y_{k+1} + a_1 y_k = u_k \ .$$

□

Example 5.5
Construct a simulation diagram to represent the difference equation

$$y_{k+2} + 2y_{k+1} + 2y_k = u_k + u_{k+1} \ . \tag{5.12}$$

The trick to solving the problem caused by the 'forward' shifted input term u_{k+1}, is to split (5.12) into two coupled equations in a similar fashion to that used in Example 5.3. This produces the equations

$$v_{k+2} + 2v_{k+1} + 2v_k = u_k \tag{5.13a}$$

$$y_k = v_k + v_{k+1} \ . \tag{5.13b}$$

Equations (5.13a) and (5.13b) then give rise to the simulation diagram of Figure 5.3.

□

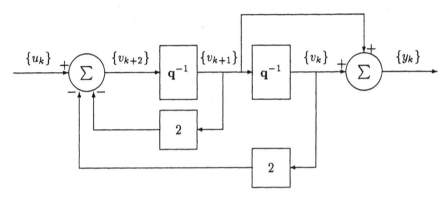

Figure 5.3: Simulation diagram for Example 5.5.

5.5   THE z-TRANSFORM

By now the reader will be familiar with the concept of a transform, and perhaps persuaded of their value. Without for the moment discussing the purpose of so doing, we define a transform for sequences. Given a sequence $\{x\} = \{x_k\}$, we define $X(z)$, the **bilateral z-transform** of $\{x\}$, by

$$\mathcal{Z}\left[\{x\}\right] = X(z) \stackrel{\text{def}}{=} \sum_{k=-\infty}^{\infty} \frac{x_k}{z^k} , \tag{5.14}$$

for those complex numbers, z, for which the sum exists.

Recall that the operation of taking the Laplace transform involved the generation of a function $F(s)$ of the complex variable s from a time signal $f(t)$. Here, we are producing a function $X(z)$ of another complex variable z, with the starting point the sequence $\{x\} = \{x_k\}$.

Example 5.6
Calculate the bilateral z-transform of the sequence

$$\{x\} \;\; = \;\; \{\dots 0, \, 1, \, 1, \, 1, \, 0, \, 0, \, 1, \, 0, \, 0, \, 1, \, 1, \, 1, \, 0, \, 0, \, \dots\} \, .$$
$$\uparrow$$

Here, we see that

$$x_k = \begin{cases} 0, & \text{for } |k| > 5 \\ 1, & \text{for } 3 \le |k| \le 5 \\ 0, & \text{for } 0 < |k| < 3 \\ 1, & \text{for } k = 0 , \end{cases}$$

and so

$$X(z) = z^5 + z^4 + z^3 + 1 + \frac{1}{z^3} + \frac{1}{z^4} + \frac{1}{z^5} \, .$$

This sum exists for all nonzero complex numbers, z.

□

In an analogous fashion to that used for the Laplace transform, we also define the **(unilateral) z-transform** by

$$\mathcal{Z}\left[\{x\}\right] = X(z) \overset{\text{def}}{=} \sum_{k=0}^{\infty} \frac{x_k}{z^k} \,, \tag{5.15}$$

for those complex numbers for which the sum exists. If the sum exists for some z then it will always exist for $|z|$ sufficiently large and, for this reason, the region of the complex plane over which $X(z)$ is defined is not usually of importance. This cannot be said for the bilateral transform. We note that, if $\{x\}$ is causal, i.e. $x_k = 0$ for $k < 0$, then the bilateral and unilateral transforms give the same result.

Example 5.7
From first principles, calculate the z-transform of the sequence

$$\{x\} = \left\{ \left(\frac{1}{2}\right)^k \right\} .$$

From the definition,

$$X(z) = \sum_{k=0}^{\infty} \frac{x_k}{z^k} = \sum_{k=0}^{\infty} \frac{1}{2^k z^k} \,.$$

This is a geometric series with common ratio $1/(2z)$ and first term 1. The sum is thus

$$\frac{1}{1 - \frac{1}{2z}} = \frac{z}{z - \frac{1}{2}} = \frac{2z}{2z - 1}$$

provided that $|1/(2z)| < 1$, i.e. $|z| > \frac{1}{2}$.

$\square$

The above example can be generalized to the sequence $\{x\} = \{a^k\}$, where a is a real or complex constant. This gives rise to

$$\mathcal{Z}\left[\{a^k\}\right] = \frac{z}{z - a} \,, \tag{5.16}$$

which exists outside the circle $|z| = |a|$ in the complex plane. The reader may spot a similarity between (5.16) and e^{at} and its Laplace transform, $1/(s - a)$.

Example 5.8
Use (5.16) to write down the z-transform of the sequence $\{x\} = \{(-\frac{1}{3})^k\}$.

At once we have

$$X(z) = \frac{z}{z + \frac{1}{3}} = \frac{3z}{3z + 1}, \qquad |z| > \frac{1}{3} \,.$$

$\square$

We use the notation

$$\{a^k\} \leftrightarrow \frac{z}{z - a} \tag{5.17}$$

to mean that $z/(z-a)$ is the z-transform of the sequence $\{a^k\}$. We now establish two more transforms based on the pair (5.17). First note that if

$$\{x_k\} \leftrightarrow X(z)$$

is a transform pair, then

$$\{kx_k\} \leftrightarrow \sum_{k=0}^{\infty} \frac{kx_k}{z^k} \; .$$

But we have

$$\sum_{k=0}^{\infty} \frac{kx_k}{z^k} = z \sum_{k=0}^{\infty} \frac{kx_k}{z^{k+1}} = z \sum_{k=0}^{\infty} x_k \frac{\mathrm{d}}{\mathrm{d}z}\left(-\frac{1}{z^k}\right)$$

$$= -z \frac{\mathrm{d}}{\mathrm{d}z} \sum_{k=0}^{\infty} \frac{x_k}{z^k} = -z \frac{\mathrm{d}}{\mathrm{d}z} X(z) \; ,$$

where we have used a standard result that power series may be differentiated term by term. In summary, we have shown

$$\{kx_k\} \leftrightarrow -z \frac{\mathrm{d}}{\mathrm{d}z} X(z) \; . \tag{5.18}$$

At once, from (5.17) and (5.18) we see that

$$\{ka^k\} \leftrightarrow -z \frac{\mathrm{d}}{\mathrm{d}z}\left(\frac{z}{z-a}\right)$$

$$= \frac{az}{(z-a)^2} \; .$$

In particular, on putting $a = 1$, we find that

$$\{k\} \leftrightarrow \frac{z}{(z-1)^2} \; . \tag{5.19}$$

5.6   PROPERTIES OF THE z-TRANSFORM

In order to apply z-transform techniques to the analysis of discrete-time systems, we must establish a few basic properties. Our treatment is formal in the sense that infinite series in the complex variable z are manipulated without a full discussion of convergence implications. Readers wishing to pursue this topic more thoroughly should consult specialist texts on the theory of functions of a complex variable, for example Page [23].

The linearity property
In §5.2, we defined addition and scalar multiplication (scaling) of sequences. Let

$\{w\} = \alpha\{x\} + \beta\{y\}$, where α and β are constants and $\{x\}$ and $\{y\}$ are sequences which have z-transforms for $|z|$ sufficiently large. We then have

$$W(z) = \sum_{k=0}^{\infty} \frac{(\alpha x_k + \beta y_k)}{z^k}$$

$$= \alpha \sum_{k=0}^{\infty} \frac{x_k}{z^k} + \beta \sum_{k=0}^{\infty} \frac{y_k}{z^k}$$

$$= \alpha X(z) + \beta Y(z) .$$

In summary, we have

$$\alpha\{x\} + \beta\{y\} \leftrightarrow \alpha X(z) + \beta Y(z) . \tag{5.20}$$

The above linearity property also holds for the bilateral transform. However, it is only possible to guarantee that $W(z)$ exists for those complex numbers z for which both $X(z)$ and $Y(z)$ exist. (This does not mean that $W(z)$ might not exist over a larger region than this.) In the unilateral case, this guarantees that $W(z)$ exists for $|z|$ sufficiently large. In the bilateral case, it is possible that the regions over which $X(z)$ and $Y(z)$ exist are disjoint and there is then no guarantee that $W(z)$ exists for any z. For example, suppose that $\{x\}$ and $\{y\}$ are given by

$$x_k = \begin{cases} 0, & \text{for } k < 0 \\ 1, & \text{for } k \geq 0 \end{cases} \quad \text{and} \quad y_k = \begin{cases} 1, & \text{for } k < 0 \\ 0, & \text{for } k \geq 0 \end{cases} .$$

The bilateral transforms of $\{x\}$ and $\{y\}$ are then given by $X(z) = z/(z-1)$, $|z| > 1$ and $Y(z) = z/(1-z)$, $|z| < 1$. In particular if we let $\{w\} = \{x\} + \{y\}$, i.e. $w_k = 1$ for all integers, k, then the bilateral transform of $\{w\}$ does not exist for any z.

Example 5.9
Let $\{x\}$ and $\{y\}$ be the sequences defined by

$$x_k = \begin{cases} 0, & \text{for } k < 0 \\ 1, & \text{for } k \geq 0 \end{cases} \quad \text{and} \quad y_k = \begin{cases} 0, & \text{for } k < 3 \\ 1, & \text{for } k \geq 3, \end{cases}$$

i.e. $\{x_k\} = \{\zeta_k\}$ and $\{y_k\} = \{\zeta_{k-3}\}$. Calculate the z-transform of the sequence $\{w\}$ defined by $\{w\} = \{x\} - \{y\}$.

We know that $X(z) = z/(z-1)$, $|z| > 1$. By definition, we have

$$Y(z) = \sum_{k=3}^{\infty} z^{-k} = \sum_{k=0}^{\infty} z^{-(k+3)} = z^{-3} X(z).$$

It follows that $Y(z) = 1/z^2(z-1)$, $|z| > 1$. Using the linearity property, we deduce that

$$W(z) = X(z) - Y(z) = \frac{z}{z-1} - \frac{1}{z^2(z-1)}$$

$$= \frac{z^3 - 1}{z^2(z-1)}$$

$$= \frac{z^2 + z + 1}{z^2} = 1 + \frac{1}{z} + \frac{1}{z^2} .$$

This result also follows directly if we note that $\{w\}$ is given by

$$w_k = \{1,\ 1,\ 1,\ 0,\ 0,\ \ldots\}.$$

We also note that, although $X(z)$ and $Y(z)$ are only defined for $|z| > 1$, $W(z)$ is well-defined over the larger region $z \neq 0$.

□

Example 5.10
Calculate the z-transform of the sequence $\{x\} = \{x_k\}$ where $x_k = 2$ for all k.

Let us first note that this sequence can be denoted by $\{2\}$. If the lack of any appearance of the index k causes any confusion, another possibility is to use the notation $\{2.(1)^k\}$. Note the use of brackets so that $2.(1)^k$ is not confused with $(2.1)^k$. We have already met the transform pair $\{1^k\} \leftrightarrow z/(z-1)$ and so it follows from the linearity property, (with $\beta = 0$), that

$$\{2\} \leftrightarrow 2.\frac{z}{z-1} = \frac{2z}{z-1}.$$

□

Example 5.11
Use the transform pair $\{a^k\} \leftrightarrow z/(z-a)$ to find the z-transform of the sequence $\{w\} = \{\sin(k\theta)\}$.

Using $\sin(k\theta) = (e^{jk\theta} - e^{-jk\theta})/(2j)$, we see that

$$\mathcal{Z}\left[\{\sin(k\theta)\}\right] = \mathcal{Z}\left[\left\{\frac{1}{2j}(e^{jk\theta} - e^{-jk\theta})\right\}\right]$$

$$= \frac{1}{2j}\mathcal{Z}\left[\{(e^{j\theta})^k\}\right] - \frac{1}{2j}\mathcal{Z}\left[\{(e^{-j\theta})^k\}\right], \qquad \text{by linearity}$$

$$= \frac{1}{2j}\frac{z}{z - e^{j\theta}} - \frac{1}{2j}\frac{z}{z - e^{-j\theta}}$$

$$= \frac{1}{2j}\left[\frac{z(e^{j\theta} - e^{-j\theta})}{z^2 - z(e^{j\theta} + e^{-j\theta}) + 1}\right]$$

$$= \frac{z\sin(\theta)}{z^2 - 2z\cos(\theta) + 1}.$$

□

Table 5.1 contains a list of some useful transform pairs for reference purposes.

The right shifting (delaying) property
Suppose the sequence $\{y\}$ is the same as a sequence $\{x\}$ that has been shifted to the right (delayed) by k_0 time steps. That is to say,

$$y_k = x_{k-k_0}.$$

$\{x_k\}$	$X(z)$
$x_k = \begin{cases} 1, & \text{for } k = 0 \\ 0, & \text{for } k \neq 0 \end{cases}$	1
$x_k = 1$	$\dfrac{z}{z-1}$
$x_k = a^k$	$\dfrac{z}{z-a}$
$x_k = ka^k$	$\dfrac{az}{(z-a)^2}$
$x_k = k$	$\dfrac{z}{(z-1)^2}$
$x_k = \sin(k\theta)$	$\dfrac{z\sin(\theta)}{z^2 - 2z\cos(\theta) + 1}$
$x_k = \cos(k\theta)$	$\dfrac{z(z - \cos(\theta))}{z^2 - 2z\cos(\theta) + 1}$

Table 5.1: Elementary z-transforms.

Let us now work out the bilateral z-transform of $\{y\}$. We have

$$Y(z) = \sum_{k=-\infty}^{\infty} \frac{x_{k-k_0}}{z^k} = \frac{1}{z^{k_0}} \sum_{k=-\infty}^{\infty} \frac{x_{k-k_0}}{z^{k-k_0}} = \frac{1}{z^{k_0}} X(z) \ .$$

For the (unilateral) z-transform, we have

$$Y(z) = \sum_{k=0}^{\infty} \frac{x_{k-k_0}}{z^k} = \frac{1}{z^{k_0}} \sum_{k=0}^{\infty} \frac{x_{k-k_0}}{z^{k-k_0}} = \frac{1}{z^{k_0}} \sum_{k=-k_0}^{\infty} \frac{x_k}{z^k} \ .$$

This gives rise to the transform pair

$$\{x_{k-k_0}\} \leftrightarrow \frac{1}{z^{k_0}} \left[X(z) + \sum_{k=-k_0}^{-1} \frac{x_k}{z^k} \right] \ .$$

In this text, we only use this delaying property with the unilateral z-transform when the sequence $\{x\}$ is causal. This then gives

$$\{x_{k-k_0}\} \leftrightarrow \frac{1}{z^{k_0}} X(z) \qquad (\{x\} \text{ causal}). \tag{5.21}$$

Example 5.12
Let $\{x\}$ be the causal sequence given by $x_k = \sin(k\theta)\zeta_k$, i.e.

$$\{x\} = \{0, \ \sin(\theta), \ \sin(2\theta), \ \sin(3\theta), \ldots\}$$

where we have used our previous convention that the above sequence starts at $k = 0$ and is padded-out with zeros for $k < 0$. Now let $\{y\}$ be given by $y_k = x_{k-2}$, i.e.

$$\{y\} = \{0, \ 0, \ 0, \ \sin(\theta), \ \sin(2\theta), \ \sin(3\theta), \ldots\} \ .$$

Find the z-transform of $\{y\}$.

It follows from (5.21) that the z-transform of $\{y\}$ is given by

$$Y(z) = \frac{1}{z^2} \frac{z\sin(\theta)}{z^2 - 2z\cos(\theta) + 1}$$

$$= \frac{\sin(\theta)}{z(z^2 - 2z\cos(\theta) + 1)} \ .$$

$\square$

Example 5.13
Find the inverse z-transform of $X(z) = \dfrac{4}{z(z-2)}$.

We can write $X(z) = \dfrac{1}{z^2} Y(z)$ where $Y(z) = \dfrac{4z}{z-2}$. From Table 5.1 and the linearity property, we deduce that $\{y\}$ is given by

$$y_k = 4.(2)^k = 2^{k+2} \qquad k \geq 0.$$

Now using the right shifting property, we see that $\{x\}$ is given by $x_k = y_{k-2}$, where $\{y\}$ must be considered to be causal. This therefore gives

$$x_k = \begin{cases} 0, & \text{for } k = 0, 1 \\ 2^k, & \text{for } k \geq 2. \end{cases}$$

$\square$

The left shifting (advancing) property

Suppose now that the sequence $\{y\}$ is found by left-shifting the sequence $\{x\}$, i.e.

$$y_k = x_{k+k_0} \; .$$

As before, let us begin by calculating the bilateral z-transform of $\{y\}$. We have

$$Y(z) = \sum_{k=-\infty}^{\infty} \frac{x_{k+k_0}}{z^k} = z^{k_0} \sum_{k=-\infty}^{\infty} \frac{x_{k+k_0}}{z^{k+k_0}} = z^{k_0} X(z) \; .$$

This time, even under the assumption that $\{x\}$ is causal, we can no longer assume $\{y\}$ is causal and so the unilateral and bilateral transforms of $\{y\}$ differ. More precisely, the unilateral transform of $\{y\}$ is given by

$$Y(z) = \sum_{k=0}^{\infty} \frac{x_{k+k_0}}{z^k} = z^{k_0} \sum_{k=0}^{\infty} \frac{x_{k+k_0}}{z^{k+k_0}} = z^{k_0} \sum_{k=k_0}^{\infty} \frac{x_k}{z^k} \; .$$

This gives rise to the transform pair

$$\{x_{k+k_0}\} \leftrightarrow z^{k_0} \left[X(z) - \sum_{k=0}^{k_0-1} \frac{x_k}{z^k} \right] \; . \tag{5.22}$$

In particular, the case $k_0 = 1$ gives

$$\{x_{k+1}\} \leftrightarrow z X(z) - z x_0 \tag{5.23}$$

and the case $k_0 = 2$ gives

$$\{x_{k+2}\} \leftrightarrow z^2 X(z) - z^2 x_0 - z x_1 \; . \tag{5.24}$$

The reader may be struck by the similarity of (5.23) and (5.24) to the results for the transforms of first and second derivatives in the Laplace transform domain. Notice, however, that the similarity is not total in view of the powers of z beyond the first term of the right-hand side in each case.

Example 5.14

Let $\{x\}$ be the causal sequence given by $x_k = (-1/2)^k$, $k \geq 0$, i.e.

$$\{x\} \quad = \quad \{\ldots 0, \, 0, \, 1, \, -\frac{1}{2}, \, \frac{1}{4}, \, -\frac{1}{8}, \, \frac{1}{16}, \ldots\}$$
$$\uparrow$$

and let $\{y\}$ be this sequence advanced by two time steps, so

$$\{y\} = \{\ldots 0,\ 0,\ 1,\ -\frac{1}{2},\ \frac{1}{4},\ -\frac{1}{8},\ \frac{1}{16},\ \ldots\} \ .$$
$$\uparrow$$

Find the bilateral and unilateral z-transforms of $\{y\}$.

The z-transform of $\{x\}$ is given by $z/(z + \frac{1}{2}) = 2z/(2z + 1)$ and so we see that the bilateral transform of $\{y\}$ is given by

$$Y(z) = z^2 \frac{2z}{2z + 1} = \frac{2z^3}{2z + 1} \ ,$$

whereas the (unilateral) transform is given by

$$Y(z) = \frac{2z^3}{2z + 1} - z^2 - z.(-\frac{1}{2}) = \frac{z}{2(2z + 1)} \ .$$

□

A scaling property

Now suppose that $\{y\}$ is given by $y_k = a^k x_k$, we then have

$$Y(z) = \sum_{k=0}^{\infty} \frac{a^k x_k}{z^k} = \sum_{k=0}^{\infty} \frac{x_k}{(z/a)^k} = X(z/a) \ ,$$

with a similar result holding in the bilateral case. So we have

$$\{a^k x_k\} \leftrightarrow X(z/a) \ . \tag{5.25}$$

Example 5.15

Find the z-transform of the sequence of sampled values of $x(t) = e^{-t}\sin(t)$ with samples taken at $t = 0,\ T,\ 2T,\ 3T, \ldots$ for some given sample period $T > 0$.

The sampled sequence is given by $\{x\} = \{x_k\}$ where $x_k = e^{-kT}\sin(kT) = (e^{-T})^k \sin(kT)$. It therefore follows that $X(z)$ can be obtained by replacing z by $z/(e^{-T}) = e^T z$ in the formula for $\mathcal{Z}\left[\{\sin(kT)\}\right]$. This gives

$$X(z) = \frac{e^T z \sin(T)}{e^{2T} z^2 - 2e^T z \cos(T) + 1} = \frac{e^{-T} z \sin(T)}{z^2 - 2e^{-T} z \cos(T) + e^{-2T}} \ .$$

□

The various properties of the z-transform are summarized in Table 5.2.

5.7 APPLICATION TO LINEAR TIME-INVARIANT SYSTEMS

In this section, we examine how the z-transform can be used to solve linear, constant coefficient difference equations by means of a few illustrative examples.

Linearity	$\alpha\{x\} + \beta\{y\} \leftrightarrow \alpha X(z) + \beta Y(z)$
Right Shift	$\{x_{k-k_0}\} \leftrightarrow \dfrac{1}{z^{k_0}} X(z)$ $\{x\}$ causal for unilateral z-transform
Left Shift	$\{x_{k+k_0}\} \leftrightarrow z^{k_0}\left[X(z) - \displaystyle\sum_{k=0}^{k_0-1}\dfrac{x_k}{z^k}\right]$ $z^{k_0}X(z)$ (Bilateral Case)
Scaling	$\{a^k x_k\} \leftrightarrow X(z/a)$
Multiplication by k	$\{k x_k\} \leftrightarrow -z\dfrac{\mathrm{d}}{\mathrm{d}z}X(z)$

Table 5.2: z-transform properties.

Example 5.16
Solve the difference equation

$$y_{k+2} + \frac{3}{4}y_{k+1} + \frac{1}{8}y_k = u_k, \quad k \in \mathbb{Z}, \tag{5.26}$$

for the case $u_k = \zeta_k$ and under the assumption that the system is initially quiescent, i.e. $y_k = 0, \ k < 0$.

Let us begin by noting that the difference equation is equivalent to the sequence equation

$$\{y_{k+2}\} + \frac{3}{4}\{y_{k+1}\} + \frac{1}{8}\{y_k\} = \{u_k\}. \tag{5.27}$$

Since we are not given values for y_0 and y_1, we can either calculate them from (5.26) by setting $k = -2$ and $k = -1$ respectively and then take z-transforms in (5.27) or, more simply, take bilateral z-transforms in (5.27). This latter approach gives

$$\left(z^2 + \frac{3}{4}z + \frac{1}{8}\right)Y(z) = U(z) = \frac{z}{z-1}.$$

It follows that

$$Y(z) = \frac{1}{z^2 + \frac{3}{4}z + \frac{1}{8}}\frac{z}{z-1} = \frac{z}{(z-1)(z+\frac{1}{2})(z+\frac{1}{4})}. \tag{5.28}$$

Determination of the sequence $\{y_k\}$ means the inversion of (5.28). The techniques for the inversion of Laplace transforms carry over to the inversion of z-transforms with one significant modification. This is in view of the factor z in the numerators of the transforms in Table 5.1. For this reason, we write (5.28) as

$$Y(z) = z \left[\frac{1}{(z-1)(z+\frac{1}{2})(z+\frac{1}{4})} \right]$$

$$= z \left[\frac{8/15}{z-1} + \frac{8/3}{z+\frac{1}{2}} - \frac{16/5}{z+\frac{1}{4}} \right]$$

$$= \frac{8}{15} \frac{z}{z-1} + \frac{8}{3} \frac{z}{z+\frac{1}{2}} - \frac{16}{5} \frac{z}{z+\frac{1}{4}}.$$

At this point, we should note that, although we have taken bilateral z-transforms in (5.27), the assumption that the system is initially quiescent means that the bilateral and unilateral transforms of $\{y\}$ are identical. This in turn means we can use (unilateral) transform tables, such as Table 5.1, to obtain

$$y_k = \frac{8}{15} + \frac{8}{3} \left(-\frac{1}{2}\right)^k - \frac{16}{5} \left(-\frac{1}{4}\right)^k \qquad k \geq 0.$$

This response is depicted in Figure 5.4. This calculation demonstrates the main steps in the analysis of discrete-time systems by transform methods. We demonstrate the technique by considering a further example.

□

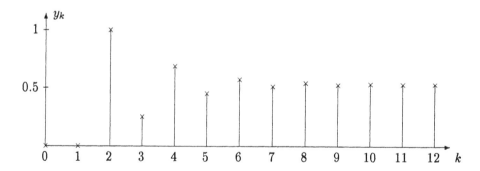

Figure 5.4: Step response for Example 5.16.

Example 5.17
Find the response of the system

$$y_{k+2} + 2y_{k+1} + 2y_k = u_k + u_{k+1}, \quad k \in \mathbb{Z}, \tag{5.29}$$

to the input $\{u_k\} = \{(1/2)^k \zeta_k\}$, assuming the system is initially quiescent.

Proceeding as in the last example, we begin by taking bilateral transforms but note that the causality of both $\{y\}$ and $\{u\}$ means that their bilateral and unilateral transforms are identical. This leads to

$$Y(z) = \frac{z+1}{z^2+2z+2} \, U(z) = \frac{z(z+1)}{(z - \frac{1}{2})(z^2+2z+2)} \, .$$

We therefore have

$$\frac{Y(z)}{z} = \frac{z+1}{(z - \frac{1}{2})(z+1+j)(z+1-j)} \, .$$

We now express $Y(z)/z$ in partial fraction form, so that

$$\frac{Y(z)}{z} = \frac{1}{13} \left[\frac{6}{z - \frac{1}{2}} + \frac{-3-2j}{z+1-j} + \frac{-3+2j}{z+1+j} \right] \, ,$$

and so

$$Y(z) = \frac{1}{13} \left[6 \frac{z}{z - \frac{1}{2}} + (-3-2j)\frac{z}{z+1-j} + (-3+2j)\frac{z}{z+1+j} \right] \, .$$

Using Table 5.1, we obtain for $k \geq 0$

$$y_k = \frac{1}{13} \left[6 \left(\frac{1}{2} \right)^k + (-3-2j)(-1+j)^k + (-3+2j)(-1-j)^k \right]$$

$$= \frac{1}{13} \left[6 \left(\frac{1}{2} \right)^k + (-3-2j) \left(\sqrt{2} \, e^{j\frac{3\pi}{4}} \right)^k + (-3+2j) \left(\sqrt{2} \, e^{-j\frac{3\pi}{4}} \right)^k \right] \, .$$

Note that the last two terms are conjugates of each other and we have, for any complex number z, $z + \bar{z} = 2\Re(z)$, where $\Re(z)$ denotes the real part of z. Therefore, it follows that

$$y_k = \frac{6}{13} \left(\frac{1}{2} \right)^k + \frac{1}{13} \cdot 2\Re \left[(-3-2j) \left(\sqrt{2} \right)^k e^{jk\frac{3\pi}{4}} \right]$$

$$= \frac{6}{13} \left(\frac{1}{2} \right)^k + \frac{2}{13} \left(\sqrt{2} \right)^k \Re \left[(-3-2j)(\cos \left(\frac{3k\pi}{4} \right) + j \sin \left(\frac{3k\pi}{4} \right)) \right]$$

$$= \frac{6}{13} \left(\frac{1}{2} \right)^k + \frac{2}{13} \left(\sqrt{2} \right)^k (-3 \cos \left(\frac{3k\pi}{4} \right) + 2 \sin \left(\frac{3k\pi}{4} \right)), \quad (k \geq 0).$$

Before leaving this example, let us compare the above approach, using the bilateral z-transform, to an approach using the unilateral transform. Setting $k = -2$ in (5.29) leads to

$$y_0 + 2y_{-1} + 2y_{-2} = u_{-2} + u_{-1} \, .$$

Since the system is assumed to have causal input and output sequences, this gives $y_0 = 0$. If we now set $k = -1$, we obtain

$$y_1 + 2y_0 + 2y_{-1} = u_{-1} + u_0 \, .$$

Since $\{u_k\} = \{(1/2)^k \zeta_k\}$, we see that $u_0 = 1$ and hence $y_1 = 1$. Therefore, taking z-transforms in (5.29) gives

$$z^2 Y(z) - z + 2zY(z) + 2Y(z) = U(z) + zU(z) - z$$

which is clearly equivalent to our previous result obtained more directly through use of the bilateral z-transform. Another common, and equivalent, approach is to rewrite (5.29) in the form

$$y_k = -2y_{k-1} - 2y_{k-2} + u_{k-1} + u_{k-2} .$$

If we now take z-transforms, using the right-shifting property of (5.21), together with the assumed causality of $\{u\}$ and $\{y\}$, we obtain

$$Y(z) = -2z^{-1}Y(z) - 2z^{-2}Y(z) + z^{-1}U(z) + z^{-2}U(z) .$$

Multiplying both sides by z^2 and re-arranging then gives

$$(z^2 + 2z + 2)Y(z) = (z + 1)U(z)$$

providing the same result as before.

$\square$

5.8   THE z-TRANSFER FUNCTION

In §1.10, we introduced the concept of a system Laplace transfer function for continuous-time systems. This idea led us to an understanding of system stability and, hence, to a frequency domain characterization of systems using the frequency response. Drawing on this experience, we wish to develop the corresponding theory for discrete-time systems.

The examples we have considered so far have concentrated on systems modelled by constant coefficient linear difference equations that can be considered as special cases of the system model

$$y_{k+n} + a_{n-1}y_{k+n-1} + \cdots + a_1 y_{k+1} + a_0 y_k =$$
$$b_m u_{k+m} + b_{m-1}u_{k+m-1} + \cdots + b_0 u_k , \quad (5.30)$$

where k is an integer variable and n, m are fixed positive integers with $n \geq m$. We say such a difference equation has order n. Equivalently, using the notation of (5.9), we can write (5.30) in the form

$$a(\mathbf{q})\{y\} = b(\mathbf{q})\{u\} ,$$

where $a(\mathbf{q})$ is an nth order polynomial in $\mathbf{q}$ and $b(\mathbf{q})$ is an mth order polynomial in $\mathbf{q}$ with $m \leq n$. If we now take bilateral z-transforms in this equation, using the appropriate 'left shift' property as given in Table 5.2, we obtain

$$(z^n + a_{n-1}z^{n-1} + \cdots + a_1 z + a_0)Y(z) = (b_m z^m + b_{m-1}z^{m-1} + \cdots + b_0)U(z)$$

or

$$\frac{Y(z)}{U(z)} = D(z) = \frac{b_m z^m + b_{m-1} z^{m-1} + \cdots + b_0}{z^n + a_{n-1} z^{n-1} + \cdots + a_1 z + a_0} = \frac{b(z)}{a(z)} \ . \tag{5.31}$$

$D(z)$ is then the **z-transfer function** of the system (5.30). It should be noted that, in many examples, the system is considered to be initially quiescent. This means we can model both $\{u\}$ and $\{y\}$ as being causal sequences. In this case the bilateral and unilateral transforms are identical and so (5.31) can be considered as an equation involving the usual (unilateral) z-transform.

Example 5.18
Find the z-transfer functions of the systems

$$(1) \quad y_{k+3} + 2y_{k+2} + y_{k+1} = u_{k+1} + u_k \quad (k \in \mathbb{Z})$$
$$(2) \quad y_k - y_{k-1} = u_{k-1} + u_{k-2} - 2u_{k-3} \quad (k \in \mathbb{Z})$$

(1)   At once, we have

$$D(z) = \frac{z+1}{z^3 + 2z^2 + z} = \frac{z+1}{z(z+1)^2} = \frac{1}{z(z+1)} \ .$$

(2)   Taking bilateral z-transforms, together with the right shift property of Table 5.2, we obtain

$$(1 - z^{-1})Y(z) = (z^{-1} + z^{-2} - 2z^{-3})U(z)$$

and so

$$D(z) = \frac{z^{-1} + z^{-2} - 2z^{-3}}{1 - z^{-1}} = \frac{z^2 + z - 2}{z^3 - z^2} = \frac{(z+2)(z-1)}{z^2(z-1)} = \frac{z+2}{z^2} \ .$$

□

5.9   A CONNECTION WITH THE LAPLACE TRANSFORM

In many situations, sequences are obtained through sampling a continuous-time signal. Let the sequence $\{x\} = \{x_k\}$ be given by

$$x_k = x(kT)$$

where $T > 0$ represents a sample period and $x(t)$ is some given continuous-time signal. In such circumstances, it is often a convenient analysis tool to associate with $\{x\}$ the impulse train given by

$$x_s(t) = \sum_{k=-\infty}^{\infty} x_k \delta(t - kT) \ .$$

(See also §3.9.) We see that $x_s(t)$ consists of a sequence of impulses at times $t = kT$ with magnitudes given by x_k. It is then natural to create a transform for the sequence $\{x\}$ by taking the Laplace transform of the function $x_s(t)$. This transform is given by

$$X_s(s) = \sum_{k=-\infty}^{\infty} x_k e^{-skT} = \sum_{k=-\infty}^{\infty} x_k \left(e^{sT}\right)^{-k} .$$

We then note that, substituting z for e^{sT}, we recover the z-transform. This technique is particularly useful in analysing **sampled-data systems**, that is to say systems containing both discrete-time and continuous-time signals, since only the one transform (Laplace) is then required.

5.10   EXERCISES

1. Calculate the z-transform of the following sequences.

   (a) $\{0,\ 0,\ 1,\ 2,\ -2,\ -1,\ 0,\ 2\}$.

   (b) $\{x_k\}$, with $x_k = 1$, $0 \le k \le 5$: $x_k = 0$, $k > 5$.

   (c) $\left\{ 5^k + 3\left(-\dfrac{1}{4}\right)^k \right\}$.

   (d) $\left\{ \left(\dfrac{1}{2}\right)^{k-3} \right\}$.

   (e) $\{5k^2\}$.

2. Following a similar procedure to that used in Example 5.11, show that the z-transform of the sequence $\{\cos(2k\theta)\}$ is given by

$$\frac{z(z - \cos(2\theta))}{z^2 - 2z\cos(2\theta) + 1} .$$

3. Invert the following z-transforms of causal sequences.

   (a) $\dfrac{2z^2 - 3z}{z^2 - 3z + 2}$.

   (b) $\dfrac{z^2 - z}{(z - 4)(z - 2)^2}$.

   (c) $\dfrac{z - 3}{z^2 - 3z + 2}$.

4. Solve the following difference equations, defined for $k \geq 0$, $k \in \mathbb{Z}$.

   (a) $2y_{k+2} - 5y_{k+1} + 2y_k = 0$, $y_0 = 1$, $y_1 = 0$.

   (b) $6y_{k+2} + 5y_{k+1} - y_k = 10$, $y_0 = 0$, $y_1 = 1$.

   (c) $12y_{k+2} - 7y_{k+1} + y_k = 20(-1)^k$, $y_0 = 0$, $y_1 = -1$.

   (d) $y(k+2) + 2y(k) = \sin(k\pi/2)$, $y(0) = 1$, $y(1) = 2$.

   For each of the above difference equations, calculate y_k, $(y(k))$ for $k = 2, 3$ and 4 directly from the difference equation and compare with the values obtained using the solution formula.

5. Find the z-transfer function for the system with input sequence $\{u_k\}$, output sequence $\{y_k\}$, and governed by the following coupled difference equations,

$$w_{k+1} = x_k + u_{k+1}$$
$$x_{k+1} = -x_k - 2y_k$$
$$y_{k+1} = w_k.$$

6. In §2.5, we considered the 'staircase' approximation

$$u(t) \approx \hat{u}(t) = \sum_k u(t_k) \left[\zeta(t - t_k) - \zeta(t - t_{k+1}) \right] .$$

Letting $t_k = kT$, take Laplace transforms to show that

$$\hat{U}(s) = \frac{1 - z^{-1}}{s} U(z),$$

where $z = e^{sT}$ and $U(z)$ is the z-transform of the sequence of sampled values, $\{u(t_k)\}$.

6

Discrete-time system responses

6.1 INTRODUCTION

As in the continuous-time setting, it is important to examine discrete-time system properties such as stability, and responses to test inputs such as steps, impulses and sinusoids. For example, quantization effects can be considered as a type of disturbance to the system input signal say, and it is important that such bounded disturbances do not produce unbounded effects in the output signal. In recent years, the replacement of analogue filters by digital ones has become more widespread and, in such areas, it is clear that calculating responses to (sampled) sinusoids is of utmost importance. In certain circumstances, as we shall see in later chapters, step and impulse responses can also be of importance in filter design. This is also the case in the related area of compensator design in the field of control engineering.

6.2 BIBO AND MARGINAL STABILITY

As in the continuous-time setting, we say that a system with transfer function, $D(z)$, is **BIBO, (Bounded Input, Bounded Output), stable** if *bounded inputs always give rise to bounded outputs*, at least under the assumption that the system is initially quiescent.

Example 6.1
Find the responses of the systems with transfer function given by

$$(1) \quad D(z) = \frac{z+1}{z^2 + 0.75z + 0.125} \qquad (2) \quad D(z) = \frac{z}{z^2 + 3z + 2}$$

when the input is a unit impulse, i.e. $\{u_k\} = \{\delta_k\}$, where $\delta_k = 1$ when $k = 0$ and $\delta_k = 0$ otherwise. In each case, assume the system is initially quiescent.

(1) Since $U(z) = 1$, the response $\{y_k\}$ has z-transform given by

$$Y(z) = D(z) = \frac{z+1}{(z+0.5)(z+0.25)}$$
$$= \frac{-2}{z+0.5} + \frac{3}{z+0.25} \ .$$

It therefore follows from the 'right shift property' of z-transforms that

$$y_k = \begin{cases} 0, & \text{for } k \leq 0 \\ (-2)(-0.5)^{k-1} + (3)(-0.25)^{k-1}, & \text{for } k \geq 1. \end{cases}$$

(2)   This time the response $\{y_k\}$ has z-transform given by

$$Y(z) = D(z) = \frac{z}{(z+1)(z+2)}$$

$$= z \left[\frac{1}{z+1} - \frac{1}{z+2} \right]$$

$$= \frac{z}{z+1} - \frac{z}{z+2}.$$

In this case, it follows that $\{y_k\}$ is given by

$$y_k = \begin{cases} 0, & \text{for } k < 0 \\ (-1)^k - (-2)^k & \text{for } k \geq 0. \end{cases}$$

$\square$

We note that the first of the above two responses decays to zero as k tends to infinity. We also see that the reason for this lies in the fact that both the poles of $D(z)$ have modulus strictly less than one. In contrast, the second response blows up in magnitude as k tends to infinity due to the pole at $z = -2$, which has magnitude strictly greater than one. It can be shown, although we do not show this here, that *a system is BIBO stable if and only if its impulse response, (from initially quiescent conditions), decays to zero* and so the first of the two systems is indeed stable. In general, we have the result that *a discrete-time system is BIBO stable if and only if all the poles of its transfer function lie in the open unit disk $|z| < 1$, $z \in \mathbb{C}$.*

A weaker concept to BIBO stability is that of **marginal stability**. As in the continuous-time setting, marginal stability can be characterized by the property that *the impulse response remains bounded but does not decay to zero.* In terms of the transfer function poles, *a discrete-time system is marginally stable if and only if all the poles lie in the closed unit disk $|z| \leq 1$ and, moreover, any poles on the unit circle $|z| = 1$, (of which there must be at least one), must be simple, i.e. not repeated.* For marginally stable systems, bounded inputs give rise to bounded outputs except in the special cases where the input $\{u_k\}$ has a z-transform $U(z)$ which has at least one pole on the unit circle that co-incides with a transfer function pole.

Example 6.2
Examine the stability of the system with transfer function given by

$$D(z) = \frac{1}{z^2 + \frac{3}{2}z + \frac{1}{2}} = \frac{1}{(z+1)(z+\frac{1}{2})}$$

and find the response to the bounded input sequence $\{u\} = \{u_k\}$ given by $u_k = (-1)^k$, $k \geq 0$, assuming input and output are both causal.

The system poles lie at $z = -1$ and $z = -\frac{1}{2}$. Both poles lie inside the closed unit disk and, moreover, the pole $z = -1$, lying on the unit circle, is simple. It follows that this system is not BIBO stable but is marginally stable. Let us now find the response of this system to the bounded input sequence $\{u\}$ defined above. We have

$$
\begin{aligned}
Y(z) &= D(z)U(z) \\
&= \frac{z}{(z+1)^2(z+\frac{1}{2})} \\
&= \frac{-2z}{(z+1)^2} - \frac{4z}{z+1} + \frac{4z}{z+\frac{1}{2}} \ .
\end{aligned}
$$

Using Table 5.1, we deduce that

$$
y_k = 2k(-1)^k - 4(-1)^k + 4\left(-\frac{1}{2}\right)^k, \qquad \text{for } k \geq 0.
$$

In particular, we see that the output is now unbounded as k tends to infinity.

□

6.3   THE IMPULSE RESPONSE

In an analogous fashion to the continuous-time case, we define the **impulse response** of a discrete-time system to be *the output that results when the input is a unit impulse sequence, $\{\delta_k\}$, and the system is initially quiescent.* If we denote the impulse response by $\{y_k^\delta\}$, we see that we have

$$
\{y_k^\delta\} = \mathcal{Z}^{-1}\left[D(z)\right] \ . \tag{6.1}
$$

Example 6.3
Calculate the impulse response for the system with transfer function

$$
D(z) = \frac{z}{(z+1)(z+\frac{1}{2})} \ .
$$

The impulse response is

$$
\begin{aligned}
\{y_k^\delta\} = \mathcal{Z}^{-1}\left[D(z)\right] &= \mathcal{Z}^{-1}\left[\frac{2z}{z+\frac{1}{2}} - \frac{2z}{z+1}\right] \\
&= \left\{2\left(-\frac{1}{2}\right)^k - 2(-1)^k\right\}, \qquad k \geq 0.
\end{aligned}
$$

A plot of this impulse response is given in Figure 6.1.

□

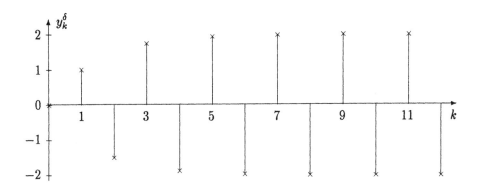

Figure 6.1: Impulse response for Example 6.3.

6.4   THE STEP RESPONSE

The **step response** for a discrete-time linear system is defined to be *the output that results when the input is a unit step sequence, $\{\zeta_k\}$, assuming the system is initially quiescent.* Denoting this response by $\{y_k^\zeta\}$, we deduce that

$$\{y_k^\zeta\} = \mathcal{Z}^{-1}\left[D(z)\,\frac{z}{z-1}\right] . \tag{6.2}$$

We recall that, in a continuous-time setting, we can consider an impulse function to be the derivative of a step function. As a consequence, the impulse response is the derivative of the step response. In the discrete-time setting, we have $\{\delta_k\} = \{\zeta_k\} - \{\zeta_{k-1}\} = (1 - \mathbf{q}^{-1})\{\zeta_k\}$, and we can deduce from this that

$$\{y_k^\delta\} = \{y_k^\zeta - y_{k-1}^\zeta\} . \tag{6.3}$$

Example 6.4
Find the step response for the system of Example 6.3 and demonstrate (6.3).

Using (6.2), we see that the step response is given by

$$\{y_k^\zeta\} = \mathcal{Z}^{-1}\left[\frac{z^2}{(z+1)(z+\frac{1}{2})(z-1)}\right]$$

$$= \mathcal{Z}^{-1}\left[\frac{1}{3}\frac{z}{z-1} + \frac{2}{3}\frac{z}{z+\frac{1}{2}} - \frac{z}{z+1}\right]$$

$$= \left\{\frac{1}{3} + \frac{2}{3}\left(-\frac{1}{2}\right)^k - (-1)^k\right\}_0^\infty .$$

Next we demonstrate (6.3). We have, for $k \geq 1$,

$$y_k^\zeta - y_{k-1}^\zeta = \left[\frac{1}{3} + \frac{2}{3}\left(-\frac{1}{2}\right)^k - (-1)^k\right] - \left[\frac{1}{3} + \frac{2}{3}\left(-\frac{1}{2}\right)^{k-1} - (-1)^{k-1}\right]$$

$$= 2 \left(-\frac{1}{2} \right)^k - 2(-1)^k .$$

For $k = 0$, we obtain $y_0^\varsigma - y_{-1}^\varsigma = y_0^\varsigma = 0$ and so the above formula is also valid for $k = 0$. This is in agreement with the formula for $\{y_k^\delta\}$ obtained in Example 6.3. A plot of this step response is given in Figure 6.2.

□

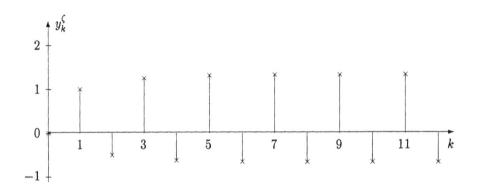

Figure 6.2: Step response for Example 6.4.

Example 6.5
Consider the analogue system with Laplace transfer function

$$H(s) = \frac{1}{s^2 + \sqrt{2}\,s + 1} .$$

Find the z-transfer function for a discrete-time system which has a step response, $\{y_k^\varsigma\}$, which matches sampled values of the step response of the above analogue system, i.e. $y_k^\varsigma = y_\varsigma(kT)$, where $y_\varsigma(t)$ is the analogue step response and $T > 0$ is the sample period.

We have

$$Y_\varsigma(s) = \frac{H(s)}{s} = \frac{1}{s(s^2 + \sqrt{2}\,s + 1)}$$

$$= \frac{1}{s} - \frac{s + \sqrt{2}}{s^2 + \sqrt{2}\,s + 1}$$

$$= \frac{1}{s} - \frac{s + \frac{1}{\sqrt{2}}}{(s + \frac{1}{\sqrt{2}})^2 + \frac{1}{2}} - \frac{\frac{1}{\sqrt{2}}}{(s + \frac{1}{\sqrt{2}})^2 + \frac{1}{2}} .$$

The analogue step response is therefore

$$y_\varsigma(t) = \left(1 - e^{-\frac{t}{\sqrt{2}}} \left[\cos \left(\frac{t}{\sqrt{2}} \right) + \sin \left(\frac{t}{\sqrt{2}} \right) \right] \right) \varsigma(t) .$$

It follows that $\{y_k^\varsigma\}$ must be given by

$$y_k^\varsigma = y_\varsigma(kT)$$

$$= \left(1 - \left(e^{-\frac{T}{\sqrt{2}}}\right)^k \left[\cos\left(\frac{kT}{\sqrt{2}}\right) + \sin\left(\frac{kT}{\sqrt{2}}\right)\right]\right)\varsigma_k \ .$$

Letting $c = \cos\left(\frac{T}{\sqrt{2}}\right)$ and $s = \sin\left(\frac{T}{\sqrt{2}}\right)$ and using the scaling property of z-transforms as given in Table 5.2, we obtain

$$Y^\varsigma(z) = \frac{z}{z-1} - \frac{e^{\frac{T}{\sqrt{2}}}z(e^{\frac{T}{\sqrt{2}}}z - c) + e^{\frac{T}{\sqrt{2}}}zs}{(e^{\frac{T}{\sqrt{2}}}z)^2 - 2(e^{\frac{T}{\sqrt{2}}}z)c + 1}$$

$$= \frac{z}{z-1} - \frac{z(z - e^{-\frac{T}{\sqrt{2}}}c) + e^{-\frac{T}{\sqrt{2}}}zs}{z^2 - 2e^{-\frac{T}{\sqrt{2}}}zc + e^{-\sqrt{2}T}} \ .$$

Since $Y^\varsigma(z) = D(z)\dfrac{z}{z-1}$, we deduce

$$D(z) = \frac{z-1}{z}Y^\varsigma(z)$$

$$= 1 - \frac{(z-1)(z - e^{-\frac{T}{\sqrt{2}}}c + e^{-\frac{T}{\sqrt{2}}}s)}{z^2 - 2e^{-\frac{T}{\sqrt{2}}}zc + e^{-\sqrt{2}T}}$$

$$= \frac{(1 - e^{-\frac{T}{\sqrt{2}}}(c+s))z + e^{-\frac{T}{\sqrt{2}}}(e^{-\frac{T}{\sqrt{2}}} - c + s)}{z^2 - 2e^{-\frac{T}{\sqrt{2}}}zc + e^{-\sqrt{2}T}} \ .$$

Later in this book, we examine further methods for creating digital systems that approximate given analogue prototypes.

$$\square$$

6.5   DISCRETE-TIME CONVOLUTION

Let $X(z)$ and $Y(z)$ represent bilateral z-transforms of sequences $\{x\}$ and $\{y\}$ respectively. Now suppose we wish to find the sequence $\{w\}$ which has the bilateral transform, $W(z) = X(z)Y(z)$. It is tempting to think the answer may be $w_k = x_k y_k$ but, as our previous work on Laplace transforms should have indicated, this is not the case. In fact, we have

$$W(z) = \sum_{k=-\infty}^{\infty} \frac{x_k}{z^k} \sum_{r=-\infty}^{\infty} \frac{y_r}{z^r}$$

$$= \sum_{k=-\infty}^{\infty} \left(\sum_{r=-\infty}^{\infty} \frac{x_k y_r}{z^{k+r}}\right) \ .$$

Letting $m = k + r$ in the sum contained within the brackets, we may write

$$W(z) = \sum_{k=-\infty}^{\infty} \sum_{m=-\infty}^{\infty} \frac{x_k y_{m-k}}{z^m} = \sum_{m=-\infty}^{\infty} \left(\sum_{k=-\infty}^{\infty} x_k y_{m-k} \right) z^{-m} .$$

But, by definition, we also have $W(z) = \sum_{m=-\infty}^{\infty} w_m z^{-m}$ and so, by equating like powers of z^{-1}, we obtain

$$w_k = \sum_{r=-\infty}^{\infty} x_r y_{k-r} = \sum_{r=-\infty}^{\infty} x_{k-r} y_r , \qquad (6.4)$$

where the latter equality follows from the commutative property of convolution. The sequence $\{w\} = \{w_k\}$ given by (6.4) is said to be the **convolution** of $\{x\}$ with $\{y\}$ and we use the notation $\{w\} = \{x\} * \{y\}$. We therefore have the transform pair

$$\{x\} * \{y\} \leftrightarrow X(z)Y(z) . \qquad (6.5)$$

We note that $X(z)Y(z) = Y(z)X(z)$, from which the commutative property of convolution clearly follows.

In the special case where both $\{x\}$ and $\{y\}$ are causal sequences, it can easily be shown that $\{w\} = \{x\} * \{y\}$ is also causal with

$$w_k = \sum_{r=0}^{k} x_{k-r} y_r , \qquad \text{for } k \geq 0. \qquad (6.6)$$

In §2.5, we spent some time discussing convolution in the time domain in order to gain some insight into the process. The main idea then was to investigate how a signal may be viewed as the superposition of a continuum of weighted impulses. We adopt a similar approach here and note that we can formally write a sequence $\{u\} = \{u_k\}$ as the infinite sum

$$\{u_k\}_{k=-\infty}^{\infty} = \sum_{r=-\infty}^{\infty} u_r \{\delta_{k-r}\}_{k=-\infty}^{\infty} , \qquad (6.7)$$

where we recall, from the definition of the impulse sequence $\{\delta_k\}$, that

$$\delta_{k-r} = \begin{cases} 1, & \text{for } k = r \\ 0, & \text{for } k \neq r. \end{cases}$$

It follows that $u_r \{\delta_{k-r}\}_{k=-\infty}^{\infty}$ is the sequence with term u_r in the rth position and 0 everywhere else. Since (6.7) contains an infinite sum of sequences, care should be taken in its interpretation. We omit any such details and content ourselves with noting that no problems occur if $\{u\}$ is finitely nonzero, since (6.7) is then only a finite sum.

Suppose we now supply the signal sequence $\{u\}$ as input to a causal linear time-invariant system, with impulse response $\{y_k^\delta\}$. Formally, we obtain

$$\{y_k\} = \sum_{r=-\infty}^{\infty} u_r \{y_{k-r}^\delta\} . \qquad (6.8)$$

Again this holds if $\{u\}$ is finitely nonzero and may be extended to infinite sequences under certain mild conditions which we omit for brevity. It follows that the discrete-time analogue, ('discretealogue'?), of Equations (2.22) and (2.23) are

$$\{u_k\} = \sum_{r=-\infty}^{\infty} u_r \{\delta_{k-r}\} \qquad \text{(Sifting Property)} \qquad (6.9)$$

$$\{y_k\} = \sum_{r=-\infty}^{\infty} u_r \{y_{k-r}^{\delta}\} \ . \qquad (6.10)$$

Equivalently, using the convolution operator

$$\{u\} = \{\delta\} * \{u\}$$
$$\{y\} = \{y^{\delta}\} * \{u\} \ .$$

Example 6.6
Consider the system with z-transfer function

$$D(z) = \frac{z}{(z+1)(z+\frac{1}{2})} \ .$$

Using the impulse response as found in Example 6.3, find the step response using a convolution sum and check the result with that obtained using z-transforms in Example 6.4.

The impulse response is

$$\{y_k^{\delta}\} = \left\{ 2 \left(-\frac{1}{2} \right)^k - 2(-1)^k \right\} \ .$$

Using (6.10) with input $\{u_k\} = \{\zeta_k\}$ to calculate the step response, we see this response sequence is generated by

$$y_k = \sum_{r=0}^{k} u_{k-r} y_r^{\delta} \qquad k \geq 0,$$

where we have used the causal form for convolution and have used the commutativity property, $\{y^{\delta}\} * \{u\} = \{u\} * \{y^{\delta}\}$. Substituting $u_{k-r} = \zeta_{k-r} = 1$, $(k \geq r)$ into the above gives rise to $(k \geq 0)$

$$y_k = \sum_{r=0}^{k} y_r^{\delta}$$
$$= \sum_{r=0}^{k} \left[2 \left(-\frac{1}{2} \right)^r - 2(-1)^r \right]$$
$$= 2 \sum_{r=0}^{k} \left(-\frac{1}{2} \right)^r - 2 \sum_{r=0}^{k} (-1)^r$$

$$= 2\left[\frac{1-(-\frac{1}{2})^{k+1}}{1+\frac{1}{2}}\right] - 2\left[\frac{1-(-1)^{k+1}}{1+1}\right]$$

$$= \frac{4}{3}\left[1-\left(-\frac{1}{2}\right)^{k+1}\right] - [1-(-1)^{k+1}]$$

$$= \frac{1}{3} - \left(\frac{4}{3}\right)\left(-\frac{1}{2}\right)\left(-\frac{1}{2}\right)^k + (-1)(-1)^k \ .$$

We therefore obtain

$$y_k = \left(\frac{1}{3} + \frac{2}{3}\left(-\frac{1}{2}\right)^k - (-1)^k\right)\zeta_k \ , \tag{6.11}$$

which agrees with the result of Example 6.4.

□

The experience gained in analysing continuous-time systems, which showed that *if a particular response is required, then a direct calculation is more efficient than the use of convolution integrals*, is clearly relevant here also. Nevertheless, *convolution again provides a solution formula for the output in the general case*, and this is its major value.

6.6   THE FREQUENCY RESPONSE

The next phase of our development moves to the frequency domain. First we examine the response of a discrete-time system to a 'pure' complex exponential input, $e^{j\omega t}$. By taking real or imaginary parts, we can then deduce the response to the real inputs, $\cos(\omega t)$ and $\sin(\omega t)$ respectively. When this has been explored, we turn our attention to a wider class of inputs and this generates the need to extend our work on Fourier analysis to the frequency domain decomposition of signal sequences.

Let us begin our analysis by examining the response of a discrete-time system with transfer function

$$D(z) = \frac{b(z)}{a(z)}$$

to a causal input sequence obtained by sampling $e^{j\omega t}$. Thus, the input sequence is $\{u_k\} = \{e^{j\omega kT}\zeta_k\} = \{(e^{j\omega T})^k\zeta_k\}$ and so, assuming the system is initially quiescent, we obtain

$$Y(z) = D(z)U(z) = D(z)\frac{z}{z-e^{j\theta}} \ , \tag{6.12}$$

where $\boxed{\theta = \omega T}$ is referred to as a **normalized frequency** variable. It therefore follows that

$$\frac{Y(z)}{z} = \frac{b(z)}{a(z)}\frac{1}{z-e^{j\theta}} \ . \tag{6.13}$$

Let us now assume that our system is BIBO stable and, for the sake of simplicity, that it has only simple poles at $z = \lambda_i$, $i = 1, 2, \ldots, n$. Performing a partial fraction expansion on (6.13) gives

$$\frac{Y(z)}{z} = \sum_{i=1}^{n} \frac{\beta_i}{z - \lambda_i} + \frac{\beta'}{z - e^{j\theta}} \ . \tag{6.14}$$

Inversion of (6.14) for $\{y_k\}$ then yields

$$\{y_k\} = \mathcal{Z}^{-1}[Y(z)] = \mathcal{Z}^{-1}\left(\sum_{i=1}^{n} \frac{\beta_i z}{z - \lambda_i} + \frac{\beta' z}{z - e^{j\theta}} \right)$$

$$= \left\{ \sum_{i=1}^{n} \beta_i (\lambda_i)^k + \beta' e^{j k\theta} \right\} \ . \tag{6.15}$$

Since we are assuming stability, we have $|\lambda_i| < 1$ for all i and all the terms in the summation decay to zero as $k \to \infty$. This means that the long-term or **steady-state response** of a stable system with input sequence $\{u_k\} = \{e^{j k\theta}\}$ is simply $\{\beta' e^{j k\theta}\} = \beta' \{e^{j k\theta}\} = \beta' \{u_k\}$. The factor β' can easily be obtained from (6.12) using the cover-up rule, or otherwise, giving

$$\beta' = D(e^{j\theta}) \ . \tag{6.16}$$

Thus, β' is just the transfer function $D(z)$ evaluated on the unit circle $z = e^{j\theta}$, the stability boundary for discrete-time systems. We refer to $D(e^{j\theta})$ as the system's **frequency response**.

Example 6.7
Calculate the frequency response of the system of Figure 6.3.

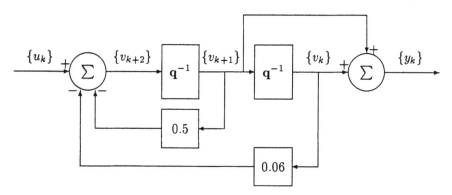

Figure 6.3: Discrete-time system of Example 6.7.

The system difference equation is calculated following the methods of §5.4 as

$$y_{k+2} + 0.5 y_{k+1} + 0.06 y_k = u_{k+1} + u_k \ .$$

Therefore, the z-transfer function is given by

$$D(z) = \frac{z + 1}{z^2 + 0.5z + 0.06} = \frac{z + 1}{(z + 0.3)(z + 0.2)} \ .$$

We see that, since the poles are at $z = -0.3$ and $z = -0.2$, which both have modulus strictly less than one, the system is stable. Therefore, the frequency response exists and is given by

$$D(e^{j\theta}) = \frac{e^{j\theta} + 1}{(e^{j\theta} + 0.3)(e^{j\theta} + 0.2)} .$$

□

We have seen that, for a stable system $D(z)$, the input signal $\{u_k\} = \{e^{jk\theta}\}$ gives rise to the steady-state response

$$\{y_k\} = D(e^{j\theta})\{e^{jk\theta}\} . \tag{6.17}$$

As the previous example illustrates, the frequency response is, in general, a complex quantity, and we can write

$$D(e^{j\theta}) = |D(e^{j\theta})|e^{j\phi_0},$$

where $\phi_0 = \arg D(e^{j\theta})$. It follows that (6.17) can be rewritten as

$$\{y_k\} = \{|D(e^{j\theta})|e^{j(k\theta + \phi_0)}\} . \tag{6.18}$$

With this in mind, we parallel the continuous-time terminology by defining $|D(e^{j\theta})|$ to be the system's **amplitude response**, (or **gain spectrum** or **magnitude response**), and defining $\arg D(e^{j\theta})$ to be the system's **phase response**, (or **phase-shift spectrum**). By taking real and imaginary parts in (6.18), we deduce that the respective steady state responses to the inputs $\{\cos(k\theta)\}$ and $\{\sin(k\theta)\}$ are given by

$$\{|D(e^{j\theta})|\cos(k\theta + \phi_0)\} \qquad \text{and} \qquad \{|D(e^{j\theta})|\sin(k\theta + \phi_0)\} .$$

Example 6.8
Calculate the amplitude and phase responses for the system of Example 6.7. Hence deduce the steady state response to a cosine wave with frequency 50 Hz., given a sampling frequency of (1) 200 Hz.   (2) 60 Hz.

In Example 6.7, we established the frequency response as

$$D(e^{j\theta}) = \frac{e^{j\theta} + 1}{(e^{j\theta} + 0.3)(e^{j\theta} + 0.2)} .$$

Let us note that, for any complex number z, we have $|z|^2 = z\bar{z}$ where $\bar{z}$ denotes the complex conjugate. In particular, it follows that if z has the form $z = e^{j\theta} + \alpha$ for some real constant α, then

$$|e^{j\theta} + \alpha|^2 = (e^{j\theta} + \alpha)(e^{-j\theta} + \alpha)$$
$$= 1 + \alpha^2 + \alpha(e^{j\theta} + e^{-j\theta})$$
$$= 1 + \alpha^2 + 2\alpha\cos\theta.$$

Using this result, we obtain

$$|D(e^{j\theta})| = \frac{|e^{j\theta} + 1|}{|e^{j\theta} + 0.3|\,|e^{j\theta} + 0.2|}$$

$$= \frac{\sqrt{2 + 2\cos\theta}}{\sqrt{1.09 + 0.6\cos\theta}\,\sqrt{1.04 + 0.4\cos\theta}} .$$

The phase response ϕ_0 is given by

$$\phi_0 = \arg(e^{j\theta} + 1) - \arg(e^{j\theta} + 0.3) - \arg(e^{j\theta} + 0.2), \qquad -\pi < \phi_0 \le \pi .$$

A plot of these two responses can be seen in Figure 6.4.

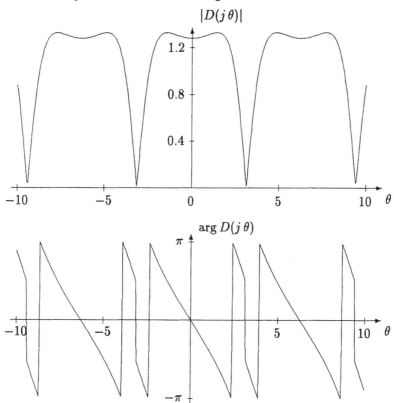

Figure 6.4: Frequency response plots for the system of Example 6.8.

For the sampling frequency of case (1), we deduce that the normalized frequency, θ, is given by $\theta = \omega T = (100\pi)/200 = \pi/2$. The corresponding system gain and phase-shift are therefore given by

$$|D(e^{j\frac{\pi}{2}})| = |D(j)| = \frac{\sqrt{2}}{\sqrt{1.09}\,\sqrt{1.04}} \approx 1.33$$

and

$$\arg(D(j)) = \frac{\pi}{4} - \tan^{-1}\left(\frac{1}{0.3}\right) - \tan^{-1}\left(\frac{1}{0.2}\right) \approx -1.87 .$$

The steady-state response is therefore given by

$$\{y_k\} \approx \{1.33 \cos(k\pi/2 - 1.87)\} .$$

For case (2), we see that the normalized frequency is now given by $\theta = (100\pi)/60 = 5\pi/3$. The corresponding gain and phase-shift are

$$|D(e^{j\frac{5\pi}{3}})| = \frac{\sqrt{3}}{\sqrt{1.39}\sqrt{1.24}} \approx 1.32$$

and

$$\arg(D(e^{j\frac{5\pi}{3}})) = -\frac{\pi}{6} + \tan^{-1}\left(\frac{5\sqrt{3}}{8}\right) + \tan^{-1}\left(\frac{5\sqrt{3}}{7}\right) \approx 1.19 .$$

The corresponding steady-state response is therefore given by

$$\{y_k\} \approx \{1.32\cos(5k\pi/3 + 1.19)\} .$$

$\square$

We now have a picture of the steady-state operation of causal, linear, time-invariant stable systems in the frequency domain. To make the most of this concept, *we must seek a decomposition of general signal sequences in the frequency domain.* This is the task of the next chapter. However, before doing this, we observe an important difference between the frequency response functions for continuous-time systems and those for discrete-time systems. This difference is apparent in Figure 6.4, where we see that $D(e^{j\theta})$ is periodic in θ with period given by

$$\theta_P = \omega_P T = 2\pi , \tag{6.19}$$

in contrast to the non-periodic responses obtained for continuous-time systems. Since the function $e^{j\theta}$ is periodic in θ with period 2π, we see that *frequency response periodicity is inevitable for discrete-time systems.* Another way of looking at this is to re-examine the sampling operation on the complex exponential signal $e^{j\omega t}$ giving the sequence $\{e^{j\omega kT}\} = \{e^{jk\theta}\}$. Now suppose we increase the frequency of the complex exponential by an integer multiple of $\frac{2\pi}{T}$ radians per second to give $e^{j(\omega+n\frac{2\pi}{T})t}$. If we again sample with sample-period, T, we obtain the sequence

$$\{e^{j(\omega+n\frac{2\pi}{T})kT}\} = \{e^{j\omega kT}\, e^{jnk2\pi}\} = \{e^{jk\theta}\}$$

since $e^{jnk2\pi} = (e^{j2\pi})^{nk} = (1)^{nk} = 1$. So *sampled values of two complex exponentials do not differ if the frequencies are integer multiples of $\frac{2\pi}{T}$ rad./sec. apart.* By inference the same conclusion must hold if complex exponentials are replaced by real sinusoids with equal phases. Since a discrete-time system acts on such sampled values, it clearly cannot distinguish between any two such sinusoids and so we again see the inevitability of periodicity in the frequency response.

This phenomenon is sometimes referred to as the **wagon-wheel effect**. This is due to the fact then when you view a wagon-wheel on a movie screen, you only see discrete snap-shots of the wheel. If the rotational speed, (frequency), of the wheel is sufficiently high then these snap-shots, (sampled values), are identical to those of a wheel with lower speed/frequency. The eye then perceives the wheel as if it is going at this lower speed—possibly even backwards—even though this is clearly a false perception. The more scientific term for this phenomenon is **aliasing**.

6.7 EXERCISES

1. Find the impulse response of the system

$$8y_{k+2} - 6y_{k+1} + y_k = u_k, \qquad k \in \mathbb{Z}.$$

   Is the system stable? Illustrate the response with a sketch.

2. The system

$$50y(k + 2) - 35y(k + 1) + 6y(k) = 25(u(k + 1) + u(k)), \qquad k \in \mathbb{Z},$$

   is initially quiescent. Find the impulse response and the response to the unit step input, $u(k) = \begin{cases} 1, & \text{for } k \geq 0, \\ 0, & \text{for } k < 0. \end{cases}$

3. Show that the system modelled by

$$2y_{k+1} + \alpha y_k = u_k, \qquad k \in \mathbb{Z}, \ \alpha \in \mathbb{R}$$

   is BIBO stable if and only if $|\alpha| < 2$. For the case $\alpha = 1$, calculate the frequency response and plot the amplitude response over the range $-2\pi \leq \theta \leq 2\pi$.

4. Find the z-transfer function for the system

$$8y(k + 2) + 2y(k + 1) - y(k) = u(k) + 2u(k + 1), \qquad k \in \mathbb{Z},$$

   and verify that it is a stable system. Plot the amplitude response over the range $-2\pi \leq \theta \leq 2\pi$.

5. Draw simulation diagrams to represent the systems of Exercises 1,2 and 4.

6. Find the frequency response of the autoregressive filter

$$y_k = ay_{k-1} + u_k, \qquad |a| < 1.$$

   Hence, or otherwise, determine the steady-state response to the input $u_k = \cos(k\theta)$.

7. Suppose the unit step response of a linear, time-invariant filter is

$$y(k) = 2\left(\frac{1}{2}\right)^k - \left(-\frac{3}{4}\right)^k, \quad k \geq 0.$$

(a) Find the transfer function for this filter. Plot the poles and zeros of the transfer function.

(b) What is the impulse response of the filter?

(c) Explain why the filter is BIBO stable.

(d) Obtain a difference equation realization for this filter.

8. Investigate BIBO stability for the linear system with transfer function

$$D(z) = \frac{1}{z^2 + \alpha z - \alpha},$$

where the parameter α is real.

7

Discrete-time Fourier analysis

7.1 INTRODUCTION

In Chapter 3, we examined how continuous-time signals could be mapped to the frequency domain via the Fourier transform. This theory was then applied to the study of analogue filters. In recent times, the replacement of such filters using digital technology has become increasingly widespread. Digital filters have the advantage that they can often be easily reprogrammed and are more robust to environmental conditions such as temperature variations. Disadvantages include errors caused by sampling and quantization, and the common need to use them in series with analogue pre-filters to reduce aliasing effects. In order to provide a proper mathematical treatment of digital filters, it is clear that a discrete-time version of Fourier methods is required and this forms the topic for this chapter.

7.2 THE DISCRETE FOURIER TRANSFORM, (DFT)

Following similar lines to that used in the continuous-time setting, let us begin our analysis with the case of **periodic sequences**. With this in mind, we are looking for a frequency domain representation for a sequence $\{u\} = \{u_k\}$, $k \in \mathbb{Z}$, with $u_{k+N} = u_k$ for all k and some $N > 0$. The smallest such value for N is referred to as the **period of the sequence**. To begin our investigation, let us consider a continuous-time periodic signal, $u(t)$, with period P. From our previous work on Fourier series, we know that we can represent $u(t)$ in the form

$$u(t) \sim \sum_{n=-\infty}^{\infty} A_n e^{j\, n\omega_0 t}\,, \qquad \omega_0 = \frac{2\pi}{P}\,. \qquad (7.1)$$

Note that we have used A_n rather than U_n for the Fourier series coefficients. This is to help avoid confusion with the discrete Fourier transform coefficients introduced later in this section.

Let us now sample $u(t)$ at times $t = kT$, where $T = P/N$. This then produces the periodic sequence, $\{u_k\} = \{u(kT)\}$, and (7.1) gives

$$u_k = \sum_{n=-\infty}^{\infty} A_n e^{j\, n\omega_0 kT} = \sum_{n=-\infty}^{\infty} A_n e^{j\, nk \frac{2\pi}{N}} \qquad (7.2)$$

where we have used, $\omega_0 = 2\pi/P = 2\pi/NT$. Let us now note that, if $\sum\limits_{n=-\infty}^{\infty} x_n$ is an absolutely convergent series, then

$$\sum_{n=-\infty}^{\infty} x_n = \cdots + x_{-2} + x_{-1} + x_0 + x_1 + x_2 + \cdots$$

$$= (\cdots + x_{-2N} + x_{-N} + x_0 + x_N + x_{2N} + \cdots)$$
$$+ (\cdots + x_{-2N+1} + x_{-N+1} + x_1 + x_{N+1} + x_{2N+1} + \cdots)$$
$$\vdots$$
$$+ (\cdots + x_{-N-1} + x_{-1} + x_{N-1} + x_{2N-1} + x_{3N-1} + \cdots).$$

It therefore follows that

$$\sum_{n=-\infty}^{\infty} x_n = \sum_{n=0}^{N-1} \sum_{r=-\infty}^{\infty} x_{n+rN} \ . \tag{7.3}$$

If we now note that $e^{j\,(n+rN)k\frac{2\pi}{N}} = e^{j\,nk\frac{2\pi}{N}}e^{j\,rk2\pi} = e^{j\,nk\frac{2\pi}{N}}$, we can rewrite (7.2), using (7.3), as

$$u_k = \sum_{n=0}^{N-1} \left(\sum_{r=-\infty}^{\infty} A_{n+rN} \right) e^{j\,nk\frac{2\pi}{N}} = \frac{1}{N} \sum_{n=0}^{N-1} U_n e^{j\,nk\frac{2\pi}{N}} \ , \tag{7.4}$$

where

$$U_n = N \sum_{r=-\infty}^{\infty} A_{n+rN} \ .$$

The factor $\frac{1}{N}$ has been introduced into (7.4) so as to be consistent with the most common definition for the Discrete Fourier transform. Equation (7.4) can now be viewed as a mapping between the time domain sequence, $\{u_k\}$, and the frequency domain sequence $\{U_n\}$. Clearly the formula $U_n = N \sum\limits_{r=-\infty}^{\infty} A_{n+rN}$ is of little use for finding $\{U_n\}$. Since $\{U_n\}$ has period N, we need only calculate U_n for $n = 0, 1, \ldots, N-1$. We can then write (7.4) in the matrix form

$$u = \frac{1}{N}MU, \tag{7.5}$$

where

$$[u]_k = u_{k-1} \qquad\qquad \text{for } 1 \leq k \leq N,$$
$$[U]_k = U_{k-1} \qquad\qquad \text{for } 1 \leq k \leq N, \text{ and}$$
$$[M]_{rs} = e^{j\,(r-1)(s-1)\frac{2\pi}{N}} \qquad \text{for } 1 \leq r, s \leq N.$$

(The notation $[x]_k$ refers to the kth entry of a (column) vector x and, similarly, $[M]_{rs}$ refers to the entry in row r and column s of a matrix M.) It follows from (7.5) that the problem of finding a direct formula for $\{U_n\}$ from $\{u_k\}$ is equivalent to that of inverting the $N \times N$ matrix $\boldsymbol{M}$. Now let $\boldsymbol{M}^*$ denote the conjugate transpose of $\boldsymbol{M}$. It then follows that

$$
\begin{aligned}
[\boldsymbol{M}^*\boldsymbol{M}]_{rs} &= \sum_{k=1}^{N} m^*_{rk} m_{ks} \\
&= \sum_{k=1}^{N} \overline{m}_{kr} m_{ks} \\
&= \sum_{k=0}^{N-1} e^{-j\,k(r-1)\frac{2\pi}{N}} e^{j\,k(s-1)\frac{2\pi}{N}} \\
&= \sum_{k=0}^{N-1} \left(e^{j\,(s-r)\frac{2\pi}{N}} \right)^k .
\end{aligned}
$$

This latter expression is a finite geometric series. Using the well-known formula for such a series we obtain

$$
[\boldsymbol{M}^*\boldsymbol{M}]_{rs} = \begin{cases} 0, & \text{for } r \neq s \\ N, & \text{for } r = s. \end{cases}
$$

This means that $\boldsymbol{M}^*\boldsymbol{M} = N\boldsymbol{I}$ and so it follows that

$$
\boldsymbol{M}^{-1} = \frac{1}{N}\boldsymbol{M}^* .
$$

Therefore, it follows from (7.5) that $\boldsymbol{U} = \boldsymbol{M}^*\boldsymbol{u}$, or equivalently

$$
U_n = \sum_{k=0}^{N-1} u_k e^{-j\,kn\frac{2\pi}{N}} . \tag{7.6}
$$

The periodic sequence $\{U_n\}$ as given by (7.6) is said to be the **Discrete Fourier Transform (DFT)** of the periodic sequence $\{u_k\}$. Both sequences $\{u_k\}$ and $\{U_n\}$ have the same period, N. The **inverse Discrete Fourier Transform, (IDFT)**, is then given by (7.4). If we let w denote the complex number $\boxed{w = e^{-j\frac{2\pi}{N}}}$, the DFT and IDFT can be written in the form

$$
\textbf{DFT} \qquad U_n = \sum_{k=0}^{N-1} u_k w^{nk} = \sum_{k=0}^{N-1} u_k (w^n)^k \tag{7.7a}
$$

$$
\textbf{IDFT} \qquad u_k = \frac{1}{N}\sum_{n=0}^{N-1} U_n w^{-nk} = \frac{1}{N}\sum_{n=0}^{N-1} U_n (w^{-k})^n . \tag{7.7b}
$$

Since a periodic sequence is uniquely specified by its values over one period, (7.7a) and (7.7b) are sometimes considered as transforms with respect to the finite sequences, $\{u_0, u_1, \ldots, u_{N-1}\}$ and $\{U_0, U_1, \ldots, U_{N-1}\}$. We use both representations interchangeably.

Example 7.1
Calculate the DFT of the finite sequence

$$\{u\} = \{1, 2, 2, 1\} \ .$$

In this case, $N = 4$, and so $w = e^{-j\frac{2\pi}{4}} = -j$. As a consequence of (7.7a), it follows that

$$U_0 = u_0 + u_1 + u_2 + u_3 = 6$$
$$U_1 = u_0 - j\,u_1 - u_2 + j\,u_3 = -1 - j$$
$$U_2 = u_0 - u_1 + u_2 - u_3 = 0$$
$$U_3 = u_0 + j\,u_1 - u_2 - j\,u_3 = -1 + j \ ,$$

and so $\{U\} = \{6, -1 - j, 0, -1 + j\}$.

□

Let us now note that

$$U_{N-n} = \sum_{k=0}^{N-1} u_k w^{(N-n)k}$$

$$= \sum_{k=0}^{N-1} u_k w^{-nk}, \qquad \text{since } w^N = 1.$$

If $\{u\}$ is a real sequence then $u_k w^{-nk}$ is the complex conjugate of $u_k w^{nk}$ and so we see that

$$U_{N-n} = \overline{U}_n \qquad \text{if } \{u\} \text{ is real.}$$

Applying this to the previous example with $n = 1$, we see that we must have $U_3 = \overline{U}_1$, which matches the obtained result. Also, letting $n = 0$ and $n = 2$ give $U_4 = \overline{U}_0$ and $U_2 = \overline{U}_2$. Since $U_4 = U_0$ by periodicity, this means both U_0 and U_2 must be real, which also matches the obtained result.

Let us also note that, when $\{u_k\} = u(kT)$ is obtained from sampling a continuous-time signal $u(t)$ with samples T secs. apart, then the Fourier coefficients are $\omega_0 = 2\pi/NT$ rads./sec. apart. In other words, the sequence $\{U_n\}$ may be viewed in the frequency domain with U_n corresponding to the frequency $n\Omega$ where $\Omega = 2\pi/NT$ rads./sec. Alternatively, if normalized frequency $\theta = \omega T$ is used, sampling is $2\pi/N$ rads. apart.

With this in mind, we refer to the function mapping the normalized frequencies, $n.2\pi/N$, to the magnitudes, $|U_n|$, as the **amplitude spectrum** of the periodic sequence, $\{u_k\}$, and the function mapping $n.2\pi/N$ to the angles, $\arg U_n$, as the corresponding **phase spectrum**.

Let us now give an example illustrating the IDFT.

Example 7.2
Calculate the inverse discrete Fourier transform of the sequence

$$\{U\} = \{6, \ -(1 + j), \ 0, \ -(1 - j)\}$$

of Example 7.1.

Using (7.7b) with $N = 4$ and hence $w = -j$, we obtain

$$u_0 = \frac{1}{4}(U_0 + U_1 + U_2 + U_3)$$

$$= \frac{1}{4}(6 - 1 - j + 0 - 1 + j) = \frac{1}{4}(4) = 1$$

$$u_1 = \frac{1}{4}(U_0 + jU_1 - U_2 - jU_3)$$

$$= \frac{1}{4}(6 - j + 1 - 0 + j + 1) = \frac{1}{4}(8) = 2$$

$$u_2 = \frac{1}{4}(U_0 - U_1 + U_2 - U_3)$$

$$= \frac{1}{4}(6 + 1 + j + 0 + 1 - j) = \frac{1}{4}(8) = 2$$

$$u_3 = \frac{1}{4}(U_0 - jU_1 - U_2 + jU_3)$$

$$= \frac{1}{4}(6 + j - 1 - 0 - j - 1) = \frac{1}{4}(4) = 1 .$$

Thus $\{u\} = \{1, 2, 2, 1\}$, which was the starting sequence of Example 7.1. We have obtained an exact inversion, as predicted.

□

The **FFT (Fast Fourier Transform)** refers to an efficient algorithm for computing the DFT. There are actually several different algorithms that are collectively referred to as FFT's. The most efficient algorithms tend to be the so-called **radix-2** algorithms which rely on N being a power of 2. In most practical applications, such a radix-2 algorithm is used but in certain circumstances N cannot be chosen to be a power of 2. In this case other algorithms, such as mixed-radix algorithms, should be used. A naïve implementation of (7.7a) is impractical except for very small values of N. Such FFT algorithms are readily available in many packages and software libraries and it is assumed that the reader can gain access to these algorithms when required. We present a brief outline of an FFT algorithm in §7.7.

Before leaving this section, let us note that conjugation of (7.7b), followed by multiplication by N, leads to

$$N\bar{u}_k = \sum_{n=0}^{N-1} \overline{U}_n w^{nk}.$$

This implies that the inverse discrete Fourier transform of a sequence $\{U_n\}$ may be found as follows:

(i) find the conjugate sequence, $\{\overline{U}_n\}$;

(ii) find the discrete Fourier transform of $\{\overline{U}_n\}$, i.e.

$$\left\{ \sum_{n=0}^{N-1} \overline{U}_n w^{kn} \right\} = \{N\bar{u}_k\}, \quad k = 0, 1, \ldots, N - 1;$$

(*iii*) divide by N and, if $\{u_k\}$ is not real-valued, conjugate to give $\{u_k\}$.

Step (*ii*) can then be performed by means of an FFT algorithm. This shows that an inverse discrete Fourier transform can be programmed using an FFT algorithm, together with the minor modifications of steps (*i*) and (*iii*) above.

7.3   THE DISCRETE-TIME FOURIER TRANSFORM, DTFT

In the previous section, we considered a transform for discrete-time periodic sequences mirrored on the Fourier series expansion for continuous-time signals. In this section, we introduce a transform for non-periodic sequences mirrored on the Fourier transform. Suppose $f(t)$ has the Fourier transform $F(j\,\omega)$ and so

$$f(t) = \frac{1}{2\pi} \int_{-\infty}^{\infty} F(j\,\omega)\, e^{j\,\omega t}\, d\omega \ . \tag{7.8}$$

If we now sample $f(t)$ at times $t = kT$, $k \in \mathbb{Z}$, we obtain the sequence $\{f_k\} = \{f(kT)\}$ and (7.8) gives

$$f_k = \frac{1}{2\pi} \int_{-\infty}^{\infty} F(j\,\omega)\, e^{j\,\omega kT}\, d\omega. \tag{7.9}$$

Splitting this infinite interval of integration into intervals of length $2\pi/T$, we obtain

$$f_k = \frac{1}{2\pi} \sum_{n=-\infty}^{\infty} \int_{n\frac{2\pi}{T}}^{(n+1)\frac{2\pi}{T}} F(j\,\omega)\, e^{j\,\omega kT}\, d\omega$$

$$= \frac{1}{2\pi} \sum_{n=-\infty}^{\infty} \int_{0}^{\frac{2\pi}{T}} F\left(j\left(\omega + n\frac{2\pi}{T}\right)\right)\, e^{j\left(\omega + n\frac{2\pi}{T}\right)kT}\, d\omega$$

$$= \frac{1}{2\pi} \int_{0}^{\frac{2\pi}{T}} \left(\sum_{n=-\infty}^{\infty} F\left(j\left(\omega + n\frac{2\pi}{T}\right)\right) \right) e^{j\,\omega kT}\, d\omega \ ,$$

since $e^{j\left(\omega+n\frac{2\pi}{T}\right)kT} = e^{j\,\omega kT} e^{j\,nk2\pi} = e^{j\,\omega kT}$. As usual, we do not attempt to give conditions under which the above interchange between an integral and an infinite sum is valid. In any case, the above is only intended as a formal procedure leading to a definition for the discrete-time Fourier transform.

If we now let θ be the normalized frequency $\theta = \omega T$ and set

$$F(e^{j\,\theta}) = \frac{1}{T} \sum_{n=-\infty}^{\infty} F\left(j\left(\frac{\theta + n2\pi}{T}\right)\right) , \tag{7.10}$$

where we note that the right-hand side has period 2π in θ, we obtain

$$f_k = \frac{1}{2\pi} \int_{0}^{2\pi} F(e^{j\,\theta})\, e^{j\,k\theta}\, d\theta \ . \tag{7.11}$$

The periodic function, $F(e^{j\theta})$, is referred to as **the discrete-time Fourier transform, (DTFT)**, of the sequence $\{f_k\}$. Equation (7.10) is unsuitable for calculation of the DTFT and so, following similar lines to §7.2, we instead use (7.11) to define the inverse DTFT and invert this in order to define the direct transform. We claim that the DTFT is, in fact, given by

$$F(e^{j\theta}) = \sum_{n=-\infty}^{\infty} f_n e^{-jn\theta} = \sum_{n=-\infty}^{\infty} f_n (e^{j\theta})^{-n} . \qquad (7.12)$$

Note that this is the same as the bilateral z-transform of $\{f\}$ evaluated at $z = e^{j\theta}$. This fact also explains the use of the notation, $F(e^{j\theta})$. To show that (7.12) is valid, we substitute into the right-hand side of (7.11) to give

$$\frac{1}{2\pi} \int_0^{2\pi} F(e^{j\theta}) e^{jk\theta} \, d\theta = \frac{1}{2\pi} \int_0^{2\pi} \sum_{n=-\infty}^{\infty} f_n e^{-jn\theta} e^{jk\theta} \, d\theta$$

$$= \sum_{n=-\infty}^{\infty} f_n \frac{1}{2\pi} \int_0^{2\pi} e^{j(k-n)\theta} \, d\theta ,$$

assuming the interchange of summation and integration is permissible. However, it is easy to see that

$$\frac{1}{2\pi} \int_0^{2\pi} e^{j(k-n)\theta} \, d\theta = \begin{cases} 0, & \text{for } k \neq n \\ 1, & \text{for } k = n \end{cases} = \delta_{k-n}$$

and so the right-hand side of (7.11) reduces to

$$\sum_{n=-\infty}^{\infty} f_n \delta_{k-n} = f_k,$$

as desired. To summarize, we have the two equations

$$\textbf{DTFT} \qquad F(e^{j\theta}) = \sum_{k=-\infty}^{\infty} f_k e^{-jk\theta}, \qquad (7.13a)$$

$$\textbf{IDTFT} \qquad f_k = \frac{1}{2\pi} \int_0^{2\pi} F(e^{j\theta}) e^{jk\theta} \, d\theta . \qquad (7.13b)$$

Example 7.3
Calculate the discrete-time Fourier transform of the finite sequence

$$\{u\} = \{1, 2, 2, 1\} .$$

Let us first note that we are no longer dealing with periodic sequences. We are therefore adopting our previous convention that the above sequence is 'padded-out' with zeros, i.e. we have $\{u\} = \{u_k\}$ where $u_0 = u_3 = 1$, $u_1 = u_2 = 2$ and $u_k = 0$ otherwise. It follows from (7.13a) that

$$U(e^{j\,\theta}) = 1 + 2e^{-j\,\theta} + 2e^{-2j\,\theta} + e^{-3j\,\theta}$$

$$= e^{-\frac{3}{2}j\,\theta}\left[(e^{\frac{3}{2}j\,\theta} + e^{-\frac{3}{2}j\,\theta}) + 2(e^{\frac{1}{2}j\,\theta} + e^{-\frac{1}{2}j\,\theta})\right]$$

$$= e^{-\frac{3}{2}j\,\theta}\left[2\cos\left(\frac{3\theta}{2}\right) + 4\cos\left(\frac{\theta}{2}\right)\right].$$

A sketch of $|U(e^{j\,\theta})| = |2\cos\left(\frac{3\theta}{2}\right) + 4\cos\left(\frac{\theta}{2}\right)|$ is given in Figure 7.1. $|U(e^{j\,\theta})|$ is called the **amplitude spectrum** of the sequence $\{u\}$. Figure 7.1 clearly shows the periodicity of $|U(e^{j\,\theta})|$, which by now is not surprising. In a similar fashion, we refer to $\arg U(e^{j\,\theta})$ as the **phase spectrum** of $\{u\}$.

$\square$

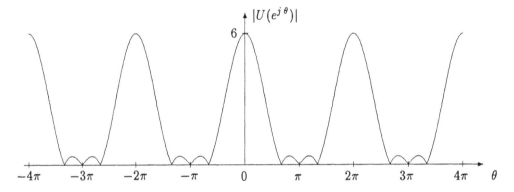

Figure 7.1: Amplitude spectrum $|U(e^{j\,\theta})|$ for Example 7.3.

7.4 ESTIMATING THE DTFT

So far we have mentioned the existence of FFT algorithms for numerically estimating the discrete Fourier transform. Let us now consider the estimation of the discrete-time Fourier transform. Since $U(e^{j\,\theta})$ is a periodic function of the continuous variable θ, it makes sense to estimate N equally spaced samples over one period, $0 \le \theta < 2\pi$. We then have

$$U(e^{j\,n\frac{2\pi}{N}}) = \sum_{k=-\infty}^{\infty} u_k e^{-j\,kn\frac{2\pi}{N}}, \quad n = 0, 1, \ldots, N-1.$$

If we now re-order this series, using the result of (7.3) for absolutely convergent series, we obtain

$$U(e^{j\,n\frac{2\pi}{N}}) = \sum_{k=0}^{N-1} \sum_{r=-\infty}^{\infty} u_{k+rN} e^{-j\,(k+rN)n\frac{2\pi}{N}}$$

$$= \sum_{k=0}^{N-1} \left(\sum_{r=-\infty}^{\infty} u_{k+rN} \right) e^{-jkn\frac{2\pi}{N}}$$

$$= \sum_{k=0}^{N-1} \hat{u}_k w^{kn}$$

where $\hat{u}_k = \sum_{r=-\infty}^{\infty} u_{k+rN}$ and $w = e^{-j\frac{2\pi}{N}}$. In other words, N equally spaced samples of $U(e^{j\theta})$ over the period, $0 \le \theta < 2\pi$, may be obtained from the discrete Fourier transform of the periodic sequence $\{\hat{u}\}$ which has period N). We now illustrate this idea with a few examples.

Example 7.4
Consider again the finite sequence of Example 7.3, namely

$$\{u\} = \{1, 2, 2, 1\} .$$

Using the DFT, calculate 8 equally spaced samples of the DTFT of $\{u\}$ over the unit circle.

Let us begin by calculating the periodic sequence $\{\hat{u}\}$ with period $N = 8$. We have

$$\begin{aligned}
\hat{u}_0 &= \ldots + u_{-16} + u_{-8} + u_0 + u_8 + u_{16} + \ldots &= u_0 &= 1 \\
\hat{u}_1 &= \ldots + u_{-15} + u_{-7} + u_1 + u_9 + u_{17} + \ldots &= u_1 &= 2.
\end{aligned}$$

In a similar fashion, we can show in general that

$$\hat{u}_k = \begin{cases} u_k, & \text{for } 0 \le k \le 3 \\ 0, & \text{for } 4 \le k \le 7 \end{cases} \quad \text{with } \hat{u}_{k+8} = \hat{u}_k \text{ for all } k.$$

The discrete Fourier transform, $\{\hat{U}\}$, of $\{\hat{u}\}$ is then given by (7.7a) with $N = 8$ and hence $w = e^{-j\frac{\pi}{4}} = \frac{1}{\sqrt{2}}(1 - j)$. We therefore obtain

$$\hat{U}_0 = u_0 + u_1 + u_2 + u_3$$
$$= 1 + 2 + 2 + 1 = 6$$
$$\hat{U}_1 = u_0 + \frac{1}{\sqrt{2}}(1 - j)u_1 - j u_2 - \frac{1}{\sqrt{2}}(1 + j)u_3$$
$$= \left(1 + \sqrt{2} - \frac{1}{\sqrt{2}}\right) + j\left(-\sqrt{2} - 2 - \frac{1}{\sqrt{2}}\right) = \left(1 + \frac{1}{\sqrt{2}}\right) - j\left(2 + \frac{3}{\sqrt{2}}\right)$$
$$\hat{U}_2 = u_0 - j u_1 - u_2 + j u_3$$
$$= 1 - j2 - 2 + j = -1 - j$$
$$\hat{U}_3 = u_0 - \frac{1}{\sqrt{2}}(1 + j)u_1 + j u_2 + \frac{1}{\sqrt{2}}(1 - j)u_3$$
$$= \left(1 - \sqrt{2} + \frac{1}{\sqrt{2}}\right) + j\left(-\sqrt{2} + 2 - \frac{1}{\sqrt{2}}\right) = \left(1 - \frac{1}{\sqrt{2}}\right) + j\left(2 - \frac{3}{\sqrt{2}}\right)$$
$$\hat{U}_4 = u_0 - u_1 + u_2 - u_3$$
$$= 1 - 2 + 2 - 1 = 0.$$

Since $\{\hat{u}\}$ is a real sequence, we can use the relation $\hat{U}_{8-n} = \overline{\hat{U}}_n$ to obtain

$$\hat{U}_5 = \overline{\hat{U}}_3 = \left(1 - \frac{1}{\sqrt{2}}\right) - j\left(2 - \frac{3}{\sqrt{2}}\right)$$

$$\hat{U}_6 = \overline{\hat{U}}_2 = -1 + j$$

$$\hat{U}_7 = \overline{\hat{U}}_1 = \left(1 + \frac{1}{\sqrt{2}}\right) + j\left(2 + \frac{3}{\sqrt{2}}\right) .$$

The reader may wish to verify that these values match those obtained from the exact formula for $U(e^{j\theta})$, as obtained in Example 7.3, with $\theta = n\pi/4$, $n = 0, 1, \ldots, 7$.

$\square$

Example 7.5
Consider the finite sequence given by

$$\{u\} = \{-2, -1, 0, 1, 2\} .$$
$$\uparrow$$

Using the DFT calculate 8 equally spaced samples of the DTFT of $\{u\}$ over the unit circle. What happens if only 4 equally spaced samples are required?

As in the previous example, the sequence $\{\hat{u}\}$ is obtained by periodically repeating $\{u\}$ with a period equal to 8. In particular, for the period $0 \le k \le 7$, we obtain

$$\{\hat{u}\} = \{0, 1, 2, 0, 0, 0, -2, -1\} .$$

This gives

$$\hat{U}_0 = \sum_{k=0}^{7} \hat{u}_k = \sum_{k=-2}^{2} u_k = 0$$

$$\hat{U}_1 = \sum_{k=0}^{7} \hat{u}_k \left(e^{-j\frac{\pi}{4}}\right)^k = \sum_{k=-2}^{2} u_k \left(e^{-j\frac{\pi}{4}}\right)^k$$

$$= -2e^{j\frac{\pi}{2}} - e^{j\frac{\pi}{4}} + 0 + e^{-j\frac{\pi}{4}} + 2e^{-j\frac{\pi}{2}}$$

$$= -2j\left[2\sin\left(\frac{\pi}{2}\right) + \sin\left(\frac{\pi}{4}\right)\right] = -j\left(4 + \sqrt{2}\right)$$

$$\hat{U}_2 = -u_{-2} + j\,u_{-1} + u_0 - j\,u_1 - u_2$$

$$= 2 - j + 0 - j - 2 = -j\,2$$

$$\hat{U}_3 = -j\,u_{-2} - \frac{1}{\sqrt{2}}(1 - j)u_{-1} + u_0 - \frac{1}{\sqrt{2}}(1 + j)u_1 + j\,u_2$$

$$= j\,2 + \frac{1}{\sqrt{2}}(1 - j) + 0 - \frac{1}{\sqrt{2}}(1 + j) + j\,2 = j\left(4 - \sqrt{2}\right)$$

$$\hat{U}_4 = u_{-2} - u_{-1} + u_0 - u_1 + u_2$$

$$= -2 + 1 + 0 - 1 + 2 = 0.$$

As before, the symmetry relation $\hat{U}_{8-n} = \overline{\hat{U}}_n$ can be used to deduce that

$$\hat{U}_5 = -j\left(4 - \sqrt{2}\right)$$
$$\hat{U}_6 = j\,2$$
$$\hat{U}_7 = j\left(4 + \sqrt{2}\right) \,.$$

As explained in the theory, the sequence $\{\hat{U}\}$ gives the desired sampled values of the DTFT of the given sequence $\{u\}$. As a simple check, the FFT of the sequence $\{\hat{u}\}$ was calculated in MATLAB using the command

```
fft([0 1 2 0 0 0 -2 -1])
```

This returned the vector

```
[0 -5.4142j -2.0000j 2.5858j 0 -2.5858j 2.0000j 5.4142j]
```

This can be seen to match the hand-calculated results. Let us now repeat the above calculation with $N = 4$. We now have

$$
\begin{array}{rclcll}
\hat{u}_0 & = & \ldots + u_{-8} + u_{-4} + u_0 + u_4 + u_8 + \ldots & = & u_0 & = & 0 \\
\hat{u}_1 & = & \ldots + u_{-7} + u_{-3} + u_1 + u_5 + u_9 + \ldots & = & u_1 & = & 1 \\
\hat{u}_2 & = & \ldots + u_{-6} + u_{-2} + u_2 + u_6 + u_{10} + \ldots & = & u_{-2} + u_2 & = & 0 \\
\hat{u}_3 & = & \ldots + u_{-5} + u_{-1} + u_3 + u_7 + u_{11} + \ldots & = & u_{-1} & = & -1.
\end{array}
$$

We now calculate the DFT of $\{\hat{u}\}$ using (7.7a) with $N = 4$ and hence $w = e^{-j\frac{\pi}{2}} = -j$. This gives rise to

$$\hat{U}_0 = \hat{u}_0 + \hat{u}_1 + \hat{u}_2 + \hat{u}_3 = 0$$
$$\hat{U}_1 = \hat{u}_0 - j\,\hat{u}_1 - \hat{u}_2 + j\,\hat{u}_3$$
$$= 0 - j - 0 - j = -j\,2$$
$$\hat{U}_2 = \hat{u}_0 - \hat{u}_1 + \hat{u}_2 - \hat{u}_3 = 0$$

and $\hat{U}_3 = \overline{\hat{U}}_1 = j\,2$. We see that these values are identical to the values of $\hat{U}_0$, $\hat{U}_2$, $\hat{U}_4$ and $\hat{U}_6$, obtained when $N = 8$, as expected. We should note, however, that the value of $\hat{u}_2$ is now the sum of u_{-2} and u_2. Therefore, it is clear that we cannot recover our original sequence values, from knowledge of $\{\hat{U}\}$ alone, by means of the IDFT. To be more precise, the inverse discrete Fourier transform of the sequence $\{\hat{U}\}$ is the sequence $\{\hat{u}\}$. If $\{u\}$ is a finite sequence of length at most N, then it is a simple matter to recover $\{u\}$ from $\{\hat{u}\}$. If $\{u\}$ has length greater than N, then we see that the values of $\{\hat{u}\}$ become corrupted. This corruption is similar to that of **aliasing**, which we introduced earlier in §6.6. This time the corruption occurs in going from the frequency domain back to the time domain.

□

Both the previous examples were concerned with finite sequences. Theoretically speaking, given an infinite sequence $\{u\}$, we can still construct the periodic sequence $\{\hat{u}\}$ for any given N and, hence, find N equally spaced estimates of the associated

DTFT over the unit circle. In practice, however, we are likely to only have access
to a truncated version of $\{u\}$. A simple approach would be to approximate $\{u\}$ as
being zero outside the given truncated range and to proceed as for finite sequences.
Unfortunately, this can introduce a sharp cut-off at the ends which is unlikely to
have been present in the sequence before truncation. This sharp cut-off in turn
produces spurious results in the calculated DTFT. One approach to ameliorate this
problem is to pointwise multiply the truncated sequence by a so-called **window
sequence**, $\{w\}$, which has the property of being close to one around the middle
and decays 'smoothly' towards zero at the two ends. The resulting product sequence
is then used in place of the truncated $\{u\}$ in the DTFT calculation. We return to
the topic of 'windowing' in §7.6 and also in Chapters 8 and 9.

7.5   ESTIMATING FOURIER SERIES

In §7.2, we saw that, if $u(t)$ has the Fourier series expansion

$$u(t) = \sum_{n=-\infty}^{\infty} A_n e^{j n \omega_0 t} ,$$

then the discrete Fourier transform of the sequence, $\{u_k\} = \{u(kT)\}$, where
$T = P/N = 2\pi/N\omega_0$, satisfies the equation

$$U_n = N \sum_{r=-\infty}^{\infty} A_{n+rN} . \tag{7.14}$$

We define the periodic sequence $\{\hat{A}_n\}$ by $\hat{A}_n = \sum_{r=-\infty}^{\infty} A_{n+rN}$, which parallels the
notation introduced in the previous section. It therefore follows that the sequence
$\{\hat{A}_n\}$ may be calculated using

$$\{\hat{A}_n\} = \frac{1}{N}\{U_n\} = \frac{1}{N} DFT\,[\{u_k\}] . \tag{7.15}$$

The sequence $\{\hat{A}_n\}$ then represents an aliased version of the desired sequence $\{A_n\}$.
If N is chosen sufficiently large then aliasing errors may prove negligibly small.
In particular, if we have $|A_n| \approx 0$ for $|n| \geq N/2$ then we see that $\hat{A}_n \approx A_n$ for
$0 \leq n < N/2$. (This should not be considered as a rigorous argument since an
infinite sum of negligibly small numbers need no longer be negligibly small. In
practical examples, the decay rate of the Fourier series coefficients may be used
to add the needed rigour.) In addition, we have $\hat{A}_n \approx A_{n-N}$ for $N/2 \leq n < N$.
For real-valued functions, $u(t)$, we have the relation $A_{-n} = \overline{A}_n$ and so these latter
values are superfluous.

Example 7.6
Let $u(t)$ be the triangular wave depicted in Figure 7.2. Calculate the Fourier series
expansion for $u(t)$ and check the result using a calculation based on the DFT as
given in (7.15).

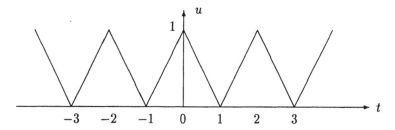

Figure 7.2: Triangular wave for Example 7.6.

From Figure 7.2, we see that $u(t)$ has period $P = 2$ and, hence, the fundamental frequency is given by $\omega_0 = \pi$. Therefore, the Fourier series expansion for u has the form

$$u(t) = \sum_{k=-\infty}^{\infty} A_k e^{j k \pi t} \, ,$$

where

$$A_k = \frac{1}{2} \int_{-1}^{1} u(t) \, e^{-j k \pi t} \, \mathrm{d}t = \int_{0}^{1} u(t) \, \cos(k\pi t) \, \mathrm{d}t \ .$$

The latter integral follows from the even property of u, i.e. $u(-t) = u(t)$. This gives

$$A_k = \int_{0}^{1} (1 - t) \cos(k\pi t) \, \mathrm{d}t = \left[(1 - t) \frac{\sin(k\pi t)}{k\pi} - \frac{\cos(k\pi t)}{k^2 \pi^2} \right]_{0}^{1} \quad k \neq 0$$

$$= \frac{1}{k^2 \pi^2} (1 - (-1)^k) = \begin{cases} 0, & \text{for } k \text{ even, } k \neq 0 \\ \dfrac{2}{k^2 \pi^2}, & \text{for } k \text{ odd.} \end{cases}$$

Since A_0 is the mean value, we can see by inspection of Figure 7.2 that $A_0 = \frac{1}{2}$. In conclusion, the Fourier series expansion for u is given by

$$u(t) = \frac{1}{2} + \frac{2}{\pi^2} \sum_{n=-\infty}^{\infty} \frac{e^{j (2n+1)\pi t}}{(2n + 1)^2} \ .$$

Next, we estimate the Fourier series expansion using (7.15). For illustrative purposes, we use $N = 256$ and the MATLAB package with the following commands:

```
u = 1/128*[[128:-1:0] [1:127]]; % sampled u values over 0<=t<2
A = fft(u)/256; A = A(1:128);
```

A plot of $|A|$ is shown in Figure 7.3 and we see that this suggests that $A_n \approx 0$ for $n \geq 128$ and so aliasing should be negligible.

The first eight estimated values, i.e. those for A_n, $0 \leq n \leq 7$ turn out to be

$$\{0.5000, \ 0.2027, \ 0, \ 0.0225, \ 0, \ 0.0081, \ 0, \ 0.0041\}.$$

which show good agreement with the hand-calculated answer.

$\square$

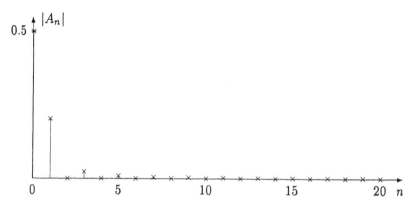

Figure 7.3: Amplitude spectrum for Example 7.6.

7.6 ESTIMATING FOURIER TRANSFORMS

In §7.3, we saw that, if $f(t)$ has the Fourier transform $F(j\omega)$, i.e.

$$f(t) = \frac{1}{2\pi} \int_{-\infty}^{\infty} F(j\omega)\, e^{j\omega t}\, d\omega \ ,$$

then the discrete-time Fourier transform, $F(e^{j\theta})$, of the sequence of sampled-values, $\{f_k\} = \{f(kT)\}$, satisfies (7.10). This gives

$$\sum_{n=-\infty}^{\infty} F\left(j\left(\omega + n\frac{2\pi}{T}\right)\right) = T.F(e^{j\omega T}) = T.F(e^{j\theta}) \ . \tag{7.16}$$

As mentioned in §7.4, the discrete-time Fourier transform is itself estimated using the DFT (FFT). Let us now put these ideas together to show how the FFT algorithm can be used to estimate sampled values of the Fourier transform of a finite-duration time signal. Let us begin by assuming that we sample the finite-duration signal, $f(t)$, at N equally-spaced samples over the time-interval, $k_1 T \le t \le k_2 T$, where $k_2 = k_1 + N - 1$. Let us also assume that this interval is chosen large enough to contain the whole duration of f. This then gives rise to the finite sequence

$$\{f\} = \{f(k_1 T), f((k_1 + 1)T), \ldots, f(k_2 T)\} \ .$$

Then, as detailed in §7.4, we can construct the periodic sequence, $\{\hat{f}\}$, with period given by N, whose discrete Fourier transform provides N equally spaced samples of $F(e^{j\theta})$ over the unit circle. If these values are then scaled by a factor of T, then, from (7.16), we obtain estimates for

$$\hat{F}(j\omega) = \sum_{n=-\infty}^{\infty} F\left(j\left(\omega + n\frac{2\pi}{T}\right)\right)$$

at the sampled frequencies $\omega = n\Omega$, $n = 0, 1, \ldots, N - 1$, where $\Omega = 2\pi/NT$. There are two importants points to note about this observation.

First, the **frequency resolution**, Ω, is inversely proportional to the product NT, i.e. the length of the time interval over which sampling is performed. Therefore, *in order to obtain a finer resolution of the Fourier transform, it is necessary to increase the size of the interval over which $f(t)$ is sampled.* (Since we are assuming $f(t)$ is of finite duration, this effectively means extra zeros are appended to the sample sequence.)

Secondly, *in order to reduce the effects of aliasing errors, it is necessary that T be sufficiently small so that $|F(j\omega)| \approx 0$ for $|\omega| \geq \pi/T$.* If this is the case, we might reasonably expect that $\hat{F}(j\,n\Omega) \approx F(j\,n\Omega)$ for $0 \leq n < N/2$ and $\hat{F}(j\,n\Omega) \approx F(j\,(n-N)\Omega)$ for $N/2 \leq n < N$. If $f(t)$ is real-valued, we have the relation $F(-j\omega) = \overline{F}(j\omega)$ and so the values for $N/2 \leq n < N$ are superfluous.

Since, in practice we are limited as to how large a value N can take, *there is a trade-off between making NT large (for fine frequency-domain sampling), and T small (to reduce aliasing errors).* Let us now illustrate these ideas by means of a simple example.

Example 7.7
Suppose $f(t)$ is as illustrated in Figure 7.4. Estimate the Fourier transform and compare with the exact values.

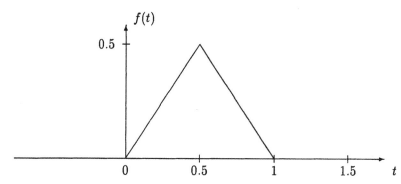

Figure 7.4: Signal for Example 7.7. A triangular pulse.

Let us begin by using $N = 8$ equally spaced samples of $f(t)$, using a sample period of $T = 0.25$ with the first sample at $t = 0$. This gives rise to the sequence

$$\{f(kT)\} = \{f(0),\ f(0.25),\ f(0.5),\ f(0.75),\ f(1.0),\ f(1.25),\ f(1.5),\ f(1.75)\}$$
$$= \{0,\ 0.25,\ 0.5,\ 0.25,\ 0,\ 0,\ 0,\ 0\}.$$

The DFT of this sequence is the sequence $\{F_n\}$ generated by

$$F_n = \sum_{k=0}^{7} f(kT)w^{nk}\ , \quad n = 0, 1, \ldots, 7,$$

where $w = e^{-j\frac{2\pi}{8}}$. Using an FFT algorithm, the following estimate for $\{F_n\}$ was obtained

$$\{F_n\} = \{1.0000,\ -j\,0.8536,\ -0.5000,\ j\,0.1464,\ 0,\ -j\,0.1464,\ -0.5000,\ j\,0.8536\}.$$

From the above theoretical discussion, this sequence needs to be scaled by a factor of $T = 0.25$ in order to provide estimates for the sampled values, $\hat{F}(j\,n\Omega)$, where $\Omega = 2\pi/NT = 2\pi/2 = \pi$. We also note that we should discard the values of F_n for $n \geq N/2 = 4$, and also that aliasing is clearly going to be a problem with this large a value for T. The exact transform is easily computed as

$$F(j\omega) = \frac{e^{-j\omega/2}}{4}\,\operatorname{sinc}^2\frac{\omega}{4}\ .$$

Sampling the exact transform at $\omega = n\pi$, for $0 \leq n \leq 3$, gives the sequence

$$\{F(j\,n\pi)\} = \left\{0.25,\ -j\,\frac{2}{\pi^2},\ -\frac{1}{\pi^2},\ j\,\frac{2}{9\pi^2}\right\}$$
$$\approx \{0.25,\ -j\,0.2026,\ -0.1013,\ j\,0.0225\}.$$

If we now scale the values computed via the FFT algorithm by $T = 0.25$, we obtain

$$\{\hat{F}(j\,n\pi)\} = \{0.2500,\ -j\,0.2134,\ -0.1250,\ j\,0.0366\}.$$

As expected, we see that the aliasing errors are non-negligible.

Let us now perform a similar estimate with $N = 512$, with T chosen such that NT still equals 2. In this case, the appropriately scaled FFT of the sampled signal values gives rise to

$$\{\hat{F}(j\,n\pi)\} \approx \{0.2500,\ -j\,0.2026,\ -0.1013,\ j\,0.0225, \ldots\}.$$

These values agree to 4 decimal places with the exact values given above. Figure 7.5 shows a plot of $\{|\hat{F}(j\,n\pi)|\}$.

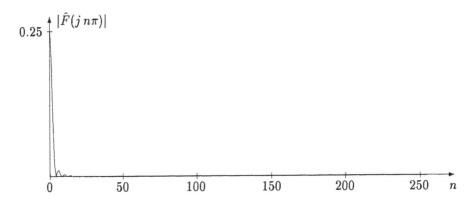

Figure 7.5: Plot of $\{|\hat{F}(j\,n\pi)|\}$ for $N = 512$ and $NT = 2$.

From this plot, we can estimate that $F(j\omega)$ is negligibly small for $\omega \geq 40\pi$ say. In order that aliasing should not prove to be a problem, we therefore require that T be chosen such that $\pi/T > 40\pi$, i.e. $T < 1/40$. So, for example, if we retain the number of samples at $N = 512$, we should be able to sample over the larger interval $0 \leq t < NT = 12$ say. The advantage of doing this is that the frequency domain sample period is now given by $\Omega = 2\pi/12 = \pi/6$ rather than π, as was the case with $NT = 2$. Figure 7.6 shows a plot of $|F(j\omega)|$ over the frequency range, $0 \le \omega \le 3\pi$, together with the estimates obtained in the following three cases:

(i) $N = 8$, $NT = 2$;
(ii) $N = 512$, $NT = 2$; and
(iii) $N = 512$, $NT = 12$.

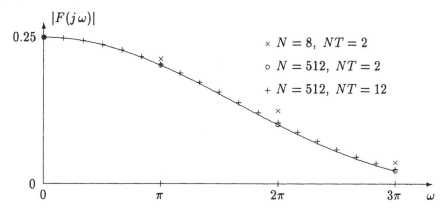

Figure 7.6: Plot of $|F(j\omega)|$ together with estimates.

For the sake of comparison with our previous estimates, we list below the estimates for $F(jn\pi)$, $n = 0, 1, 2, 3$, as obtained using case (iii).

$$\{0.2499, \; -0.0001 - j\,0.2026, \; -0.1013, \; -0.0001 + j\,0.0225\}.$$

We see evidence of a small aliasing error. If this error is unacceptable then we need to decrease the value of T.

□

In the next example, we illustrate the concept of windowing for signals of infinite duration.

Example 7.8
Let $f(t)$ be the signal defined by

$$f(t) = \begin{cases} e^t, & \text{for } -\infty < t < 0 \\ e^{-t}, & \text{for } 0 \le t < \infty. \end{cases}$$

It is easy to show that this signal has the Fourier transform

$$F(j\omega) = \frac{2}{1 + \omega^2} \; .$$

Now suppose we are only given the truncated signal over the time interval $-2 \le t \le 2$. Let $g(t)$ be the finite duration signal given by

$$g(t) = \begin{cases} f(t), & \text{for } -2 \le t \le 2 \\ 0, & \text{otherwise.} \end{cases}$$

1. Estimate $G(j\,\omega)$ for $\omega = n\pi/4$ using a time sampling frequency of $\frac{1}{T} = 64$ Hz. and compare this with $F(j\,\omega)$.

2. Repeat the above but now replacing $g(t)$ by the 'windowed' signal $h(t) = w(t)g(t)$, where $w(t)$ is the Hanning window defined by
   $w(t) = 0.5(1 + \cos(\pi t/2))$.

(1) We are given that $T = 1/64$ and for a frequency resolution of $\Omega = 2\pi/NT = \pi/4$, we have $NT = 8$. This in turn leads to the choice $N = 512$. Since our time signal, $g(t)$, is not causal, we must either use the 'time-delay' property of the Fourier transform or equivalently, we must sample the periodic signal, $\hat{g}(t)$, over $0 \leq t < 8$, where $\hat{g}(t)$ is given by the finite-duration pulse $g(t)$ repeated every 8 secs. A plot of $\hat{g}(t)$ over the period $0 \leq t < 8$ is given in Figure 7.7.

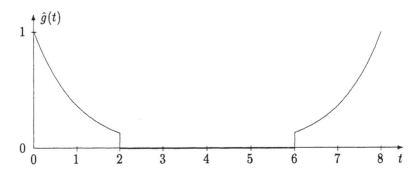

Figure 7.7: Plot of $\hat{g}(t)$ for Example 7.8.

Proceeding as in the previous examples, we then obtain sampled values of $G(j\,\omega)$ at $\omega = n\pi/4$, $n = 0, 1, \ldots, 255$, by taking the FFT of the sampled values of $\hat{g}(t)$ and then scaling by a factor of $T = 1/64$. In Figure 7.8, a plot is given of $|F(j\omega)|$ over the frequency range $0 \leq \omega \leq 5\pi$, together with the estimates, $|G(j\,n\pi/4)|$, $0 \leq n \leq 20$. In particular, we note the characteristic ripples in $|G(j\,n\pi/4)|$ caused by the sharp cut-off induced by straight truncation.

(2) We now repeat the above using $h(t) = 0.5(1 + \cos(\pi t/2))g(t)$ in place of $g(t)$. Figure 7.9 shows a plot of $|F(j\omega)|$ over the frequency range $0 \leq \omega \leq 5\pi$, together with the estimates, $|H(j\,n\pi/4)|$, $0 \leq n \leq 20$. We see that the previous ripples have now been effectively eliminated at the expense of a poorer fit at low frequency.

□

7.7   THE FAST FOURIER TRANSFORM

In previous sections, we have made reference to FFT (fast Fourier transform) algorithms for computing the discrete Fourier transform. There are in fact many

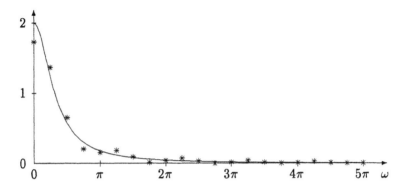

Figure 7.8: Plot of $|F(j\,\omega)|$ and $|G(j\,n\pi/4)|$, $0 \le n \le 20$.

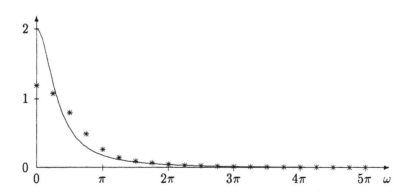

Figure 7.9: Plot of $|F(j\,\omega)|$ and $|H(j\,n\pi/4)|$, $0 \le n \le 20$.

different algorithms that fall under the umbrella of being called FFTs. In this section, we present a brief introduction to one such highly efficient algorithm. The algorithm we discuss was given by Cooley & Tukey [4] in 1965. The reader who wishes to see a fuller discussion should consult Brigham [2].

From (7.7a), we know that the discrete Fourier transform sequence, $\{X_n\}$, $n = 0, 1, \ldots, N - 1$ of the sequence, $\{x_n\}$, $n = 0, 1, \ldots, N - 1$, is given by

$$X_n = \sum_{k=0}^{N-1} x_k w_N^{nk} \qquad \text{where } w_N = e^{-j\frac{2\pi}{N}}. \tag{7.17}$$

Now, let us suppose that N can be factored as the product $N = LM$, with $L, M \geq 2$. The sequence $\{x\}$ can then be entered into an $L \times M$ matrix given by

$$M_x = \begin{bmatrix} x_0 & x_L & \cdots & x_{(M-1)L} \\ x_1 & x_{L+1} & \cdots & x_{(M-1)L+1} \\ \vdots & \vdots & & \vdots \\ x_{L-1} & x_{2L-1} & \cdots & x_{N-1} \end{bmatrix}.$$

We write x_{rs}, $0 \leq r \leq L - 1$, $0 \leq s \leq M - 1$, to represent the entry in row $r + 1$ and column $s + 1$. In other words, we have

$$x_{rs} = x_{sL+r}.$$

Similarly, we can enter the sequence $\{X\}$ into an $L \times M$ matrix. For reasons that should become apparent, $\{X\}$ is entered row-wise to give

$$M_X = \begin{bmatrix} X_0 & X_1 & \cdots & X_{M-1} \\ X_M & X_{M+1} & \cdots & X_{2M-1} \\ \vdots & \vdots & & \vdots \\ X_{(L-1)M} & X_{(L-1)M+1} & \cdots & X_{N-1} \end{bmatrix}.$$

We write X_{uv}, $0 \leq u \leq L-1$, $0 \leq v \leq M-1$ for the entry in row $u+1$ and column $v+1$, i.e.

$$X_{uv} = X_{uM+v}.$$

Using (7.17), we see that

$$X_{uv} = \sum_{r=0}^{L-1} \sum_{s=0}^{M-1} x_{rs} w_N^{(uM+v)(sL+r)}. \tag{7.18}$$

Consider the factor $w_N^{(uM+v)(sL+r)}$ in (7.18); now

$$w_N^{(uM+v)(sL+r)} = w_N^{usN} w_N^{vsL} w_N^{urM} w_N^{vr}. \tag{7.19}$$

But $w_N^N = \left(e^{-j\frac{2\pi}{N}}\right)^N = e^{-j\,2\pi} = 1$ and so $w_N^{usN} = 1^{us} = 1$. (This is the reason behind entering one sequence column-wise and the other row-wise.)

Also $w_N^L = \left(e^{-j\frac{2\pi}{LM}}\right)^L = e^{-j\frac{2\pi}{M}} = w_M$ and similarly $w_N^M = w_L$. Therefore, it follows from (7.19) that

$$w_N^{(uM+v)(sL+r)} = w_M^{vs} w_L^{ur} w_N^{vr}.$$

Substituting this back into (7.18) gives

$$X_{uv} = \sum_{r=0}^{L-1} w_L^{ur} \left\{ w_N^{vr} \sum_{s=0}^{M-1} x_{rs} w_M^{vs} \right\}. \qquad (7.20)$$

We now split this calculation into three stages.

Stage 1

For each fixed value of r, $0 \le r \le L-1$, calculate the sequence $\{y'_{rv}\}$, $v = 0, 1, \ldots, M-1$, given by

$$y'_{rv} = \sum_{s=0}^{M-1} x_{rs} w_M^{vs}.$$

In other words, $\{y'_{rv}\}$, $v = 0, 1, \ldots, M-1$ is the discrete Fourier transform of $\{x_{rs}\}$, $s = 0, 1, \ldots, M-1$ for each fixed r, $0 \le r \le L-1$.

If y'_{rv} is stored as the entry in row $r+1$ and column $v+1$ of an $L \times M$ matrix

$$\boldsymbol{M}_{y'} = \begin{bmatrix} y'_{00} & y'_{01} & \cdots & y'_{0,M-1} \\ \vdots & \vdots & & \vdots \\ y'_{L-1,0} & y'_{L-1,1} & \cdots & y'_{L-1,M-1} \end{bmatrix}$$

then *each row of* $\boldsymbol{M}_{y'}$ *is obtained by taking the discrete Fourier transform of the analogous row of* $\boldsymbol{M}_x$.

Stage 2

For each fixed value of r, $0 \le r \le L-1$, calculate the sequence $\{y_{rv}\}$, $v = 0, 1, \ldots, M-1$, given by

$$y_{rv} = w_N^{rv} y'_{rv}.$$

That is to say that the matrix $\boldsymbol{M}_y$ can be constructed, using the componentwise multiplication denoted by $.^*$, as

$$\boldsymbol{M}_y = \begin{bmatrix} 1 & 1 & 1 & \cdots & 1 \\ 1 & w_N & w_N^2 & \cdots & w_N^{M-1} \\ 1 & w_N^2 & w_N^4 & \cdots & w_N^{2(M-1)} \\ \vdots & \vdots & \vdots & & \vdots \\ 1 & w_N^{L-1} & w_N^{2(L-1)} & \cdots & w_N^{(L-1)(M-1)} \end{bmatrix} .^* \boldsymbol{M}_{y'} \ .$$

The complex factors w_N^{rv} are sometimes referred to as **twiddle factors** and **Stage 2** is then referred to as **twiddling**.

Stage 3
From (7.20), we see that

$$X_{uv} = \sum_{r=0}^{L-1} y_{rv} w_L^{ur}$$

and so, for each v, $0 \le v \le M-1$, the sequence $\{X_{uv}\}$, $u = 0, 1, \ldots, L-1$, is given by taking the discrete Fourier transform of the sequence $\{y_{rv}\}$, $r = 0, 1, \ldots, L-1$. This simply says that *the matrix M_X is given column-by-column by taking discrete Fourier transforms of the analogous columns in the matrix M_y.*

This completes the algorithm based upon the factorization $N = LM$. We can always assume L is a prime factor but that M may possibly be factored further $M = PQ$ say, (with P prime), so $N = LPQ$. Using this factorization, the discrete Fourier transforms of **Stage 1** can each be split into three stages via an identical procedure. This can be continued until N is written as a product of prime factors.

When this algorithm is programmed, see Brigham [2] for more details, it becomes particularly efficient when N is a power of 2 and so every prime factor of N equals 2. In this case it is known as a radix-2 algorithm. These computationally efficient radix-2 algorithms are those that are most widely used in practice. This restriction on N is not often troublesome since, in many applications, sequence lengths are selected rather than imposed, and the selection is then made of the smallest number of terms of the form $N = 2^\gamma$ consistent with sampling constraints. Let us now look in more detail at the cases $N = 2, 4$ and 8.

The Case $N = 2$
For $N = 2$, we cannot factor N and so we use the discrete Fourier transform as it stands. We see that

$$X_n = \sum_{k=0}^{1} x_k w_2^{nk}, \quad n = 0, 1,$$

where $w_2 = e^{-j\frac{2\pi}{2}} = -1$. In matrix form, we have

$$\begin{bmatrix} X_0 \\ X_1 \end{bmatrix} = \begin{bmatrix} 1 & 1 \\ 1 & -1 \end{bmatrix} \begin{bmatrix} x_0 \\ x_1 \end{bmatrix},$$

i.e. $X_0 = x_0 + x_1$ and $X_1 = x_0 - x_1$.

The Case $N = 4$
For implementation as an in-place serial algorithm, the matrices $M_x, M_{y'}, M_y$ and M_X are usually stored as vectors with row 1 entries followed by row 2 entries etc. The matrix

$$M_x = \begin{bmatrix} x_0 & x_2 \\ x_1 & x_3 \end{bmatrix}$$

is therefore stored in the vector $\begin{bmatrix} x_0 & x_2 & x_1 & x_3 \end{bmatrix}$. This initial scrambling of the data vector is best seen in the following light, with a view to generalization. If the data had been entered row-wise, then the entry in row $r+1$ and column $s+1$ would have a subscript given in binary as the number rs, where r and s are known as

bits. Since the data were in fact entered column-wise, the subscript is bit-reversed. Therefore, if we take the binary numbers from 00 to 11 and perform bit-reversal to give $00, 10, 01, 11$, we obtain the binary representation for the scrambled data subscripts.

Example 7.9
Calculate the discrete Fourier transform of the sequence $\{x\} = \{1, 2, 2, 1\}$ using the 3-stage algorithm with $L = M = 2$.

The matrix M_x is given by entering the data $\{1, 2, 2, 1\}$ column-wise into a 2×2 matrix to give

$$M_x = \begin{bmatrix} 1 & 2 \\ 2 & 1 \end{bmatrix}.$$

The matrix $M_{y'}$ is then found by taking the discrete Fourier transform for each row of M_x in turn to give

$$M_{y'} = \begin{bmatrix} 3 & -1 \\ 3 & 1 \end{bmatrix}.$$

The matrix M_y is found by pointwise multiplication by the matrix of 'twiddle factors'

$$\begin{bmatrix} 1 & 1 \\ 1 & w_N \end{bmatrix}$$

where $w_N = w_4 = e^{-j\frac{2\pi}{4}} = -j$. This gives

$$M_y = \begin{bmatrix} 3 & -1 \\ 3 & -j \end{bmatrix}.$$

We then take discrete Fourier transforms for each column of M_y in turn to give

$$M_X = \begin{bmatrix} 6 & -1-j \\ 0 & -1+j \end{bmatrix}.$$

Recalling that $\{X\}$ is entered row-wise in this matrix, we finally see that the required discrete Fourier transform sequence is given by

$$\{X\} = \{6, -1-j, 0, -1+j\}$$

in agreement with Example 7.1.

□

The Case $N = 8$
Taking $L = 2$ and $M = 4$, we see that the matrix M_x is given by

$$M_x = \begin{bmatrix} x_0 & x_2 & x_4 & x_6 \\ x_1 & x_3 & x_5 & x_7 \end{bmatrix}.$$

When performing the discrete Fourier transforms on the rows of M_x and using the algorithm with $M = PQ$ with $P = Q = 2$, we use the matrices

$$M_{x_1} = \begin{bmatrix} x_0 & x_4 \\ x_2 & x_6 \end{bmatrix} \quad \text{and} \quad M_{x_2} = \begin{bmatrix} x_1 & x_5 \\ x_3 & x_7 \end{bmatrix}.$$

Again, for algorithmic purposes, the data are usually held in a vector,
i.e. $\begin{bmatrix} x_0 & x_4 & x_2 & x_6 & x_1 & x_5 & x_3 & x_7 \end{bmatrix}$. As for the case $N = 4$, the subscripts
can be obtained by taking the binary numbers 000 to 111 and invoking bit-reversal.
Indeed, scrambling is equivalent to bit-reversal in the general radix-2 algorithm.

Example 7.10
Calculate the discrete Fourier transform of the sequence

$$\{x\} = \{0, 1, 2, 0, 0, 0, -2, -1\}.$$

The matrices $\boldsymbol{M}_{x_1}$ and $\boldsymbol{M}_{x_2}$ are given by

$$\boldsymbol{M}_{x_1} = \begin{bmatrix} 0 & 0 \\ 2 & -2 \end{bmatrix} \quad \text{and} \quad \boldsymbol{M}_{x_2} = \begin{bmatrix} 1 & 0 \\ 0 & -1 \end{bmatrix}.$$

Taking row discrete Fourier transforms gives

$$\boldsymbol{M}_{y_1'} = \begin{bmatrix} 0 & 0 \\ 0 & 4 \end{bmatrix} \quad \text{and} \quad \boldsymbol{M}_{y_2'} = \begin{bmatrix} 1 & 1 \\ -1 & 1 \end{bmatrix}.$$

After 'twiddling', we obtain

$$\boldsymbol{M}_{y_1} = \begin{bmatrix} 0 & 0 \\ 0 & -j4 \end{bmatrix} \quad \text{and} \quad \boldsymbol{M}_{y_2} = \begin{bmatrix} 1 & 1 \\ -1 & -j \end{bmatrix}.$$

Now taking discrete Fourier transforms along the columns gives

$$\boldsymbol{M}_{X_1} = \begin{bmatrix} 0 & -j4 \\ 0 & j4 \end{bmatrix} \quad \text{and} \quad \boldsymbol{M}_{X_2} = \begin{bmatrix} 0 & 1-j \\ 2 & 1+j \end{bmatrix}.$$

Putting these together completes the first stage of the $L = 2$, $M = 4$ algorithm and
gives

$$\boldsymbol{M}_{y'} = \begin{bmatrix} 0 & -j4 & 0 & j4 \\ 0 & 1-j & 2 & 1+j \end{bmatrix}.$$

The matrix of 'twiddle factors' is now given by

$$\begin{bmatrix} 1 & 1 & 1 & 1 \\ 1 & w_8 & w_8^2 & w_8^3 \end{bmatrix}$$

where $w_8 = e^{-j\frac{\pi}{4}} = \dfrac{1}{\sqrt{2}}(1 - j)$. This gives

$$\boldsymbol{M}_y = \begin{bmatrix} 1 & 1 & 1 & 1 \\ 1 & \frac{1}{\sqrt{2}}(1-j) & -j & -\frac{1}{\sqrt{2}}(1+j) \end{bmatrix} * \begin{bmatrix} 0 & -j4 & 0 & j4 \\ 0 & 1-j & 2 & 1+j \end{bmatrix}$$

$$= \begin{bmatrix} 0 & -j4 & 0 & j4 \\ 0 & -j\sqrt{2} & -j2 & -j\sqrt{2} \end{bmatrix}.$$

Taking column discrete Fourier transforms gives

$$\boldsymbol{M}_X = \begin{bmatrix} 0 & -j(4+\sqrt{2}) & -j2 & j(4-\sqrt{2}) \\ 0 & -j(4-\sqrt{2}) & j2 & j(4+\sqrt{2}) \end{bmatrix}$$

and so

$$\{X\} = \left\{0, -j\left(4 + \sqrt{2}\right), -j\,2, j\left(4 - \sqrt{2}\right), 0, -j\left(4 - \sqrt{2}\right), j\,2, j\left(4 + \sqrt{2}\right)\right\}$$

in agreement with the result obtained in Example 7.5.

$\square$

The real power of the FFT algorithm outlined above can be seen when we compare the number of complex additions and multiplications involved with those obtained via a direct implementation of the DFT. It is possible to show that for general γ, (with $N = 2^\gamma$), the FFT algorithm involves $N\gamma$ complex additions and $N\gamma/2$ complex multiplications, whereas a direct implementation involves $N(N-1)$ additions and N^2 multiplications. For example, if $N = 2^8 = 256$, direct evaluation of the DFT would require

$$N(N-1) = 256 \times 255 = 65,280 \text{ additions and}$$
$$N^2 = 256^2 = 65,536 \text{ multiplications;}$$

whereas the FFT needs only

$$N\gamma = 256 \times 8 = 2,048 \text{ additions and}$$
$$N\gamma/2 = 1024 \text{ multiplications.}$$

The algorithm we have presented with $\{x\}$ entered column-wise and $\{X\}$ row-wise leads to a so-called **decimation-in-time** algorithm which involves bit-reversal for the 'time-sequence', $\{x\}$. It is simple to construct an equivalent algorithm with $\{x\}$ entered row-wise and $\{X\}$ entered column-wise leading to bit-reversal for the 'frequency sequence', $\{X\}$. This latter algorithm is then referred to as a **decimation-in-frequency** algorithm. As mentioned at the end of §7.2, FFT algorithms can be used, with a minor modification, for finding inverse discrete Fourier transforms.

7.8   EXERCISES

1. Calculate the discrete-time Fourier transform of the sequence

$$\{1, 0, 2, 1, 2, 0, 1\}\ .$$

   Sketch the amplitude spectrum of the transform.

2. Find the DFT of the finite sequence $\{x_k\}$, $k = 0, 1, \ldots, 31$, where

$$x_k = \begin{cases} e^{-0.3k}, & \text{for } 0 \le k \le 9 \\ 0, & \text{for } 10 \le k \le 31. \end{cases}$$

3. A signal, $f(t)$, has a Fourier transform $F(j\omega)$ that is assumed to be negligibly small for frequencies higher than 4kHz, i.e. $\omega \geq 8000\pi$ rads./sec.

   (a) Determine the minimum sampling frequency, $1/T$ Hz, to avoid aliasing.

   (b) Suppose $F(j\omega)$ is to be estimated using an FFT algorithm with a frequency resolution that is finer than 10Hz. Find the smallest time interval for sampling $f(t)$ that is compatible with this requirement.

   (c) If a radix-2 algorithm is to be used, what is the smallest number of samples, N, that satisfies the above requirements?

4. Let $\{u_k\} = \left\{1, \dfrac{1}{2}, -\dfrac{1}{2}, 1\right\}$. Find its discrete Fourier transform, $\{U_n\}$. Demonstrate Parseval's relation:

$$\sum_{k=0}^{3} |u_k|^2 = \frac{1}{4}\sum_{n=0}^{3} |U_n|^2 .$$

5. (a) Using the formula for summing a finite geometric series, show that

$$\sum_{k=0}^{N-1} e^{j\,(r-n)k2\pi/N} = \begin{cases} N, & \text{for } (r-n) = mN, \ m = 0, \pm 1, \pm 2, \ldots \\ 0, & \text{otherwise.} \end{cases}$$

   Hence determine the discrete Fourier transform of a von Hann window given by

$$w(k) = 0.5[1 - \cos(2\pi k/N)], \quad k = 0, 1, \ldots, N-1.$$

   (b) If a discrete-time signal, $\{u_k\}$, is passed through a von Hann window, the discrete Fourier transform of the output signal, $\{y_k\}$, is given by

$$Y_n = -0.25U_{n-1} + 0.5U_n - 0.25U_{n+1} .$$

   For a von Hann window of length 4, find $\{Y_n\}$ when $\{u_k\}$ is given by $\{1, 0, 2, -1\}$ and hence plot the amplitude and phase spectra of the output signal.

6. The first five DFT coefficients of a real 8-point periodic sequence are:
   $\{0.5, 2 + j, 3 + 2j, j, 3\}$.
   What are the remaining three coefficients?

7. Let $\{u_k\} = \{1, 2, 1, 1, 3, 2, 1, 2\}$. Using the radix-2 FFT algorithm described in §7.7, calculate by hand the DFT coefficients, $\{U_n\}$, $n = 0, 1, \ldots 7$.

8. Let $\{U_n\} = \{5, 2 + j, -5, 2 - j\}$, be the discrete Fourier transform of $\{u_k\}$. Use the inverse discrete Fourier transform to find $\{u_k\}$, $k = 0, 1, 2, 3$.

9. Let $f(t)$ be the half-rectified sine wave defined by

$$f(t) = \begin{cases} \sin t, & \text{for } 0 \leq t < \pi, \\ 0, & \text{for } \pi \leq t < 2\pi, \end{cases} \qquad \text{with } f(t + 2\pi) = f(t).$$

Determine the complex exponential Fourier series coefficients, $\{A_n\}$ say, for $f(t)$ and compare these exact values with estimated values, for $n = 0, 1, \ldots 8$, using the method of §7.5.

10. Let $f(t)$ be the finite pulse defined by

$$f(t) = \begin{cases} |t|, & \text{for } -1 \leq t \leq 1, \\ 0, & \text{otherwise.} \end{cases}$$

Show that the Fourier transform is given by $F(j\omega) = 2\text{sinc}(\omega) - \text{sinc}^2\left(\dfrac{\omega}{2}\right)$.
Sketch $F(j\omega)$ for $-5\pi \leq \omega \leq 5\pi$. Using the method of §7.6, find an estimate for $F(j\omega)$ over the same frequency range and sketch this for comparison. (Note: If the estimate contains some small imaginary parts, due to aliasing, these should be removed so as to produce a real sketch.)

8

The design of digital filters

8.1   INTRODUCTION

In this chapter, we bring together many of the ideas of the earlier chapters to discuss the process of design of digital filters. We regard digital filters as discrete-time systems which perform similar tasks to the analogue filter designs of Chapter 4. In essence, we are laying the foundations for the software replacement of such analogue devices, although a discussion of the effects of signal quantization would be required for a full treatment. Obviously, such software processes samples drawn from continuous-time signals, and we are well aware from Chapters 3 and 5 of the implications of the sampling process on signal spectra, particularly in terms of the appearance of periodic effects.

This book is designed to illustrate the mathematical aspects of communication theory. For this reason, this chapter sets out only to explain some underlying ideas in the theory, and does not attempt to give an exhaustive treatment of current engineering practice. It is hoped that the reader finds the material in this chapter a sufficient introduction to the more advanced modern engineering texts on this topic. A secondary aim is to highlight the place of applied mathematics as an aid in the design or synthesis task. Thus, in this chapter, our principal objective is to produce digital filter designs, and readers are encouraged to extend the work to produce and experiment with their own designs. Access to a personal computer or microcomputer is valuable in this respect, although not essential, however those who do have such access will quickly appreciate the ease of production of the necessary code to test designs.

8.2   THE IMPULSE INVARIANT METHOD

In this section, we explore a method of obtaining a discrete-time system which emulates, in a particular sense, the behaviour of a continuous-time prototype system. For any method that produces digital filters based upon analogue prototypes, there are two major questions of importance:

1. (**Stability**) Will the digital filter always be stable, given a stable analogue prototype?

2. (**Accuracy**) Does the digital frequency response approximately match the analogue frequency response over the frequency range of interest?

In §6.6, we saw that digital frequency responses are always periodic, with period given by $2\pi/T$ rads./sec., where T is the sample period. With this in mind, T should always be chosen such that the frequency range of interest is contained in the interval $-\pi/T \leq \omega \leq \pi/T$. In practice, it is often necessary to use a crude analogue pre-filter to filter out high frequencies before sampling takes place.

Let us begin by considering an analogue prototype filter with transfer function, $H(s)$, and corresponding impulse response, $h(t) = \mathcal{L}^{-1}[H(s)]$. The response $y(t)$ to a given input $u(t)$, as shown in §2.5, is then given by

$$y(t) = (h * u)(t) = \int_0^t h(t - \tau)u(\tau)\,\mathrm{d}\tau.$$

If we now sample $y(t)$ at $t = kT$, $k = 0, 1, 2, \ldots$, we obtain

$$y(kT) = \int_0^{kT} h(kT - \tau)u(\tau)\,\mathrm{d}\tau = \sum_{r=0}^{k-1} \int_{rT}^{(r+1)T} h(kT - \tau)u(\tau)\,\mathrm{d}\tau. \qquad (8.1)$$

Let us now make the crude approximations,

$$u(t) \approx u(rT) \quad \text{and} \quad h(kT - t) \approx h([k - r - 1]T), \quad \text{for } rT \leq t < (r + 1)T.$$

Substituting these approximations into (8.1) gives

$$y(kT) \approx \sum_{r=0}^{k-1} Th([k - r - 1]T)u(rT). \qquad (8.2)$$

Now consider a discrete-time system with transfer function, $\tilde{H}(z)$, defined by

$$\tilde{H}(z) = \mathcal{Z}[\{h(kT)\}]. \qquad (8.3)$$

(We have elected to use the notation $\tilde{H}(z)$, as opposed to $H(z)$, since the latter notation may be misunderstood as meaning $H(s)$ with s replaced by z.) If this system has an input sequence, $\{u_k\}$, then the corresponding output sequence is given by

$$\{\tilde{y}_k\} = \sum_{r=0}^{k} \tilde{h}_{k-r}u_r, \qquad (8.4)$$

where we have $\tilde{h}_k = h(kT)$. If we now assume that $\{u_k\} = \{u(kT)\}$, then a comparison between (8.4) and (8.2) yields

$$y(kT) \approx T\tilde{y}_{k-1}.$$

This suggests that a natural choice for the digital filter transfer function is to scale $\tilde{H}(z)$ by a factor of T and then divide by z. The division by z does not affect the amplitude response and can be omitted from the design method, if so desired. In practice, the fact that digital computations are not instantaneous means that the discrete-time transfer function should be chosen to be *strictly* proper rational and so such a division by z may prove necessary for implementation. To summarize, the impulse invariant method consists of the following steps.

Impulse Invariant Method

Step 1 Find the impulse response, $h(t)$, for the analogue prototype $H(s)$, i.e. $h(t) = \mathcal{L}^{-1}[H(s)]$.

Step 2 Find the z-transform of the sequence of sampled values of $h(t)$, i.e. find $\tilde{H}(z) = \mathcal{Z}[\{h(kT)\}]$.

Step 3 Scale $\tilde{H}(z)$ to obtain the final digital filter transfer function $D(z)$. As seen above, a natural scaling is to use the sample time period T, i.e. $D(z) = T\tilde{H}(z)$. In practice, other scalings may be used. As discussed above, a final division by z, to produce a strictly proper z-transfer function, may be necessary.

The fact that *the impulse response of the digital filter is simply a scaled version of the sampled analogue impulse response* explains the name, 'Impulse Invariant Method'. Having introduced this discretization method, we must now examine the issues of **Stability** and **Accuracy**.

8.2.1   *Stability of the Impulse Invariant Method*

In sections §2.2 and §6.2, we saw that a system (analogue or digital) is BIBO stable if and only if its impulse response decays to zero. Since the analogue prototype is assumed to be BIBO stable and the associated digital filter has an impulse response given by scaled, sampled values of this response, *the BIBO stability of the digital filter is ensured.* For an alternative proof, see Exercise 2.

8.2.2   *Accuracy of the Impulse Invariant Method*

Since the impulse response of the digital filter is given by $\{Th(kT)\}$, it follows that the digital transfer function satisfies

$$D(z) = \mathcal{Z}[\{Th(kT)\}] = \sum_{k=0}^{\infty} Th(kT)z^{-k}.$$

We know $D(z)$ is BIBO stable and so the discrete-time frequency response is given by

$$D(e^{j\,\omega T}) = \sum_{k=0}^{\infty} Th(kT)e^{-j\,\omega kT}.$$

But we already know that $e^{-j\,\omega kT}$ is equal to the Fourier transform of the signal $\delta(t - kT)$, see Equation (3.23), and so, assuming we can interchange the infinite sum with the operation of taking the Fourier transform, we obtain

$$D(e^{j\,\omega T}) = \mathcal{F}\left\{ \sum_{k=0}^{\infty} Th(kT)\delta(t - kT) \right\}. \tag{8.5}$$

In §3.9, we saw that, given any analogue signal $u(t)$, we can construct the corresponding signal $u_s(t)$ defined by

$$u_s(t) = \sum_{k=-\infty}^{\infty} u(kT)\delta(t - kT) = u(t) \sum_{k=-\infty}^{\infty} \delta(t - kT).$$

It was also shown in §3.9 that the Fourier transforms of $u(t)$ and $u_s(t)$ are related by

$$U_s(j\omega) = \frac{1}{T} \sum_{k=-\infty}^{\infty} U(j(\omega - 2\pi k/T)).$$

If this result is now applied to (8.5), using $h(t)$ in place of $u(t)$, we obtain the result

$$D(e^{j\omega T}) = \sum_{k=-\infty}^{\infty} H(j(\omega - 2\pi k/T)). \tag{8.6}$$

Equation (8.6) gives the desired relationship between the analogue and digital frequency responses. We see that *the frequency response of the digital filter consists of a series of repetitions of the frequency response of the analogue prototype.* As seen in §3.9, if T is not chosen sufficiently small then the digital filter will suffer from distortion due to 'aliasing'. This means that we require

$$H(j\omega) \approx 0 \quad \text{for } |\omega| \geq \pi/T.$$

As discussed above, T should be chosen so that the frequency range of interest lies in the interval, $-\pi/T \leq \omega \leq \pi/T$. It is therefore likely, assuming that $H(s)$ is lowpass or band-pass and has a sufficiently steep cut-off, that $H(j\omega)$ will be negligibly small outside this range of interest. For further details, the reader is referred to the book by Jackson [11]. To summarize: *we should choose the sample period T to be sufficiently small so that $T \leq \pi/\omega_m$, where ω_m is the highest frequency in the input signal, (which may well be reduced by means of analogue pre-filtering), and so that $H(j\omega) \approx 0$ for $|\omega| \geq \pi/T$ to reduce the distortion due to the aliasing of the impulse response as given by (8.6).*

We now apply this technique to the discretization of a first-order Butterworth filter. Of course, in practice this technique would not be applied to such a low-order prototype, but such an example provides an easy-to-understand introduction to this particular design method.

Example 8.1
Use the impulse invariant method to design a discrete-time replacement for the first order Butterworth filter with cut-off frequency ω_c.

From Chapter 4, the transfer function of the first order Butterworth filter is

$$H(s) = \frac{\omega_c}{s + \omega_c}.$$

At once we obtain the impulse response as $h(t) = \omega_c e^{-\omega_c t}$, $t \geq 0$, and so the impulse response sequence is $\{h(kT)\} = \{\omega_c e^{-k\omega_c T}\}$, $k \geq 0$. Now

$$\tilde{H}(z) = \mathcal{Z}\left[\{\omega_c e^{-k\omega_c T}\}\right] = \omega_c \mathcal{Z}\left[\{(e^{-\omega_c T})^k\}\right] = \omega_c \frac{z}{z - e^{-\omega_c T}}$$

and, therefore, we set

$$D(z) = T\tilde{H}(z) = T\omega_c \frac{z}{z - e^{-\omega_c T}} \tag{8.7}$$

as the transfer function of the discrete-time system.

From the transform domain representation, we can derive the time domain difference equation which represents the system. From (8.7), we obtain

$$\left(z - e^{-\omega_c T}\right) Y(z) = T\omega_c z U(z)$$

and, inverting, we obtain the difference equation

$$y([k+1]T) - e^{-\omega_c T} y(kT) = T\omega_c u([k+1]T) ,$$

or,

$$y(kT) - e^{-\omega_c T} y([k-1]T) = T\omega_c u(kT) , \quad k \geq 1 .$$

This design can be implemented by the system of Figure 8.1. Finally, we note that this design has produced a digital filter that is not strictly proper rational and, in particular, $y(kT)$ depends on $u(kT)$ rather than on past inputs only. As mentioned in the previous discussion, the introduction of a unit delay, (i.e. a delay of T secs.), through dividing $D(z)$ by z, can circumvent this problem.

□

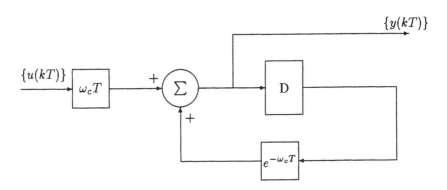

Figure 8.1: Time domain block diagram for the system of Example 8.1.

8.3   THE STEP INVARIANT METHOD

A second indirect method of filter design, based on emulating a specific response of an analogue prototype, is the so-called step invariant method. As the name implies, in this method we generate *a discrete-time design which produces, as its step response sequence, a sequence identical with a sequence of samples drawn from the step response of the analogue prototype at the appropriate instants.*

In the last section, we obtained the result that sampled values of an analogue filter's response to an input $u(t)$ are given by

$$y(kT) = \sum_{r=0}^{k-1} \int_{rT}^{(r+1)T} h(kT - \tau)u(\tau)\,d\tau.$$

For the impulse invariant method, we adopted a piecewise constant approximation to the above integrand. An alternative approach is to only approximate the input in this manner and integrate the impulse response. This leads to the approximation

$$y(kT) \approx \sum_{r=0}^{k-1} \int_{rT}^{(r+1)T} h(kT - \tau)\,d\tau\, u(rT). \tag{8.8}$$

Let $y_\zeta(t)$ be the analogue filter's step response. We then have $h(t) = \dot{y}_\zeta(t)$ and so (8.8) becomes

$$y(kT) \approx \sum_{r=0}^{k-1} \Big[- y_\zeta(kT - \tau) \Big]_{rT}^{(r+1)T} u(rT)$$

$$= \sum_{r=0}^{k-1} \Big[y_\zeta([k-r]T) - y_\zeta([k-r-1]T) \Big] u(rT). \tag{8.9}$$

Now consider a discrete-time system with transfer function given by

$$D(z) = \sum_{k=0}^{\infty} d_k z^{-k}.$$

Given any input sequence $\{u_k\}$, $k = 0, 1, 2, \ldots$, we can write

$$\{u_k\} = u_0\{\zeta_k - \zeta_{k-1}\} + u_1\{\zeta_{k-1} - \zeta_{k-2}\} + \cdots$$

$$= \sum_{r=0}^{\infty} u_r\{\zeta_{k-r} - \zeta_{k-r-1}\}.$$

Assuming the linearity property (which applies only to finite sums) can be extended to such an infinite sum, we obtain the response

$$\{y_k\} = \sum_{r=0}^{\infty} u_r\{y_{k-r}^\zeta - y_{k-r-1}^\zeta\} = \left\{ \sum_{r=0}^{k-1} u_r(y_{k-r}^\zeta - y_{k-r-1}^\zeta) \right\}, \tag{8.10}$$

where we have assumed that the discrete-time step response satisfies $y_k^\zeta = 0$ for $k = 0$, as would be the case with a strictly proper design. Comparison of (8.10) with (8.9) shows that we obtain $y(kT) \approx \{y_k\}$ if we choose $\{u_k\} = \{u(kT)\}$ and choose $D(z)$ so that $y_k^\zeta = y_\zeta(kT)$. This motivates the method of designing $D(z)$ so that its step response matches appropriate sample values of the analogue prototype's step response. Since

$$Y^\zeta(z) = D(z)\frac{z}{z-1} \iff D(z) = \frac{z-1}{z}Y^\zeta(z),$$

we obtain the following design method.

Step Invariant Method

Step 1 Find the analogue filter's step response, $y_\varsigma(t) = \mathcal{L}^{-1}\left[\dfrac{H(s)}{s}\right]$.

Step 2 Find the z-transform of the sequence of sampled values, $\{y_k^\varsigma\} = \{y_\varsigma(kT)\}$.

Step 3 Set the digital filter's transfer function to have $\{y_k^\varsigma\}$ as its step response,

i.e. $D(z) = \dfrac{z-1}{z}Y^\varsigma(z)$.

This still leaves the issues of *stability* and *accuracy*.

8.3.1   Stability of the Step Invariant Method

Let us suppose that $D(z)$ was not BIBO stable. This means that a bounded input sequence $\{u_k\}$ exists that gives rise to an unbounded response $\{y_k\}$. If we now let $u(t)$ be the bounded piecewise constant input given by

$$u(t) = u_k \quad \text{for } kT \le t < (k+1)T,$$

then it follows that the analogue filter's response to this input satisfies $y(kT) = y_k$ and, hence, is unbounded. This contradicts the assumed BIBO stability of the analogue prototype. Therefore, *the stability of the digital filter is ensured*.

8.3.2   Accuracy of the Step Invariant Method

A similar analysis to that performed for the impulse invariant method may be used. The mathematics is however a little more involved and so we elect to omit the details. The formula corresponding to that of (8.6) can be shown to be given by

$$D(e^{j\omega T}) = e^{-j\omega T/2}\sum_{k=-\infty}^{\infty}(-1)^k \operatorname{sinc}\left(\frac{\omega T}{2} - k\pi\right)H(j(\omega - k2\pi/T)). \qquad (8.11)$$

The term $e^{-j\omega T/2}$ represents a delay of $T/2$ secs. Let us also note that

$$\lim_{z\to\infty} D(z) = \lim_{z\to\infty}\frac{z-1}{z}Y^\varsigma(z) = \lim_{z\to\infty}Y^\varsigma(z)$$

$$= \lim_{z\to\infty} y_0^\varsigma + \frac{y_1^\varsigma}{z} + \frac{y_2^\varsigma}{z^2} + \cdots = y_0^\varsigma.$$

From this we can see that $D(z)$ tends to zero as z tends to infinity, i.e. $D(z)$ is strictly proper rational, whenever y_0^ς equals zero. Since the discrete-time step response is designed to match sampled values of the analogue step response, we deduce that $D(z)$ will be strictly proper rational whenever the analogue step response does not have a jump discontinuity at $t = 0$. This will be the situation whenever the analogue transfer function $H(s)$ is strictly proper rational. This is not necessarily the case with the impulse invariant method, *c.f.* Example 8.1 where the digital filter is only proper rational.

We note that, when $\omega = 0$ in (8.11), every term vanishes in the sum except the $k = 0$ term. This shows that the d.c. (zero frequency) terms match exactly, as would be expected using a step invariant approach. Let us also note that if we choose T so that

$$H(j\omega) \approx 0 \quad \text{for } |\omega| \geq \pi/T,$$

then, assuming a sufficiently rapid decay rate for $H(j\omega)$, we can ignore all terms in the above sum for $k \neq 0$ over the frequency range of interest, assumed to lie in the interval $-\pi/T \leq \omega \leq \pi/T$. Over this interval, we then have the approximation

$$D(e^{j\omega T}) \approx e^{-j\omega T/2}\text{sinc}\left(\frac{\omega T}{2}\right) H(j\omega).$$

Ignoring the delay term, $e^{-j\omega T/2}$, we see that $D(e^{j\omega T}) \approx H(j\omega)$ for ωT sufficiently small. As ω approaches π/T, we note that $\text{sinc}\left(\dfrac{\omega T}{2}\right)$ approaches $2/\pi$ with a resultant loss of match between the analogue and digital frequency responses. This is not necessarily a problem since, typically, $H(j\omega) \approx 0$ for $|\omega| \geq \omega_u$ say, with $\omega_u \ll \pi/T$. In any case, a mismatch between the digital and analogue frequency responses does not mean that the digital filter does not perform the desired filtering task, since the analogue prototype is itself only an approximation to an ideal case.

We demonstrate the design procedure in Example 8.2.

Example 8.2
Determine the step invariant design based on the first order Butterworth filter of Example 8.1.

The transfer function of the prototype analogue design is again

$$H(s) = \frac{\omega_c}{s + \omega_c}$$

and, thus, the transform of the step response is $Y_\zeta(s)$, where

$$Y_\zeta(s) = \frac{\omega_c}{s(s + \omega_c)} .$$

We can invert this to obtain

$$y_\zeta(t) = 1 - e^{-\omega_c t}, \quad t \geq 0,$$

and sampling this response, at intervals T, produces the sequence $\{y_\zeta(kT)\}$, where $y_\zeta(kT) = (1 - e^{-\omega_c kT})$, $k \geq 0$. Now take the z-transform of this sequence to obtain

$$\mathcal{Z}\left[\{y_\zeta(kT)\}\right] = \frac{z}{z - 1} - \frac{z}{z - e^{-\omega_c T}}$$

$$= Y^\zeta(z) .$$

The transfer function $D(z)$ is now selected as

$$D(z) = \frac{z - 1}{z}Y^\zeta(z)$$

$$= 1 - \frac{z - 1}{z - e^{-\omega_c T}}$$

$$= \frac{1 - e^{-\omega_c T}}{z - e^{-\omega_c T}} . \tag{8.12}$$

Working from this transfer function, we can obtain a difference equation which represents our design. Writing $Y(z) = D(z)U(z)$, then, from (8.12), we have

$$\left(z - e^{-\omega_c T}\right) Y(z) = (1 - e^{-\omega_c T})U(z) \ ,$$

and, inverting, we see that this produces the difference equation

$$y([k+1]T) - e^{-\omega_c T} y(kT) = \left(1 - e^{-\omega_c T}\right) u(kT) \ , \quad k \geq 0 \ .$$

$\square$

A similar analysis for the second-order Butterworth low-pass filter, with cut-off frequency $\omega_c = 1$, may be found in Example 6.5.

8.4 THE BILINEAR TRANSFORM METHOD

Also known as **Tustin's method** or the **trapezoidal integration technique**, this is the final indirect method of design based on an analogue prototype that we consider. The approach used is somewhat different from the two previous methods, in that we do not attempt to match specified responses of the analogue prototype. Rather, we note that multiplication by $1/s$ in the Laplace transform domain represents integration in the time domain. This operation is then emulated in the z-domain, using a representation of the trapezoidal approximation to integration. Figure 8.2 shows a schematic representation of the integration operation in the Laplace transform domain. In Figure 8.2, we see that $Q(s) = \dfrac{1}{s}P(s)$, or

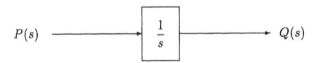

$$P(s) \longrightarrow \boxed{\dfrac{1}{s}} \longrightarrow Q(s)$$

Figure 8.2: Integration in the transform domain.

$sQ(s) = P(s)$. In the time domain, this corresponds to the relationship

$$\frac{\mathrm{d}q}{\mathrm{d}t}(t) = p(t) \tag{8.13}$$

and, integrating (8.13) between $\tau = (k-1)T$ and $\tau = kT$, we obtain

$$q(kT) - q([k-1]T) = \int_{(k-1)T}^{kT} p(\tau)\,\mathrm{d}\tau \ .$$

The integral on the right hand side is now approximated using the trapezoidal scheme, as shown in Figure 8.3. This procedure leads to the estimate

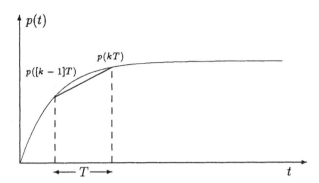

Figure 8.3: The trapezoidal integration scheme.

$$\int_{(k-1)T}^{kT} p(\tau)\,d\tau \approx \frac{T}{2}\left(p(kT) + p([k-1]T)\right),$$

and, thus, we deduce that

$$q(kT) - q([k-1]T) \approx \frac{T}{2}\left(p(kT) + p([k-1]T)\right), \quad k > 0. \qquad (8.14)$$

Defining $\mathcal{Z}\left[\{q(kT)\}\right] = \tilde{Q}(z)$ and $\mathcal{Z}\left[\{p(kT)\}\right] = \tilde{P}(z)$, we can take z-transforms in (8.14) to obtain

$$\left(1 - \frac{1}{z}\right)\tilde{Q}(z) \approx \frac{T}{2}\left(1 + \frac{1}{z}\right)\tilde{P}(z)$$

or

$$\frac{\tilde{P}(z)}{\tilde{Q}(z)} \approx \frac{2}{T}\left(\frac{z-1}{z+1}\right).$$

Recall that in the Laplace transform domain, for the continuous-time integration process, we found that

$$\frac{P(s)}{Q(s)} = s$$

and, thus, we are led to propose a discretization scheme based on replacing s in the transfer function of the continuous-time prototype by

$$\frac{2}{T}\left(\frac{z-1}{z+1}\right).$$

This operation yields the z-transfer function of a discrete-time system and the relationship between the performance of this design and that of a continuous-time prototype must now be examined. It may be enlightening to consider this method as being equivalent to taking a simulation diagram realization for the analogue prototype and then replacing all the integrators by numerical integrators based upon the trapezoidal approximation. The resulting diagram then represents our digital replacement filter. Other numerical integration schemes can be adopted, for example see Exercise 11.

8.4.1 Stability of the Bilinear Transform Method

It can be shown that this method always gives rise to a stable digital filter, given a stable analogue prototype. To prove this, it is necessary to show that the poles of the digital filter are related to the poles of the analogue prototype through the bilinear transform

$$s \mapsto \frac{2}{T}\left(\frac{z-1}{z+1}\right).$$

It can then be shown that any complex number in the s-domain with negative real part maps to a complex number in the z-domain with modulus less than one (and vice-versa). The proof of this is left to the exercises.

8.4.2 Accuracy of the Bilinear Transform Method

Since $D(z)$ is formed by replacing s by $(2/T)(z-1)/(z+1)$, it follows that

$$D(e^{j\,\omega T}) = \Big[H(s)\Big]_{s \to \frac{2}{T}\left(\frac{e^{j\,\omega T}-1}{e^{j\,\omega T}+1}\right)}.$$

Now,

$$\frac{e^{j\,\omega T}-1}{e^{j\,\omega T}+1} = \frac{\left(e^{j\,\frac{\omega T}{2}}\right)\left(e^{j\,\frac{\omega T}{2}}-e^{-j\,\frac{\omega T}{2}}\right)}{\left(e^{j\,\frac{\omega T}{2}}\right)\left(e^{j\,\frac{\omega T}{2}}+e^{-j\,\frac{\omega T}{2}}\right)} = \frac{2j\,\sin(\omega T/2)}{2\cos(\omega T/2)} = j\tan\left(\frac{\omega T}{2}\right),$$

and so we have

$$D(e^{j\,\omega T}) = H(j\,\phi) \qquad \text{with} \qquad \phi = \frac{2}{T}\tan\left(\frac{\omega T}{2}\right). \qquad (8.15)$$

Equation (8.15) provides the desired relationship between the frequency responses of the analogue prototype and digital replacement filters. We note that, for $\theta = \omega T$ sufficiently small, we have $\tan(\theta/2) \approx \theta/2$ and hence $\phi \approx \omega$. As discussed earlier, the highest frequency in the digital filter's input, i.e. immediately prior to sampling and after any analogue pre-filtering, should be less than π/T rads./sec. In terms of normalized frequency θ, the frequency range of interest is therefore contained in $-\pi < \theta < \pi$. In many applications, the analogue prototype $H(s)$ satisfies $|H(j\omega)| \approx 0$ for $|\omega| > \omega_\mu$ with $\omega_\mu \ll \pi/T$. In this case, the approximation $\tan(\theta/2) \approx \theta/2$ may well be acceptable over the whole range $-\omega_\mu < \omega < \omega_\mu$ and, hence, the analogue and digital frequency responses are well-matched over the whole frequency range of interest.

In general, we see from (8.15) that the digital frequency response at frequency ω exactly matches the analogue frequency response at frequency ϕ. Now suppose that $H(s)$ has a cut-off frequency given by $\omega = \omega_c$, which we understand as meaning

$$|H(j\,\omega_c)| = \frac{1}{\sqrt{2}} \approx 0.707.$$

If we now let ω_d be the cut-off frequency for $D(z)$, then we see that ω_d is given through solving

$$|D(e^{j\,\omega_d T})| = \frac{1}{\sqrt{2}} \iff |H(j\,\phi_d)| = \frac{1}{\sqrt{2}} \iff \phi_d \overset{\text{def}}{=} \frac{2}{T}\tan\left(\frac{\omega_d T}{2}\right) = \omega_c.$$

The unique solution for ω_d, in the range $0 \le \omega_d < \pi/T$, is therefore given by

$$\omega_d = \frac{2}{T}\tan^{-1}\left(\frac{\omega_c T}{2}\right). \tag{8.16}$$

Letting $\theta_d = \omega_d T$ and $\theta_c = \omega_c T$ be normalized frequencies, we can rewrite (8.16) in the form

$$\theta_d = 2\tan^{-1}\left(\frac{\theta_c}{2}\right). \tag{8.17}$$

In many applications, $\theta_c \ll \pi$ and we have $\theta_d \approx \theta_c$ as already discussed. However, we can specify a desired cut-off frequency θ_d for $D(z)$ by appropriately choosing the analogue cut-off frequency θ_c. This technique is known as **frequency pre-warping**, i.e. we 'warp' the analogue cut-off frequency so as to produce a desired digital cut-off frequency. From (8.17), we see that the appropriate pre-warping transformation is given by

$$\theta_c = 2\tan\left(\frac{\theta_d}{2}\right) \qquad \textbf{frequency pre-warping.} \tag{8.18}$$

In the above discussion, pre-warping was performed so as to give rise to a desired digital cut-off frequency. Clearly, we can replace $1/\sqrt{2}$ by a different value and hence pre-warp so that the digital filter has a desired amplitude at a desired frequency. Apart from the exact match at zero frequency, which always holds as can be seen from (8.15), we can only use this technique to produce a match at one further positive frequency. We now illustrate these ideas in the following two examples.

Example 8.3
Use the trapezoidal integration technique to construct a discrete-time filter based on the second-order Butterworth low-pass filter with normalized cut-off frequency $\theta_c = 1$ radian.

From Chapter 4, we recall that the transfer function of the prototype is

$$H(s) = \frac{\omega_c^2}{s^2 + \sqrt{2}\omega_c s + \omega_c^2}$$

and, making the transformation

$$s \mapsto \frac{2}{T}\left(\frac{z-1}{z+1}\right),$$

we obtain

$$D(z) = H\left(\frac{2}{T}\left(\frac{z-1}{z+1}\right)\right)$$

$$= \frac{\omega_c^2}{\dfrac{4(z-1)^2}{T^2(z+1)^2} + \dfrac{2\sqrt{2}\omega_c(z-1)}{T(z+1)} + \omega_c^2}$$

$$= \frac{T^2\omega_c^2(z+1)^2}{z^2(4 + 2\sqrt{2}T\omega_c + T^2\omega_c^2) + z(2T^2\omega_c^2 - 8) + (4 - 2\sqrt{2}T\omega_c + T^2\omega_c^2)}.$$

Using the fact that $\theta_c = \omega_c T = 1$, this reduces to

$$D(z) = \frac{(z+1)^2}{(5+2\sqrt{2})z^2 - 6z + (5-2\sqrt{2})} \approx \frac{0.1277(z+1)^2}{z^2 - 0.7664z + 0.2774}.$$

A plot of the amplitude response of this system is shown in Figure 8.4.

□

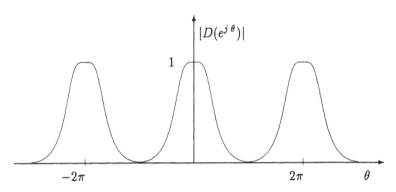

Figure 8.4: The amplitude response of the system of Example 8.3.

Example 8.4
Using as prototype the second-order Butterworth filter of Example 8.3, use frequency pre-warping to design a low-pass digital filter with normalized cut-off frequency $\theta_d = \pi/2$ rads.

From (8.18), we see that the analogue prototype needs to have a normalized cut-off frequency given by

$$\theta_c = 2\tan\left(\frac{\theta_d}{2}\right) = 2.$$

The appropriate digital replacement filter is then given from Example 8.3 as

$$D(z) = \frac{\theta_c^2(z+1)^2}{z^2(4+2\sqrt{2}\,\theta_c + \theta_c^2) + z(2\theta_c^2 - 8) + (4 - 2\sqrt{2}\,\theta_c + \theta_c^2)}$$

$$= \frac{4(z+1)^2}{4(2+\sqrt{2})z^2 + 4(2-\sqrt{2})}$$

$$\approx \frac{0.2929(z+1)^2}{z^2 + 0.1716}.$$

□

Other formulations of this technique are available; for example, see the text by Poularikis and Seeley [26].

8.5 A 'DIRECT' DESIGN METHOD:
THE FOURIER SERIES APPROACH

The methods we have studied so far have all been based on emulating the response of a prototype continuous-time filter. We now examine an entirely different approach to the design problem. Suppose that $D(z)$ is the transfer function of a stable discrete-time system, then, we can write as usual,

$$Y(z) = D(z)U(z) .$$

If the input sequence is $\{u_k\} = \{\delta_k\} = \{1, 0, 0, 0, \ldots\}$, the unit impulse sequence with z-transform $U(z) = 1$, then the transform of the output sequence, namely the impulse response sequence, is

$$Y_\delta(z) = D(z) = \sum_{n=0}^{\infty} d_n z^{-n} .$$

Since the system is stable, by assumption, there is a frequency response which is obtained by taking the DTFT of the impulse response sequence. This is achieved by replacing z by $e^{j\omega t}$ in $D(z)$ to obtain

$$D\left(e^{j\omega T}\right) = D\left(e^{j\theta}\right) = \sum_{n=0}^{\infty} d_n e^{-jn\theta} , \qquad (8.19)$$

where $\theta = \omega T$.

Now (8.19) can be interpreted as the Fourier expansion of $D\left(e^{j\theta}\right)$, using as basis functions the orthogonal set $\{e^{-jn\theta}\}$. It is then easy to show that the Fourier coefficients relative to this base are given by

$$d_n = \frac{1}{2\pi} \int_{-\pi}^{\pi} D\left(e^{j\theta}\right) e^{jn\theta} \, d\theta .$$

We now set $D\left(e^{j\theta}\right)$ to the desired ideal frequency response function and calculate the resulting Fourier coefficients, $\{h_d(n)\}$ say. It should be noted that, at this stage, we can no longer restrict to $n \geq 0$, i.e. $h_d(n)$, as defined above, is not causal and hence does not correspond with the impulse response of any *realizable* system. If a filter is to be realized using a finite number of delay elements, some form of truncation must take place. It is helpful to think of this truncation being performed by the application of a **window**, defined by a window weighting function $w(n)$. The simplest window is the **rectangular window**, with weighting function $w(n)$ defined by

$$w(n) = \left\{ \begin{array}{ll} 1 , & -n_1 \leq n \leq n_2 \\ 0 , & \text{otherwise} . \end{array} \right.$$

Using this window, we actually form

$$\sum_{n=-\infty}^{\infty} w(n)h_d(n)e^{-jn\theta} = \sum_{n=-n_1}^{n_2} h_d(n)e^{-jn\theta} = \tilde{D}\left(e^{j\theta}\right) ,$$

where, if n_1 and n_2 are sufficiently large, $\tilde{D}\left(e^{j\,\theta}\right)$ will be an adequate approximation to $D\left(e^{j\,\theta}\right)$, the desired frequency response. It is important to note that the **filter length**, that is the number of delay elements or terms in the difference equation, depends on the choice of n_1 and n_2. This means that some accuracy will always have to be sacrificed in order to produce an acceptable design.

We explore this technique by designing a low-pass filter in the following example.

Example 8.5
Use the Fourier Series, or direct design method, to produce a low-pass digital filter with cut-off frequency $f_c = 1\,\text{kHz}$, when the sampling frequency is $f_s = 5\,\text{kHz}$.

We wish to make use of the non-dimensional frequency variable θ and, since $T = 1/f_s = 1/5000$, we have

$$\theta = \omega T = 2\pi fT = \frac{2\pi f}{5000}.$$

The cut-off frequency is then $\theta_c = 2\pi f_c/5000 = 2\pi/5$ and the ideal frequency response $D\left(e^{j\,\theta}\right)$ is now defined by

$$D\left(e^{j\,\theta}\right) = \left\{ \begin{array}{ll} 1, & |\theta| \leq 2\pi/5 \\ 0, & |\theta| > 2\pi/5 \,. \end{array} \right.$$

We now calculate the coefficients $h_d(n)$ as

$$h_d(n) = \frac{1}{2\pi} \int_{-\pi}^{\pi} D\left(e^{j\,\theta}\right) e^{j\,n\theta}\, d\theta,$$

$$= \frac{1}{2\pi} \int_{-2\pi/5}^{2\pi/5} e^{j\,n\theta}\, d\theta,$$

$$= \frac{1}{n\pi} \sin\left(\frac{2n\pi}{5}\right), \quad \text{for } n \neq 0,$$

$$= \frac{2}{5}\text{sinc}\left(\frac{2n\pi}{5}\right), \quad \text{(also valid for } n = 0\text{)}.$$

At this stage, we have to choose the length of the filter. By now, we know that a 'long' filter is likely to produce superior results in terms of frequency domain performance. However, experience again tells us that there will be penalties in some form or other. Let us choose a filter of length 9, with the coefficients selected for simplicity as symmetric about $n = 0$. As already discussed, this choice leads to a non-causal system, but we deal with this problem when it arises. This scheme is equivalent to specifying the use of a rectangular window defined by

$$w(n) = \left\{ \begin{array}{ll} 1, & -4 \leq n \leq 4 \\ 0, & \text{otherwise} \,. \end{array} \right.$$

We now calculate the coefficients $h_d(-4)$, $h_d(-3), \ldots h_d(0), \ldots h_d(4)$, which are tabulated in Table 8.1.

$h_d(\pm 4)$	$h_d(\pm 3)$	$h_d(\pm 2)$	$h_d(\pm 1)$	$h_d(0)$
-0.07568	-0.06237	0.09355	0.30273	0.40000

Table 8.1: Coefficients $h_d(k)$, for $k = -4, -3, \ldots, 4$.

The transfer function of the digital filter is then $\tilde{D}(z)$, where

$$\tilde{D}(z) = \sum_{n=-4}^{4} h_d(n) z^{-n}$$

$$= -0.07568 z^{-4} - 0.06237 z^{-3} + 0.09355 z^{-2} + 0.30273 z^{-1} + 0.40000$$
$$+ 0.30273 z + 0.09355 z^{2} - 0.06237 z^{3} - 0.07568 z^{4} .$$

Although this system is indeed non-causal, since its impulse response sequence contains terms in positive powers of z, we can calculate the frequency response as

$$\tilde{D}\left(e^{j\theta}\right) = -0.15137 \cos(4\theta) - 0.12473 \cos(3\theta) + 0.18710 \cos(2\theta)$$
$$+ 0.60546 \cos(\theta) + 0.40000 .$$

Figure 8.5 illustrates the corresponding amplitude response.

□

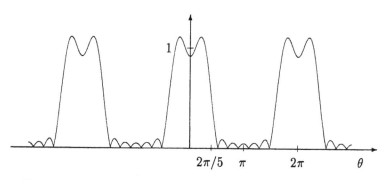

Figure 8.5: Amplitude response of the non-causal filter of Example 8.5.

Figure 8.5, of Example 8.5, shows us that the amplitude response of our filter is a reasonable approximation to the design specification. We do, however, notice that there are some oscillations in both pass- and stop-bands. These are due to the abrupt cut-off of the rectangular window function and the effect is known as **Gibb's phenomenon**. Thus, the window function generates additional spectral components, which are referred to as **spectral leakage**. A way of improving the performance in this respect is discussed in §8.6. The immediate problem is the realization of this non-causal design. To see how we can circumvent the difficulty, we proceed as follows.

The transfer function we have derived is of the general form

$$\tilde{D}(z) = \sum_{k=-N}^{N} h_d(k)z^{-k}$$
$$= z^N \left[h_d(-N) + h_d(-N+1)z^{-1} + \ldots + h_d(0)z^{-N} + \ldots + h_d(N)z^{-2N} \right] .$$

Suppose that we implement the system with transfer function

$$\hat{D}(z) = z^{-N}\tilde{D}(z) ,$$

which is a causal system. First we notice that, on setting $z = e^{j\omega T}$, the amplitude response $\left| \hat{D}\left(e^{j\omega T}\right) \right|$ is given by

$$\left| \hat{D}\left(e^{j\omega T}\right) \right| = \left| e^{-j\omega NT} \right| \left| \tilde{D}\left(e^{j\omega T}\right) \right|$$
$$= \left| \tilde{D}\left(e^{j\omega T}\right) \right| ,$$

that is, it is identical with that of the desired design. Furthermore,

$$\arg\left\{ \hat{D}\left(e^{j\omega T}\right) \right\} = \arg\left\{ \tilde{D}\left(e^{j\omega T}\right) \right\} - N\omega T ,$$

indicating a pure delay of amount NT in the response of the second system. This means that, assuming we are prepared to accept this delay, our design objective can be met by the system with transfer function $\hat{D}(z)$ given by

$$\hat{D}(z) = \left[-0.07568 - 0.06237z^{-1} + 0.09355z^{-2} + 0.30273z^{-3} \right.$$
$$\left. + 0.40000z^{-4} + 0.30273z^{-5} + 0.09355z^{-6} - 0.06237z^{-7} - 0.07568z^{-8} \right] .$$

It is evident from Figure 8.6 that the filter designed in Example 8.5 differs from the previous designs. The nature of this difference is the absence of feedback paths in the block diagram realization of Figure 8.6. One effect of this is that the impulse response sequence is finite, a fact which we already know, since the design method involved truncating the impulse response sequence. Filters of this type are known as **finite impulse response (FIR)** designs and may always be implemented using structures not involving feedback loops. Another name used for such structures is **non-recursive**, but it is not correct to assume that the only possible realization of an FIR filter is by use of a non-recursive structure; for details see Jong [12].

8.6   WINDOWS

In this concluding section, we consider the problem identified in Example 8.5 above in connection with the sharp cut-off of the rectangular window function.

8.6.1   *The rectangular window*

The rectangular window sequence, illustrated in Figure 8.7, is defined by

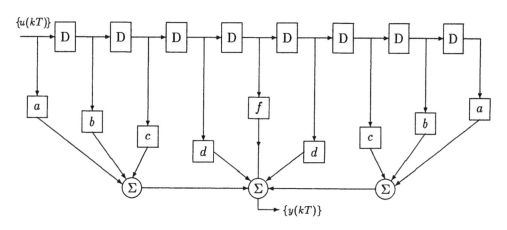

$a = -0.07568, \; b = -0.06237, \; c = 0.09355, \; d = 0.30273, \; f = 0.40000.$

Figure 8.6: A realization of the final system of Example 8.5.

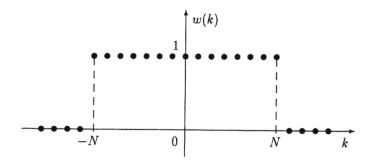

Figure 8.7: Rectangular window sequence.

$$w(k) = \begin{cases} 1, & |k| \leq N \\ 0, & \text{otherwise} \end{cases},$$

which can be expressed in the form

$$w(k) = \zeta(k + N) - \zeta(k - (N + 1)) .$$

Since

$$W(z) = \left(z^N - z^{-(N+1)}\right)\left(\frac{z}{z-1}\right) = \frac{z^{N+\frac{1}{2}} - z^{-(N+\frac{1}{2})}}{z^{\frac{1}{2}} - z^{-\frac{1}{2}}} ,$$

the DTFT of the sequence $\{w(k)\}$ is

$$W\left(e^{j\theta}\right) = \frac{\sin\left(\frac{1}{2}(2N + 1)\theta\right)}{\sin\left(\frac{1}{2}\theta\right)} \quad \text{for } \theta \neq 0$$

$$= \frac{(2N + 1)\text{sinc}\left(\frac{1}{2}(2N + 1)\theta\right)}{\text{sinc}\left(\frac{1}{2}\theta\right)} .$$

It is easy to see that $W\left(e^{j0}\right) = W(1) = \sum_{n=-N}^{N} w(n) = 2N + 1$ and so the above formula, using the sinc function, is valid for all θ, including $\theta = 0$. The graph of this function is illustrated in Figure 8.8. The first positive (negative) zero in its

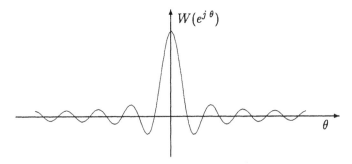

Figure 8.8: DTFT of the rectangular window sequence.

spectrum is the positive (negative) value of θ closest to zero such that $W(e^{j\theta}) = 0$. The **main lobe** of the window function is that part of the graph of $W(e^{j\theta})$ that lies between the first positive and first negative zero in $W(e^{j\theta})$. The **main lobe width** is the distance between the first positive and negative zeros in $W(e^{j\theta})$. As the length of the window increases, the main lobe narrows and its peak value rises and, in some sense, $W(e^{j\theta})$ approaches an impulse, which is desirable. However, the main disadvantage is that the amplitudes of the side lobes also increase.

The use of any window leads to distortion of the spectrum of the original signal caused by the size of the side lobes in the window spectrum and the width of the window's main spectral lobe, producing oscillations in the filter response. The window function can be selected so that the amplitudes of the sides lobes are relatively small, with the result that the size of the oscillations are reduced; however, in general, the main lobe width does not decrease. Thus, in choosing a window, it is important to know the trade-off between having narrow main lobe and low side lobes in the window spectrum.

8.6.2   Other popular windows

A considerable amount of research has been carried out, aimed at determining suitable alternative window functions which smooth out straight truncation and thus reduce the Gibb's phenomena effects observed in the amplitude response of Figure 8.5. To minimize the effect of spectral leakage, windows which approach zero smoothly at either end of the sampled signal are used. We do not discuss the derivation of the various window functions, rather we tabulate, in Table 8.2, some of the more popular examples in a form suitable for symmetric filters of length $2N+1$. For a more detailed discussion on windows and their properties, see, for example, the texts by Ifeachor and Jervis [10], Oppenheim and Schafer [21], and Stearns and Hush [29].

Window name	$w(k)$	
Bartlett	$w(k) = \begin{cases} (k+N)/N\,, \\ (N-k)/N\,, \end{cases}$	$\begin{aligned} -N \le k < 0 \\ 0 \le k \le N \end{aligned}$
von Hann or Hanning	$w(k) = 0.5 + 0.5\cos(\pi k/(N+1))\,,$	$-N \le k \le N$
Hamming	$w(k) = 0.54 + 0.46\cos(\pi k/N)\,,$	$-N \le k \le N$
Blackman	$w(k) = 0.42 + 0.5\cos(\pi k/N)$	
	$\quad\quad +0.08\cos(2\pi k/N)\,,$	$-N \le k \le N$

In each case, $w(k) = 0$ for k outside the range $[-N,\ N]$.

Table 8.2: Some popular window functions.

Note: Slight variations on the above definitions may be found in various texts. These tend to involve switching between 'division by N', 'division by $N + \frac{1}{2}$' and 'division by $N + 1$'. For example, the von Hann or Hanning window is variously defined by $w(k) = 0.5(1 + \cos(\pi k/N))$ or $w(k) = 0.5(1 + \cos(2\pi k/(2N + 1)))$ or $w(k) = 0.5(1 + \cos(\pi k/(N + 1)))$ for $|k| \le N$ with $w(k) = 0$ for $|k| > N$. The Bartlett window, or one of its variations, is sometimes referred to as a **triangular window**. It should also be noted that both the Bartlett window and the Blackman window, as defined in Table 8.2, satisfy $w(-N) = w(N) = 0$ and hence give rise to difference equations of order $2N - 2$ rather than $2N$.

Formulations for other configurations can easily be deduced, or may be found in, for example, Jackson [11] or Ziemer *et al.* [31]. The section closes with an example of the application to the design of Example 8.5 above.

Example 8.6
Plot the amplitude response for the filter design of Example 8.5, using (a) the Hamming window, and (b) the Blackman window.

(a) The transfer function coefficients are now given by $h_d(k)w_H(k)$, where $w_H(k)$ are the Hamming window coefficients, calculated with $N = 4$ and $-4 \le k \le 4$. The Hamming window coefficients are tabulated in Table 8.3.

N	± 4	± 3	± 2	± 1	0
	0.08000	0.21473	0.54000	0.86527	1.00000

Table 8.3: Hamming window coefficients for $-4 \le k \le 4$.

The transfer function then becomes

$$\hat{D}_H(z) = [-0.00605 - 0.01339z^{-1} + 0.05052z^{-2}$$
$$+ 0.26194z^{-3} + 0.40000z^{-4} + 0.26194z^{-5} + 0.05052z^{-6}$$
$$- 0.01339z^{-7} - 0.00605z^{-8}] .$$

The frequency response is then obtained by writing $z = e^{j\theta}$, as

$$\hat{D}_H\left(e^{j\theta}\right) = e^{-j\,4\theta}\left(-0.01211\cos(4\theta) - 0.02678\cos(3\theta)\right.$$
$$\left. + 0.10103\cos(2\theta) + 0.52389\cos(\theta) + 0.40000\right).$$

Figure 8.9 illustrates the magnitude of this response and the reduction of oscillations in both the pass- and stop-band is striking. The penalty is the lack of sharpness near the cut-off frequency, although the stop-band characteristics close to $\theta = \pi$ are quite good.

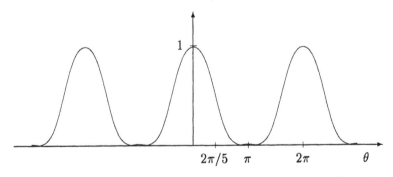

Figure 8.9: Amplitude response of the filter of Example 8.5, with Hamming window.

(b) Proceeding as in case (a), we calculate the Blackman window coefficients as shown in Table 8.4. The Blackman windowed transfer function is thus

$$\hat{D}_B(z) = -0.00414 + 0.03181z^{-1} + 0.23418z^{-2} + 0.40000z^{-3}$$
$$+ 0.23418z^{-4} + 0.03181z^{-5} - 0.00414z^{-6} ,$$

N	±4	±3	±2	±1	0
	0.00000	0.06645	0.34000	0.77355	1.00000

Table 8.4: Blackman window coefficients for $-4 \leq k \leq 4$.

and the frequency response is found as

$$\hat{D}\left(e^{j\,\theta}\right) = e^{-j\,3\theta}\Big(-0.00829\cos(3\theta) + 0.06361\cos(2\theta)$$

$$+ 0.46836\cos(\theta) + 0.40000\Big).$$

The amplitude response is shown in Figure 8.10 and this design again suffers from a relatively poor performance in terms of sharpness of cut-off. The

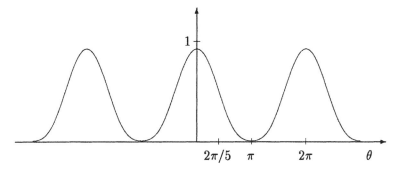

Figure 8.10: Amplitude response of the filter of Example 8.5, with Blackman window.

ripples observed in the pass- and stop-bands with the rectangular window have been removed as before. However, the 'flat' characteristic of the Hamming design close to $\theta = \pi$ is not evident when using the Blackman window for this particular filter.   □

This chapter has given a brief introduction to some methods of digital filter design. Space does not permit a longer discussion and a number of important ideas have been omitted; for example, we have not considered the possibility of direct filtering in the frequency domain, using the FFT. However, our aim has been to encourage further study.

8.7 EXERCISES

1. Draw a block diagram representing an implementation of the digital filter designed in Example 8.4.

2. Show that, under the mapping $z = e^{sT}$, points in the left half of the complex s-plane are mapped into the unit circle in the z-plane. Hence, deduce that the impulse invariant design method produces stable filter designs, provided that the continuous-time prototype is a stable design.

3. Design, using the impulse invariant technique, a digital filter corresponding to the analogue low-pass filter with transfer function

$$H(s) = \frac{1}{(s+2)^2} \, ,$$

assuming the sampling rate is 20 Hz.

Write down the frequency response for both the digital and analogue filters and, in each case, find the dc gain. Explain briefly why the dc gains differ.

4. Determine the difference equation realization of a digital filter that approximates the second-order Butterworth low-pass filter

$$H(s) = \frac{2a^2}{s^2 + 2as + 2a^2} \, , \quad a > 0 \, ,$$

using the impulse invariant method and assuming sampling period T seconds.

5. Use the impulse invariant method to design, in the form of a difference equation, a digital replacement for the second-order Butterworth low-pass filter, with cut-off frequency $\omega_c = 10$ rads./sec., given that the sampling frequency is to be five times the cut-off frequency. Determine the sampling interval T and plot the amplitude response.

6. Repeat the above exercise, this time using the step invariant technique.

7. Determine the transfer function of the digital filter that approximates an analogue filter with transfer function

$$H(s) = \frac{2s+1}{s^2 + 3s + 2}$$

using the impulse invariant technique and assuming the sampling frequency $\frac{1}{T}$ Hz.

Find the dc gain of both the analogue filter and the approximating digital filter, $D_1(z)$.

Suppose it is required that the digital filter has exactly the same dc gain as the analogue filter. By modifying the gain of $D_1(z)$, write down a transfer function, $D_2(z)$, of an approximating digital filter that satisfies this requirement.

8. By examining the bilinear mapping which defines the trapezoidal integration method of digital filter design, show that filters designed by this method are stable provided that the continuous-time prototype is stable.

9. Use the bilinear transform method, with pre-warping, to construct a digital low-pass filter with cut-off frequency ω_d, based upon the third-order Butterworth filter. If the sampling frequency ω_s is to be ten times the analogue cut-off frequency, draw up a table showing the values of ω_d and ω_s obtained when the sampling interval T takes the values $1\,\text{s}$, $0.1\,\text{s}$ and $0.0001\,\text{s}$.

10. (a) Using the bilinear transform method with pre-warping, design a second-order digital low-pass filter, with normalized cut-off frequency $\theta_d = \frac{\pi}{3}$, corresponding to the analogue low-pass Butterworth filter with frequency response

$$H(j\omega) = \frac{\omega_c^2}{(j\omega)^2 + j\sqrt{2}\omega_c\,\omega + \omega_c^2} \; .$$

Evaluate the dc gain for both the analogue and digital filters.

(b) The corresponding digital filter obtained using the impulse invariant method produces

$$D_1(z) = \frac{2ke^{-k}z\sin(k)}{z^2 - 2e^{-k}z\cos(k) + e^{-2k}} \; ,$$

where $k = \dfrac{\theta_d}{\sqrt{2}} = \dfrac{\pi}{3\sqrt{2}}$. Evaluate the corresponding dc gain.

Use MATLAB, or any other appropriate software, to plot the gain against normalized frequency for $0 \le \theta < \pi$ for all three filters, i.e. the analogue filter

$$H(j\omega) = \frac{\omega_c^2}{(j\omega)^2 + j\sqrt{2}\omega_c\,\omega + \omega_c^2} \, ,$$

where $\omega_c = \dfrac{\theta_d}{T} = \dfrac{100\pi\sqrt{6}}{3}$, together with the two digital filters, $D(z)$ and $D_1(z)$.

What happens if the frequency range is increased to $0 \le \theta < 2\pi$?

Repeat the above plots, but now assume the filter, $D_1(z)$ is to be replaced by $D_2(z)$, where extra gain is used to make the dc gain equal to unity; that is

$$D_2(z) = \frac{(1 - 2e^{-k}\cos(k) + e^{-2k})z}{z^2 - 2e^{-k}z\cos(k) + e^{-2k}} \, .$$

11. Develop a digital filter design method based upon Simpson's rule for numerical integration and design a digital replacement for the first-order Butterworth low-pass filter with cut-off frequency ω_c. Can you say anything about the stability of designs so produced?

12. Use the direct (Fourier series) design method with a rectangular window of length 11 to produce a causal low-pass filter with non-dimensional cut-off frequency $\theta_c = \pi/2$. Plot the amplitude response.

13. Repeat the above exercise, this time using the Hamming window.

14. Follow the method of Example 8.4 to obtain a 2nd-order difference equation which represents a digital low-pass filter with cut-off frequency $\omega_d = \frac{1}{10}\omega_s$. Sketch the amplitude response.

9

Aspects of speech processing

9.1 INTRODUCTION

There are many aspects of speech processing such as, for example: speech production, modelling of speech, speech analysis, speech recognition, speech enhancement, synthesis of speech, etc. All these aspects have numerous applications, including speech coding for communications, speech recognition systems in the robotic industry or world of finance (to name but a few), and speech synthesis, the technology for which has been incorporated in a number of modern educational toys for example. In this chapter, we introduce some models for the processing of speech signals, discuss some analysis techniques and consider some specific applications of speech processing.

9.2 SPEECH PRODUCTION MODEL

In this section, an appropriate system function is suggested for modelling the **vocal** and **nasal tracts** as shown in Figure 9.1 (for more details, see Deller *et al* [5] and Rabiner and Schafer [27]). The vocal tract produces a sound when the lungs contract, forcing air up through an air-passage known as the **trachea**. The increase in air pressure causes the **glottis** (a type of valve) to burst open, momentarily, before quickly closing. When this is repeatedly performed, a very nearly periodic waveform is produced. The resulting flow of air is then perturbed by the vocal tract (or vocal and nasal tracts) to produce a sound. Some factors that can affect the waveform are: tension in the vocal chords, lung pressure, and the shape of the vocal cavities. The mouth, nose and throat are the primary resonating cavities. Their resonant characteristics can be changed (thereby altering the speech waveform) by moving the lips, tongue, jaw, and soft palate at the back of the mouth.

Speech sounds can be classified into a number of distinct classes. Three important classes are:

(a) **voiced sounds**: produced when quasi-periodic (that is approximately periodic) pulses of air excite the vocal tract (for example 'u', 'i', 'd', 'w', etc.);

(b) **unvoiced sounds** (or **fricatives**): generated by forming a constriction in the vocal tract and forcing air through the constriction. This creates a noise source which excites the vocal tract (for example 'sh', 'f', 's');

(c) **plosive sounds**: obtained by making a complete closure of the vocal tract (with lips), building up pressure behind the closure, and abruptly releasing it (for example 'p', 'b').

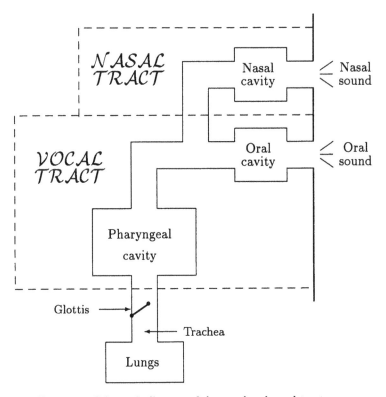

Figure 9.1: Schematic diagram of the vocal and nasal tracts.

Resonances in the vocal tract are characterized by resonant frequencies in the vocal tract spectrum, which are called **formant** frequencies (or simply **formants**). Tension of the vocal chords can affect **pitch**, that is an increase in tension causes the chords to vibrate at a higher frequency. The time between successive openings of the glottis valve, producing the quasi-periodic pulses of air, is called the **fundamental period**. By convention, see Deller et al [5] for more details, the fundamental period is often referred to as the **pitch period**. *Estimation of pitch periods and formants are two very important problems in speech processing.*

In order to analyse a speech (continuous-time) signal, the signal is sampled to obtain a discrete-time signal, $\{s(m)\}$. It is assumed that speech production can be modelled by a linear time-varying filter that is excited by an input, $\{u(m)\}$, consisting of a quasi-periodic train of impulses, in the case of voiced sounds, or a random noise source, in the case of unvoiced sounds. Usually during excitation, vocal and nasal tract properties of a speech signal change relatively slowly with time and, for most speech signals, we may assume that these properties remain fixed over short time periods, such as 10-20 msec. Thus, in these short time periods, speech production is assumed to be modelled by *time-invariant* filters. A linear time-invariant model for speech production is usually taken to be of the form:

$$S(z) = G(z)V(z)L(z)U(z) \ ,$$

where $S(z) = \mathcal{Z}[\{s(m)\}]$, $U(z) = \mathcal{Z}[\{u(m)\}]$, $G(z)$ is the **glottal shaping model**, $V(z)$ a **vocal tract model**, and $L(z)$ a **lip radiation model**. Experimentally, it is found that a reasonable approximation to the speech production system function has the form:

$$H(z) = \frac{S(z)}{U(z)} = G(z)V(z)L(z) \approx \frac{G_s \left[1 - \sum_{i=1}^{q} b_i z^{-i} \right]}{\left[1 - \sum_{k=1}^{p} a_k z^{-k} \right]} \ ,$$

where the constant G_s is known as the **speech gain parameter**. Therefore, the model

$$S(z) = H(z)U(z) = \frac{G_s \left[1 - \sum_{i=1}^{q} b_i z^{-i} \right]}{\left[1 - \sum_{k=1}^{p} a_k z^{-k} \right]} U(z)$$

is often used for the synthesis of speech. For non-nasal, voiced sounds an **all-pole** (namely **autoregressive**) filter, that is $b_j = 0$ for all j, is a good approximation for the system function. Thus, a further simplification of the speech production model (for non-nasal, voiced sounds) is:

$$S(z) = G_s \frac{U(z)}{A(z)} \ ,$$

where

$$A(z) \stackrel{\text{def}}{=} 1 - \sum_{k=1}^{p} a_k z^{-k}$$

and, in the literature, is known as the **inverse filter**.

9.3 CORRELATION FUNCTIONS

Correlation functions give a measure of 'correlation' between two signals, namely the dependence of the present value of a signal on the values of a second signal. Such correlation functions can be defined for both deterministic and stochastic (that is random) signals. Here, we consider only deterministic signals. Given two continuous-time signals $f(t)$ and $g(t)$ the correlation function is defined as

$$R_{fg}(\tau) \stackrel{\text{def}}{=} \lim_{T \to \infty} \frac{1}{T} \int_{-\frac{T}{2}}^{\frac{T}{2}} f(t)g(t+\tau)\, \mathrm{d}t \; , \tag{9.1}$$

and is a function of the time shift between two signals. If f and g have finite energy (see Chapter 3, §3.6), $R_{fg}(\tau)$ is zero and, in this case, (9.1) is not very useful. To overcome this problem, an alternative definition is given for the correlation function of two finite energy signals f and g as

$$R_{fg}(\tau) \stackrel{\text{def}}{=} \int_{-\infty}^{\infty} f(t)g(t+\tau)\, \mathrm{d}t \; . \tag{9.2}$$

Therefore, *in determining the correlation function of two signals, (9.1) is used when the signals have finite, non-zero average power, whilst for signals with finite energy, the correlation function is evaluated using (9.2).* If the two signals are sampled, the time shift, τ, is an integer multiple of the sampling period and, thus, the correlation function becomes a discrete (or 'sampled') sequence. The concepts of (total) energy and average power, introduced in Chapter 3, can easily be extended to discrete-time signals, $\{x(m)\}$, as

$$E = \sum_{m=-\infty}^{\infty} |x(m)|^2 \tag{9.3}$$

$$P_{\text{av}} = \lim_{M \to \infty} \frac{1}{2M+1} \sum_{m=-M}^{M} |x(m)|^2 \; . \tag{9.4}$$

For discrete-time signals $\{x(m)\}$ and $\{y(m)\}$ the correlation function is defined by

(*i*) for signals with finite, non-zero average power,

$$R_{xy}(k) \stackrel{\text{def}}{=} \lim_{M \to \infty} \frac{1}{2M+1} \sum_{m=-M}^{M} x(m)y(m+k) \; ; \tag{9.5}$$

(*ii*) for signals with finite energy,

$$R_{xy}(k) \stackrel{\text{def}}{=} \sum_{m=-\infty}^{\infty} x(m)y(m+k) \; . \tag{9.6}$$

Only discrete-time signals are considered in the remaining part of this chapter.

In the case when x and y are different signals, $R_{xy}(k)$ (as defined in (9.5) and (9.6)) is known as the **cross-correlation** function, whilst, when the signals are the same,

(*i*) for signals with finite, non-zero average power,

$$R_{xx}(k) \stackrel{\text{def}}{=} \lim_{M \to \infty} \frac{1}{2M+1} \sum_{m=-M}^{M} x(m)x(m+k) \; ; \tag{9.7}$$

(*ii*) for signals with finite energy,

$$R_{xx}(k) \stackrel{\text{def}}{=} \sum_{m=-\infty}^{\infty} x(m)x(m+k) , \qquad (9.8)$$

which is known as the **autocorrelation** function.

9.3.1   The autocorrelation function

Properties of the autocorrelation function are given below (for more details see, for example, Stearns and Hush [29]):

Properties

For signals with non-zero, finite average power or finite energy:

(a) $R_{xx}(-k) = R_{xx}(k)$; R_{xx} is an even function.

(b) $|R_{xx}(k)| \le R_{xx}(0)$;

For signals with non-zero, finite average power:

(c) if $\{x(m)\}$ contains a periodic component, with period M, then $R_{xx}(k)$ also contains a periodic component, with period M.

(d) if $R_{xx}(k)$ has no periodic component, then

$$\lim_{k \to \infty} R_{xx}(k) = \mu_x^2 , \quad \text{where } \mu_x \stackrel{\text{def}}{=} \lim_{M \to \infty} \frac{1}{2M+1} \sum_{m=-M}^{M} x(m)$$

denotes the average value for the sequence $\{x(m)\}$;

(e) defining $\sigma_x^2 \stackrel{\text{def}}{=} \left[\lim_{M \to \infty} \frac{1}{2M+1} \sum_{m=-M}^{M} x^2(m) \right] - \mu_x^2$, then $R_{xx}(0) = \mu_x^2 + \sigma_x^2$

and

$$-\sigma_x^2 + \mu_x^2 \le R_{xx}(k) \le \sigma_x^2 + \mu_x^2 .$$

The term μ_x^2 is known as the **dc component** of $R_{xx}(k)$.

As a consequence of (9.7) and in view of (9.4),

$$\boldsymbol{P}_{\text{av}} = R_{xx}(0) .$$

Similarly, for a discrete-time, finite energy signal $\{x(m)\}$,

$$\boldsymbol{E} = R_{xx}(0) ,$$

in view of (9.3) and (9.8). A typical graph of an autocorrelation sequence is illustrated in Figure 9.2.

Example 9.1

Using the result: $\displaystyle\sum_{r=0}^{\infty} w^r = \frac{1}{1-w}$,   for $|w| < 1$,

and given $x(k) = a^k \zeta(k)$, $|a| < 1$:

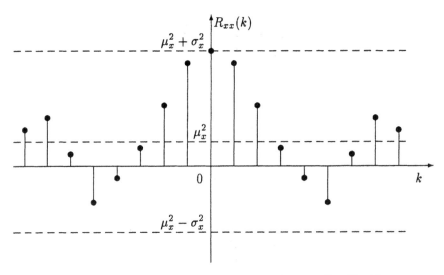

Figure 9.2: Typical graph of an autocorrelation function $R_{xx}(k)$ with no periodic component.

(a) show that the sequence $\{x(k)\}$ has finite energy;

(b) determine the autocorrelation sequence for $\{x(k)\}$.

(a) $\displaystyle\sum_{k=-\infty}^{\infty} x^2(k) = \sum_{k=0}^{\infty} a^{2k} = \frac{1}{1-a^2}$, since $|a| < 1$,

and, hence, the sequence has finite energy.

(b) Suppose $k \geq 0$, then

$$R_{xx}(k) = \sum_{m=-\infty}^{\infty} x(m)x(m+k)$$

$$= a^k \sum_{m=0}^{\infty} a^{2m}$$

$$= \frac{a^k}{1-a^2} \; .$$

Using the even symmetry property of $\{R_{xx}(k)\}$,

$$R_{xx}(k) = \frac{a^{-k}}{1-a^2} \; , \qquad \text{when } k < 0 \; .$$

Thus, for all k, $R_{xx}(k) = \dfrac{a^{|k|}}{1-a^2} .$

□

9.3.2 *The cross-correlation function*

The cross-correlation function, defined in (9.5) and (9.6), also has some important properties, which are listed below.

Properties

For signals with non-zero, finite average power or finite energy:

(a) $R_{xy}(-k) = R_{yx}(k)$

(b) $|R_{xy}(k)| \leq \{R_{xx}(0)R_{yy}(0)\}^{\frac{1}{2}} \leq \frac{1}{2}\{R_{xx}(0) + R_{yy}(0)\}$

(c) If both $\{x(k)\}$ and $\{y(k)\}$ contain a periodic component, with period M, then both $R_{xy}(k)$ and $R_{yx}(k)$ contain a periodic component, with period M.

This function is very useful for certain aspects of system identification, as shown in, for example, Ljung [16].

9.3.3 *An application of the autocorrelation function*

The autocorrelation function is very useful for estimating periodicities in signals (see Property (c) in §9.3.1), including speech. In practice, a segment of speech $\{s(m); \quad m = 0, 1, \ldots, M-1\}$ is available for analysis. This sequence has finite energy. The **'short-time' autocorrelation function** (which is an estimate of the true autocorrelation function) is determined by

$$r_{ss}(k) = \sum_{m=0}^{M-1-k} s(m)s(m+k) , \qquad (9.9)$$

for $k = 0, 1, \ldots, M-1$. It can be shown that r_{ss} has the even symmetry property (see Property (a) in §9.3.1), if the speech segment is assumed to be identically zero outside the interval $0 \leq m \leq M-1$. Let $\{\tilde{s}\}$ denote the sequence

$$\{\ldots 0, \ 0, \ s(0), \ s(1), \ldots, \ s(M-1), \ 0, \ 0, \ldots\} ,$$
$$\uparrow$$

then

$$r_{ss}(-k) = \sum_{m=0}^{M-1+k} \tilde{s}(m)\tilde{s}(m-k)$$

$$= \sum_{i=-k}^{M-1} \tilde{s}(i+k)\tilde{s}(i) , \quad \text{with} \ \ i = m-k ,$$

$$= \sum_{i=0}^{M-1-k} \tilde{s}(i+k)\tilde{s}(i) , \quad \text{since } \tilde{s}(i) = 0 \text{ for } i < 0 \text{ and } i \geq M ,$$

$$= \sum_{i=0}^{M-1-k} s(i+k)s(i)$$

$$= r_{ss}(k) .$$

For voiced speech, peaks occur in $r_{ss}(k)$ approximately at multiples of the pitch period. Hence, *the pitch period of a speech signal can be estimated by finding the location of the first maximum in the autocorrelation function.* If there are no strong periodicity peaks, then this indicates a lack of periodicity in the waveform, implying that the section of the speech signal is unvoiced. This technique is illustrated for a segment $\{x(m)\}$ of a non-nasal, voiced speech signal from the spoken word 'should'. The speech signal for the word 'should' is sampled at a rate of 13.16 kHz (sampling period 0.076 msec. approximately) and its graph is plotted in Figure 9.3. The autocorrelation function is obtained using MATLAB with the command **xcorr**.

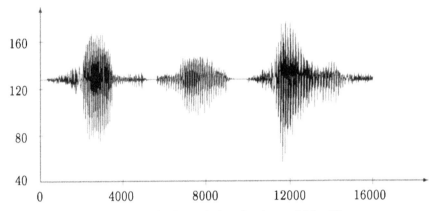

Figure 9.3: Graph of speech data for the word 'should'.

This calculates

$$\sum_{m=0}^{M-1-|k|} x(m)x(m+k) , \quad k = -(M-1), \ldots, M-1, \tag{9.10}$$

which is identical to the short-time autocorrelation function (9.9) when the time-lag k satisfies $k \geq 0$. However, the sequence is constructed so that the time-lag $k = 0$ for this sequence occurs in the middle and, thus, some adjustment is necessary. Finally, each value in the autocorrelation sequence is divided by the energy in the sequence, that is the sequence is **normalized**. To illustrate this, a segment of speech, with 512 data values (its graph is plotted in Figure 9.4), is selected from the speech data for the word 'should' and its autocorrelation function, plotted in Figure 9.5, is obtained using the MATLAB commands:

```
r=xcorr(x);        % autocorrelation sequence evaluation
l=length(r);
a=r((l+1)/2:l);    % zero th time-lag occurs at a(1)
na=a/a(1);         % normalized sequence
plot(na);
```

As can be seen in Figure 9.5, the application of (9.10) using the given data generates a decreasing function curve which masks any periodicity. To overcome this problem

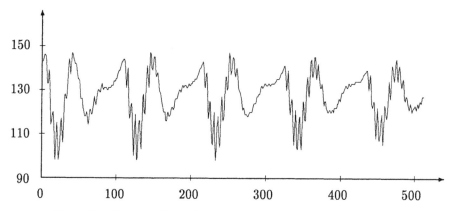

Figure 9.4: Segment of speech, 512 data values, for the word 'should'.

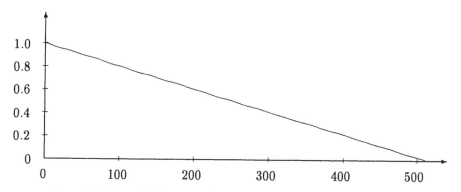

Figure 9.5: Normalized autocorrelation function for the speech segment.

the mean value of the data set is subtracted from each data value to form a new sequence. The autocorrelation function for this new sequence is shown in Figure 9.6. It is clear that the graph has dominant peaks at approximately 110, 220 and 330

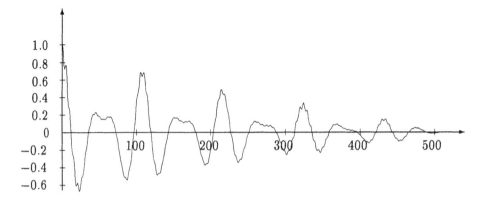

Figure 9.6: Normalized autocorrelation function for the speech data with zero mean.

which suggests that the pitch period is approximately $110 \times 0.076\,\text{msec.} \approx 8.4\,\text{msec.}$ and the pitch frequency is approximately $13160 \div 110 \approx 120\,\text{Hz}$ (or $754\,\text{rads./sec.}$). One may ask the question "Why do we need to use this procedure, when the approximate periodicity of the signal can be seen graphically in Figure 9.4?". The main point, here, is that the use of the autocorrelation function together with a 'peak picking' algorithm can be implemented quite easily electronically and, hence, the process of estimating the pitch period (or frequency) of a non-nasal voiced speech signal can be automated.

9.4   LINEAR PREDICTIVE FILTERS

The method of **linear prediction** (or **linear predictive coding**, as it is sometimes known) is based on the classical least squares method and utilizes the concept of **autocorrelation matrices**. An informative tutorial paper on this technique has been presented by Makhoul [17], and a detailed text has been written by Markel and Gray [18] (see, also, Haddad and Parsons [8], Parsons [25], Rabiner and Schafer [27] and Saito and Nakata [28]). Based on this method and properties of the autocorrelation function, discussed in §9.3.1 and §9.3.3, some filtering techniques can be used to analyse speech signals.

Let us suppose that an appropriate *linear* model for speech production is the **autoregressive** (AR) model

$$s(m) = G_s u(m) + \sum_{k=1}^{p} a_k s(m - k) \, ,$$

where a_k are the model parameters, $p \geq 1$, $u(m)$ is the model input, G_s is the speech gain parameter and $s(m)$ is the model output. Our main objective is to estimate the model parameters a_k. We introduce the concept of a **linear predictor**, which is a system whose output sequence $\{\tilde{s}\}$ is given by

$$\tilde{s}(m) = \sum_{k=1}^{p} \alpha_k s(m - k) ,$$

corresponding to the input sequence $\{s\}$. This system may be used to 'predict' $\{s(m)\}$, in the sense that, if $s(m - p)$, $s(m - (p - 1))$, ..., $s(m - 1)$ are given, $\tilde{s}(m)$ can be determined and used as an approximation to $s(m)$. Here, α_k are known as the **predictor coefficients**, p is the order of the filter, and $\tilde{s}(m)$ is the predicted value of $s(m)$. Suppose we wish to perform an analysis with a finite-length data set $\{s(m); \ m = 0, \ldots, \ M - 1\}$, where M is the number of signal samples. With sequences $\{s\}$ and $\{\tilde{s}\}$, we can define an error sequence $\{e\}$, where

$$e(m) = s(m) - \tilde{s}(m) .$$

For the present, it is assumed that $s(m) = 0$ if $m < 0$ or $m \geq M$, in which case $e(m) = 0$ when $m < 0$ or $m \geq p + M$. The predictor coefficients can be determined by minimizing the mean square error

$$e_{\mathrm{ms}} = \sum_{m=0}^{p+M-1} |e(m)|^2 .$$

A necessary condition for minimizing the mean square error, with respect to the predictor coefficients, is

$$\frac{\partial e_{\mathrm{ms}}}{\partial \alpha_i} = 0$$

$$\Longleftrightarrow \quad 2 \sum_{m=0}^{p+M-1} (s(m) - \tilde{s}(m)) \frac{\partial}{\partial \alpha_i} (s(m) - \tilde{s}(m)) \ = \ 0 .$$

But

$$\frac{\partial}{\partial \alpha_i} (s(m) - \tilde{s}(m)) = -s(m - i)$$

and therefore

$$\sum_{m=0}^{p+M-1} (\tilde{s}(m) - s(m)) s(m - i) = 0 , \quad i = 1, \ldots, p .$$

This produces the relationship

$$\sum_{m=0}^{p+M-1} \sum_{k=1}^{p} \alpha_k s(m - k) s(m - i) = \sum_{m=0}^{p+M-1} s(m) s(m - i) ,$$

or, since $s(m - k) = 0$ when $m > M - 1 + k$ and $s(m) = 0$ when $m > M - 1$,

$$\sum_{k=1}^{p} \alpha_k \sum_{m=0}^{M-1+k} s(m - k) s(m - i) = \sum_{m=0}^{M-1} s(m) s(m - i) , \quad i = 1, \ldots, p . \quad (9.11)$$

With $n = m - i$, and since $s(n) = 0$ when $n < 0$, we have

$$\sum_{m=0}^{M-1+k} s(m-k)s(m-i) = \sum_{n=-i}^{M-1+k-i} s(n)s(n+i-k)$$

$$= \sum_{n=0}^{M-1-(i-k)} s(n)s(n+i-k) , \quad i, k = 1, \ldots, p ,$$

which is the short-time autocorrelation function evaluated at $(i - k)$ (see (9.9)). Define

$$r(i - k) \stackrel{\text{def}}{=} \sum_{m=0}^{M-1+k} s(m-k)s(m-i) , \tag{9.12}$$

where, for the sake of clarity, we have omitted the double subscript on the short-time autocorrelation function r. As a consequence of (9.12),

$$r(i) = \sum_{m=0}^{M-1} s(m)s(m-i) .$$

Hence, (9.11) may be rewritten in the form

$$\sum_{k=1}^{p} \alpha_k r(i-k) = r(i) , \quad i = 1, \ldots, p .$$

As shown in §9.3.3, since $s(m)$ is zero outside the interval $0 \le m \le M - 1$, r is an even function. Thus, $r(k - i) = r(i - k)$ and so we may write

$$\sum_{k=1}^{p} \alpha_k r(|i-k|) = r(i) , \quad i = 1, \ldots, p , \tag{9.13}$$

which are known as the **normal** equations. The matrix,

$$\begin{bmatrix} r(0) & r(1) & r(2) & \cdots & r(p-1) \\ r(1) & r(0) & r(1) & \cdots & r(p-2) \\ r(2) & r(1) & r(0) & \cdots & r(p-3) \\ \vdots & \vdots & \vdots & \vdots & \vdots \\ r(p-1) & r(p-2) & r(p-3) & \cdots & r(0) \end{bmatrix} ,$$

known as the **autocorrelation matrix**, is a **Toeplitz** matrix, that is the matrix is symmetric and the elements along diagonals, sloping downwards from left to right, are all equal. An efficient recursive method for solving Equation (9.13) is **Durbin's algorithm**, which, sometimes, is attributed to Levinson or both Durbin and Levinson, see Rabiner and Schafer [27]. The algorithm is as follows:

1. *Initialization*:

$$e^{(0)} = r(0)$$
$$k_1 = r(1)/e^{(0)}$$
$$\alpha_1^{(1)} = k_1$$
$$e^{(1)} = (1 - k_1^2)e^{(0)}$$

2. (*For $2 \le i \le p$*)

$$k_i = \left\{ r(i) - \sum_{n=1}^{i-1} \alpha_n^{(i-1)} r(i-n) \right\} / e^{(i-1)}$$

$$\alpha_i^{(i)} = k_i$$

$$\alpha_n^{(i)} = \alpha_n^{(i-1)} - k_i \alpha_{i-n}^{(i-1)}, \qquad\qquad 1 \le n \le i-1$$

$$e^{(i)} = (1 - k_i^2) e^{(i-1)}.$$

The predictor coefficients are then given by $\alpha_k^{(p)}$, $\quad k = 1, \ldots, p$.

Remarks:

(a) The algorithm also generates predictor coefficients for predictors of order less than p, that is in the process of generating the prediction coefficients for a pth order predictor, at the ith stage ($i < p$), predictor coefficients are generated for an ith order predictor.

(b) Note that $e^{(i)}$ denotes the prediction error for a predictor of order i and, thus, the prediction error can be monitored at each stage (as discussed in Rabiner and Schafer [27], Chapter 8 §8.3.2).

(c) It is known (see, for example, Parsons [25]) that $|k_i| \le 1$, for every i, and that the linear predictive filter is stable.

(d) A MATLAB function m-file for this algorithm is presented in Appendix B.1.

Example 9.2

Show that Durbin's algorithm, for solving the normal equations, gives the exact values for the predictor coefficients of a second order predictor with autocorrelation matrix

$$\begin{bmatrix} 4 & 2 & 1 \\ 2 & 4 & 2 \\ 1 & 2 & 4 \end{bmatrix}.$$

We require to solve

$$r(0)\alpha_1 + r(1)\alpha_2 = r(1)$$
$$r(1)\alpha_1 + r(0)\alpha_2 = r(2)$$

with $r(0) = 4$, $r(1) = 2$, $r(2) = 1$, that is

$$4\alpha_1 + 2\alpha_2 = 2$$
$$2\alpha_1 + 4\alpha_2 = 1 .$$

This system of equations has the exact solution $\alpha_1 = \frac{1}{2}$, $\alpha_2 = 0$. Using Durbin's algorithm:

$$e^{(0)} = r(0) = 4$$

$$k_1 = r(1)/e^{(0)} = \tfrac{1}{2}$$

$$\alpha_1^{(1)} = k_1 = \tfrac{1}{2}$$

$$e^{(1)} = (1 - k_1^2)e^{(0)} = 3$$

$$k_2 = \left(r(2) - \alpha_1^{(1)} r(1) \right) / e^{(1)} = 0$$

$$\alpha_2^{(2)} = k_2 = 0$$

$$\alpha_1^{(2)} = \alpha_1^{(1)} - k_2 \alpha_1^{(1)} = \tfrac{1}{2}$$

$$e^{(2)} = (1 - k_2^2)e^{(1)} = 3 \ .$$

Thus, the predictor coefficients are $\alpha_1 = \alpha_1^{(2)} = \tfrac{1}{2}$, $\alpha_2 = \alpha_2^{(2)} = 0$.

$\square$

In speech processing, the above method is known as the **autocorrelation method**. Another method, known as the **covariance method**, is based on the assumption that some signal values outside the interval $0 \le m \le M-1$ are available for analysis, specifically, $s(m)$ for $-p \le m < 0$. For this case, the mean square error is minimized over the interval $0 \le m \le M - 1$, that is we minimize

$$e_{\text{ms}} = \sum_{m=0}^{M-1} [s(m) - \tilde{s}(m)]^2 \ ,$$

which leads to

$$\sum_{k=1}^{p} \alpha_k \sum_{m=0}^{M-1} s(m-k)s(m-i) = \sum_{m=0}^{M-1} s(m)s(m-i) \ . \tag{9.14}$$

Let

$$\rho(k,i) = \sum_{m=0}^{M-1} s(m-k)s(m-i)$$

then

$$\rho(0,i) = \sum_{m=0}^{M-1} s(m)s(m-i)$$

and, therefore, (9.14) becomes

$$\sum_{k=1}^{p} \alpha_k \rho(k,i) = \rho(0,i) \ . \tag{9.15}$$

The equations given by (9.15) are also known as the **normal equations**. Since we can use values of $s(m)$ in the interval $-p \le m \le M - 1$,

$$\rho(k,i) = \sum_{m=0}^{M-1} s(m-k)s(m-i)$$

$$= \sum_{n=-i}^{M-1-i} s(n)s(n+i-k) \ , \qquad \text{with} \quad n = m-i,$$

for $1 \leq i \leq p$ and $0 \leq k \leq p$. The **covariance matrix**

$$
\begin{bmatrix}
\rho(1,1) & \rho(1,2) & \cdots & \rho(1,p) \\
\rho(2,1) & \rho(2,2) & \cdots & \rho(2,p) \\
\vdots & \vdots & \vdots & \vdots \\
\rho(p,1) & \rho(p,2) & \cdots & \rho(p,p)
\end{bmatrix} ,
$$

is symmetric, since $\rho(k,i) = \rho(i,k)$, but it is not Toeplitz. An efficient, recursive method for solving Equation (9.15) is the **Cholesky decomposition algorithm** (see Rabiner and Schafer [27]):

1. *(For $i = 1, \ldots, p$)*

$$
\rho^{(1)}(i,k) = \rho(i,k) \qquad\qquad k = i, \ldots, p
$$

2. *(For $k = 1, \ldots, p$)*

$$
w(k,k) = \sqrt{\rho^{(k)}(k,k)}
$$
$$
w(k,n) = \rho^{(k)}(k,n)/w(k,k) \qquad n = k+1, \ldots, p
$$

For $n = k+1, \ldots, p$

$$
\rho^{(k+1)}(n,i) = \rho^{(k)}(n,i) - w(k,i)w(k,n) \quad i = n, \ldots, p
$$

3. *(For $k = 1, \ldots, p$)*

$$
a_k = \left(\rho(0,k) - \sum_{i=1}^{k-1} a_i w(i,k) \right) / w(k,k)
$$

4.

$$
\alpha_p = a_p / w(p,p)
$$
$$
\alpha_k = \left(a_k - \sum_{i=k+1}^{p} \alpha_i w(k,i) \right) / w(k,k) \quad k = p-1, \ldots, 1.
$$

A MATLAB function m-file for this algorithm is presented in Appendix B.2. For the Cholesky decomposition algorithm there is no guarantee that the linear predictive filter is stable, as was the case for the autocorrelation method.

Example 9.3
Show that the Cholesky decomposition algorithm, for solving the normal equations, gives the exact values for the predictor coefficients of a second order predictor with covariance matrix

$$
\begin{bmatrix}
10 & 4 \\
4 & 7
\end{bmatrix}
$$

and given $\rho(0,1) = 4$, $\rho(0,2) = 0$.

We require to solve

$$\rho(1,1)\alpha_1 + \rho(2,1)\alpha_2 = \rho(0,1)$$
$$\rho(1,2)\alpha_1 + \rho(2,2)\alpha_2 = \rho(0,2) \ ,$$

that is

$$10\alpha_1 + 4\alpha_2 = 4$$
$$4\alpha_1 + 7\alpha_2 = 0 \ .$$

This system of equations has the exact solution $\alpha_1 = \frac{14}{27}$, $\alpha_2 = -\frac{8}{27}$. Using the Cholesky decomposition algorithm, with $p = 2$, we obtain:

Step 1: $i = 1$, $\rho^{(1)}(1,1) = \rho(1,1) = 10$
$\rho^{(1)}(1,2) = \rho(1,2) = 4$

$\quad\quad\quad i = 2$, $\rho^{(1)}(2,2) = \rho(2,2) = 7$

Step 2: $k = 1$, $w(1,1) = \sqrt{\rho^{(1)}(1,1)} = \sqrt{10}$
$w(1,2) = \rho^{(1)}(1,2)/w(1,1) = \frac{4}{\sqrt{10}}$

$\quad\quad\quad n = 2$, $\rho^{(2)}(2,2) = \rho^{(1)}(2,2) - w^2(1,2) = \frac{27}{5}$

$\quad\quad\quad k = 2$, $w(2,2) = \sqrt{\rho^{(2)}(2,2)} = \sqrt{\frac{27}{5}}$

Step 3: $k = 1$, $a_1 = \rho(0,1)/w(1,1) = \frac{4}{\sqrt{10}}$

$\quad\quad\quad k = 2$, $a_2 = [\rho(0,2) - a_1 w(1,2)]/w(2,2) = -\frac{8}{3\sqrt{15}}$

Step 4: $\alpha_2 = a_2/w(2,2) = -\frac{8}{27}$
$\alpha_1 = (a_1 - w(1,2)\alpha_2)/w(1,1) = \frac{14}{27}$.

$\square$

As indicated in §9.2 for speech production, an all-pole (AR) model is an appropriate model for non-nasal voiced sounds. The composite effects of lip radiation, vocal tract and glottal excitation can be represented by a time-varying digital filter, which can be approximated by a linear time-invariant system over short intervals of time whose system function has the form:

$$H(z) = \frac{G_s}{1 - \sum_{k=1}^{p} a_k z^{-k}} \ ,$$

where G_s is the speech gain parameter. This model gives a good representation of speech sounds provided p is large enough. The filter is assumed to be excited by a train of impulses:

$$u(m) = \sum_{n=n_1}^{n_2} \delta(m-n) = \begin{cases} 1, & \text{if } n_1 \leq m \leq n_2 , \\ 0, & \text{otherwise} , \end{cases}$$

for voiced speech. For this filter, the speech samples $\{s(m)\}$ are related to the excitation sequence $\{u(m)\}$ by the difference equation

$$s(m) = G_s u(m) + \sum_{k=1}^{p} a_k s(m-k) \ . \tag{9.16}$$

Using a linear predictor:

$$\tilde{s}(m) = \sum_{k=1}^{p} \alpha_k s(m-k)$$

for this system, the prediction error is given by

$$e(m) = s(m) - \sum_{k=1}^{p} \alpha_k s(m-k) \ .$$

A rough rule of thumb, as described in Rabiner and Schafer [27] and Parsons [25], is that the order of predictor satisfies

$$p = f_s/1000 + q,$$

where f_s is the sampling frequency in hertz and q is typically 2 or 3.

If the speech signal fits the model (9.16) exactly and if $\alpha_k = a_k$, then $e(m) = G_s u(m)$, that is for voiced speech, $\{e(m)\}$ would consist of a train of impulses. This suggests a method for the determination of the pitch period of a voiced speech segment using the method of linear prediction. First, the autocorrelation or covariance method may be used to find the predictor coefficients. The error signal $\{e(m)\}$ is then determined and the difference between consecutive peaks in r_{ee} is an estimate of the pitch period. To illustrate this, the autocorrelation method, using Durbin's algorithm and with 20 prediction parameters, is performed on the same speech data used in §9.3. Graphs of the error signal and the autocorrelation function for the error signal are illustrated in Figure 9.7 and Figure 9.8, respectively. The graph,

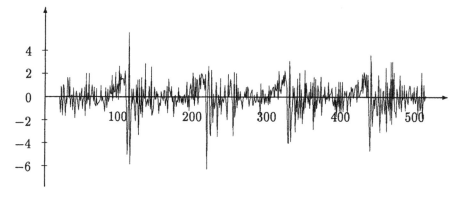

Figure 9.7: Error signal.

shown in Figure 9.8, has dominant peaks at approximately 110, 220 and 330 which confirms that the pitch period is approximately 8.4 msec. (see §9.3).

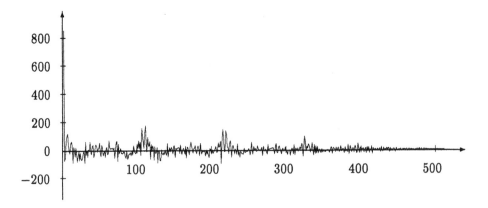

Figure 9.8: Autocorrelation function for the error signal.

9.5   ESTIMATION OF THE SPEECH GAIN PARAMETER

Suppose the excitation signal is a unit impulse at $m = 0$ and the interval for analysis is approximately the length of a pitch period, then the total energy in the error signal is G_s^2 and (9.16) holds with $u(m) = \delta(m)$. For voiced speech, since the input may be approximated by a train of impulses, then, when the interval is approximately the length of a pitch period and assuming $\alpha_k = a_k$ for all k, the impulse response sequence for the filter, $\{h(m)\}$, is described by the relationship

$$h(m) = G_s\delta(m) + \sum_{k=1}^{p} \alpha_k h(m - k) \; , \qquad (9.17)$$

and $h(m) \approx s(m)$ in the analysis interval. Assuming the filter is causal, the short-time autocorrelation function for $h(m)$ can be approximated by

$$\tilde{r}(m) = \sum_{n=0}^{M-1-m} h(n)h(m + n) \; , \quad m = 0,\, 1, \ldots \qquad (9.18)$$

$$\approx \sum_{n=0}^{M-1-m} s(n)s(m + n)$$

$$= r(m) \; , \quad m = 0,\, 1, \ldots,\, M - 1 \; . \qquad (9.19)$$

In addition, using (9.17) and (9.18),

$$\tilde{r}(0) = \sum_{n=0}^{M-1} h^2(n)$$

$$= \sum_{n=0}^{M-1} h(n) \left[G_s\delta(n) + \sum_{k=1}^{p} \alpha_k h(n - k) \right]$$

$$= G_s h(0) + \sum_{k=1}^{p} \alpha_k \sum_{n=0}^{M-1} h(n)h(n - k)$$

$$= G_s h(0) + \sum_{k=1}^{p} \alpha_k \sum_{i=-k}^{M-1-k} h(i)h(i+k) , \quad \text{with } i = n-k,$$

$$= G_s h(0) + \sum_{k=1}^{p} \alpha_k \sum_{i=0}^{M-1-k} h(i)h(i+k) , \quad \text{since } h(i) = 0 \text{ for } i < 0,$$

$$= G_s^2 + \sum_{k=1}^{p} \alpha_k \tilde{r}(k) , \quad \text{since } h(0) = G_s .$$

By matching the energy in the signal, $r(0)$, with the energy in the predicted samples, $\tilde{r}(0)$, and utilizing (9.19), the speech gain parameter can be estimated from

$$G_s^2 \approx r(0) - \sum_{k=1}^{p} \alpha_k r(k) . \tag{9.20}$$

9.6   FORMANT ESTIMATION

As stated earlier, a simple model for voiced speech is excitations produced by the vocal chords and filtered by the vocal tract. For an adult male, the resonances produced by the vocal tract are approximately 1 kHz apart. However, the location of these resonances (or formants) can be changed by moving the tongue and lips. Normally, only the first three formants are of importance (in practice). Formant estimation is often performed using frequency-domain analysis. The segment of speech, obtained by windowing, must be long enough that the individual harmonics can be resolved, but short enough that the speech signal can assumed to be modelled by a linear time-invariant system. Normally, a window is chosen to be at least two or three pitch periods long. The pitch period varies from a minimum of approximately 80 Hz in adult males to a maximum of approximately 400 Hz in adult females. For example, three periods of 100 Hz voiced speech is approximately 30 msec. The signal sampling frequency is usually in the range $8 - 15$ kHz, and the number of data samples is chosen to be a power of two. In the spectrum, the narrow peaks are the harmonics of the excitation, whilst the peaks in the envelope of the peaks in the spectrum indicate the formants due to the vocal tract. To understand more clearly what is meant by an envelope, study the dashed-curve illustrated in the diagram of Figure 9.10.

Linear prediction may be used to determine the spectrum. Assuming an all-pole model,

$$H(z) = \frac{G_s}{1 - \displaystyle\sum_{k=1}^{p} a_k z^{-k}} ,$$

the model parameters can be estimated using the method of linear prediction and the speech gain parameter G_s can be estimated as indicated in §9.5. Hence, the spectrum can be obtained by evaluating $H(z)$ on the unit circle $z = e^{j\theta}$. One method of achieving this is to construct a sequence $\{\tilde{a}\}$ which is composed of the sequence $\{1, -a_1, -a_2, \ldots, -a_p\}$ padded with zeros so that the length of the sequence, say

N, is a power of two, that is $\{\tilde{a}\} = \{1, -a_1, -a_2, \ldots, -a_p, 0, 0, 0, \ldots\}$, and divide G_s by each element of the DFT of the sequence $\{\tilde{a}\}$ to produce the sequence

$$\left\{ G_s \left[1 - \sum_{k=1}^{p} a_k e^{-j\frac{2\pi kn}{N}} \right]^{-1} , \quad n = 0, 1, \ldots, N-1 \right\}.$$

An FFT algorithm can be used to determine the DFT of the sequence $\{\tilde{a}\}$. For sufficiently large N, the magnitude of the resulting sequence yields a high-resolution representation of the vocal tract amplitude spectrum

$$|H(e^{j\theta})| = \frac{G_s}{\left| 1 - \sum_{k=1}^{p} a_k e^{-j\theta k} \right|}.$$

The formants are estimated by detecting the peaks in the spectrum envelope.

To illustrate the above procedure, we apply the technique to the voiced speech data used in §9.3, using a 512-point FFT algorithm. In addition, to identify the formants more easily, the graph of the logarithm of the vocal tract amplitude spectrum is plotted. The graph of $20 \log_{10} |H(e^{j\theta})|$, with units measured in decibels, plotted against the frequency index n is shown in Figure 9.9. The peak at approximately

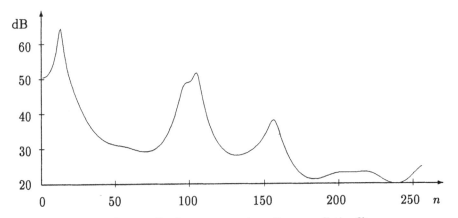

Figure 9.9: Log-amplitude spectrum using a linear predictive filter.

$n = 16$ corresponds to the first formant, denoted by F1. To obtain the frequency in Hertz, corresponding to values of the frequency index n, we recall from §6.6 that samples are $2\pi/(NT)$rads/sec. $= (NT)^{-1}$Hz apart, and so

$$f = \frac{n}{NT},$$

where T is the sampling period and N denotes the number of data values used for the FFT. Thus, since $T = 0.076$ msec. and $N = 512$, F1 has a frequency of approximately 386 Hz. Two other formants can easily be identified in Figure 9.9, and their frequencies are calculated to be approximately 2699 Hz and 4048 Hz.

The identification and estimation of formants for voiced speech can be made easier by *pre-emphasizing* the speech signal using a filter with impulse response $\delta(k) - a\delta(k-1)$, with $a \approx 1$ (see Deller *et al* [5]; Chapter 5). Such a filter has similar characteristics to those filters used to model lip radiation. If two formants are close to each other, the peaks in the log-amplitude spectrum can be accentuated by evaluating the spectrum on a circle with radius $\rho < 1$. This modification can be achieved quite easily by premultiplying the linear prediction parameters by ρ before computing the DFT. For more details see, for example, Deller *et al* [5] and Parsons [25].

The graphs of the logarithm of the spectrum of the speech signal and its envelope are shown in Figure 9.10. Comparing the graph of the envelope of the logarithm of the spectrum with Figure 9.9, we see that the peaks in the envelope correspond, approximately, to the peaks in the vocal tract amplitude spectrum and, therefore, formants can be found using Figure 9.10. The **discrete power spectrum** is also

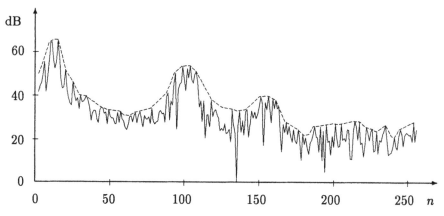

Figure 9.10: Log-amplitude spectrum and envelope (shown dashed) for the speech segment.

useful in locating formants. The power spectrum of a sequence, which is a function of frequency, is the DTFT of the autocorrelation function for the given sequence (see Stearns and Hush [29] for a more detailed discussion). A graph of the logarithm of the magnitude of the power spectrum for the zero-mean segment of speech, considered in §9.3, is shown in Figure 9.11. Comparing this graph with those illustrated in Figures 9.9 and 9.10, we notice that the two sets of peaks in the graph of the log-magnitude of the power spectrum correspond, approximately, to the first two peaks in Figure 9.9 and the first two peaks in the envelope illustrated in Figure 9.10. Thus, Figure 9.11 can also be used to identify the formant frequencies of the vocal tract spectrum.

9.7 LATTICE FILTER FORMULATION

In the ith stage of Durbin's algorithm, let $\{a_n^{(i)},\ n = 1, 2, \ldots,\ i\}$ denote the coefficients of the ith order linear predictor. The prediction error, in this case, is

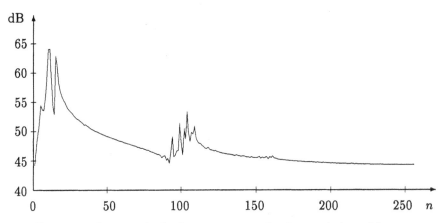

Figure 9.11: Log-magnitude of power spectrum for the speech data with zero mean.

then

$$f^{(i)}(m) = s(m) - \sum_{k=1}^{i} a_k^{(i)} s(m-k) . \qquad (9.21)$$

This can be interpreted as predicting $s(m)$ using the past i samples $\{s(m-k), \ k = 1, 2, \ldots, i\}$. Since (9.21) represents a *forward* prediction using an ith order predictor, it is generally known as a **forward prediction error sequence**. Suppose, now, the i future samples $\{s(m-i+k), \ k = 0, \ldots, i-1\}$ are used to predict $s(m-i)$. The sequence

$$b^{(i)}(m) = s(m-i) - \sum_{k=1}^{i} a_k^{(i)} s(m+k-i) , \qquad (9.22)$$

associated with the prediction of $s(m-i)$, is usually known as the **backward prediction error sequence**. Unfortunately, in this text, there is not sufficient space to elaborate on why this sequence does in fact represent an *error sequence*. For the reader interested in this particular aspect, it is suggested that the texts by Deller *et al* [5], Haddad and Parsons [8] and Rabiner and Schafer [27] be consulted. The concepts of forward and backward prediction are illustrated in Figure 9.12. It can be shown that (9.21) and (9.22) may be rewritten in the form:

$$f^{(i)}(m) = f^{(i-1)}(m) - k_i b^{(i-1)}(m-1) \qquad (9.23)$$
$$b^{(i)}(m) = b^{(i-1)}(m-1) - k_i f^{(i-1)}(m) . \qquad (9.24)$$

Initially, $f^{(0)}(m) = b^{(0)}(m) \overset{\text{def}}{=} s(m)$. The formulation (9.23)–(9.24) is known as a **lattice filter**, which can be illustrated diagrammatically as a lattice (as shown in Figure 9.13, where D represents delay), and the coefficients k_i are known as **reflection** or **PARCOR** (partial correlation) **coefficients**. Lattice filters are used in many signal processing applications and are discussed in many signal processing texts, for example, Haddad and Parsons [8].

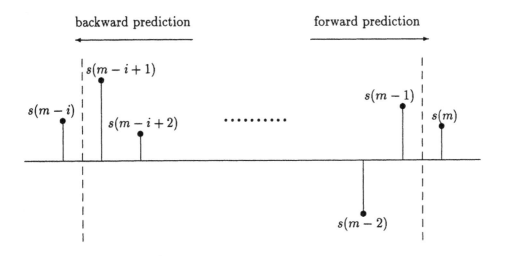

Figure 9.12: Forward and backward predictions.

This structure is known to be stable provided $|k_i| \leq 1$. Using this structure $e(m) = f^{(p)}(m)$. In particular, Burg used this lattice formulation, to develop a

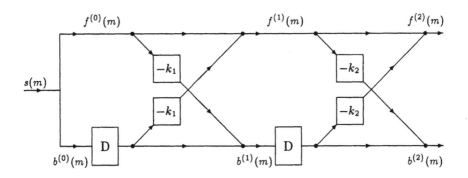

Figure 9.13: Lattice structure.

procedure based upon minimizing the sum of the mean square forward and backward prediction errors, that is Burg's method minimized

$$\sum_{m=0}^{M-1} [\{f^{(i)}(m)\}^2 + \{b^{(i)}(m)\}^2] .$$

In this case, the reflection (or PARCOR) coefficients are given by

$$k_i = \frac{2 \sum\limits_{m=i}^{M-1} f^{(i-1)}(m) b^{(i-1)}(m-1)}{\sum\limits_{m=i}^{M-1} [\{f^{(i-1)}(m)\}^2 + \{b^{(i-1)}(m-1)\}^2]} .$$

Burg's algorithm, sometimes known as the **maximum entropy method** (see Parsons [25]), is as follows:

1. *Initialization*:

$$f^{(0)}(m) = b^{(0)}(m) = s(m), \qquad m = 0, \ldots, M-1$$

$$k_1 = \frac{2 \sum\limits_{m=1}^{M-1} f^{(0)}(m) b^{(0)}(m-1)}{\sum\limits_{m=1}^{M-1} [\{f^{(0)}(m)\}^2 + \{b^{(0)}(m-1)\}^2]}$$

$$\alpha_1^{(1)} = k_1$$

2. *(For* $1 \le i \le p-1$)

$$\left. \begin{array}{rcl} f^{(i)}(m) &=& f^{(i-1)}(m) - k_i b^{(i-1)}(m-1) \\ b^{(i)}(m) &=& b^{(i-1)}(m-1) - k_i f^{(i-1)}(m) \end{array} \right\} \qquad i \le m \le M-1$$

$$k_{i+1} = \frac{2 \sum\limits_{m=i+1}^{M-1} f^{(i)}(m) b^{(i)}(m-1)}{\sum\limits_{m=i+1}^{M-1} [\{f^{(i)}(m)\}^2 + \{b^{(i)}(m-1)\}^2]}$$

$$\alpha_{i+1}^{(i+1)} = k_{i+1}$$

$$\alpha_n^{(i+1)} = \alpha_n^{(i)} - k_{i+1} \alpha_{i+1-n}^{(i)}, \qquad 1 \le n \le i .$$

As in Durbin's method, the predictor coefficients are given by $\alpha_k^{(p)}, \quad k = 1, \ldots, p$.

Example 9.4
Given the data set $\{1, 2, 2, 1\}$, apply Burg's algorithm to determine the predictor coefficients for a second order predictor and compare these coefficients with those obtained using the autocorrelation method.

Step 1: With $s(0) = 1$, $s(1) = 2$, $s(2) = 2$, $s(3) = 1$,

$$f^{(0)}(0) = b^{(0)}(0) = 1 \qquad f^{(0)}(1) = b^{(0)}(1) = 2$$

$$f^{(0)}(2) = b^{(0)}(2) = 2 \qquad f^{(0)}(3) = b^{(0)}(3) = 1$$

$$k_1 = \frac{2(f^{(0)}(1)b^{(0)}(0) + f^{(0)}(2)b^{(0)}(1) + f^{(0)}(3)b^{(0)}(2))}{([f^{(0)}(1)]^2 + [f^{(0)}(2)]^2 + [f^{(0)}(3)]^2) + ([b^{(0)}(0)]^2 + [b^{(0)}(1)]^2 + [b^{(0)}(2)]^2)} = \frac{8}{9}$$

and so $\alpha_1^{(1)} = k_1 = \frac{8}{9}$.

Step 2: $i = 1$

$$f^{(1)}(1) = f^{(0)}(1) - k_1 b^{(0)}(0) = \frac{10}{9} \qquad b^{(1)}(1) = b^{(0)}(0) - k_1 f^{(0)}(1) = -\frac{7}{9}$$
$$f^{(1)}(2) = f^{(0)}(2) - k_1 b^{(0)}(1) = \frac{2}{9} \qquad b^{(1)}(2) = b^{(0)}(1) - k_1 f^{(0)}(2) = \frac{2}{9}$$
$$f^{(1)}(3) = f^{(0)}(3) - k_1 b^{(0)}(2) = -\frac{7}{9} \qquad b^{(1)}(3) = b^{(0)}(2) - k_1 f^{(0)}(3) = \frac{10}{9}$$

$$k_2 = \frac{2(f^{(1)}(2)b^{(1)}(1) + f^{(1)}(3)b^{(1)}(2))}{([f^{(1)}(2)]^2 + [f^{(1)}(3)]^2) + ([b^{(1)}(1)]^2 + [b^{(1)}(2)]^2)} = -\frac{28}{53}$$

and so $\alpha_2^{(2)} = k_2 = -\frac{28}{53}$. Finally, $\alpha_1^{(2)} = \alpha_1^{(1)} - k_2\alpha_1^{(1)} = \frac{72}{53}$.

Hence, the predictor coefficients are $\alpha_1^{(2)} = \frac{72}{53} \approx 1.3585$ and $\alpha_2^{(2)} = -\frac{28}{53} \approx -0.5283$.

Using the autocorrelation method,

$$r(0) = \sum_{m=0}^{3} s^2(m) = 10, \; r(1) = \sum_{m=0}^{2} s(m)s(m+1) = 8, \; r(2) = \sum_{m=0}^{1} s(m)s(m+2) = 4 \,.$$

Solving
$$\begin{aligned} 10\alpha_1 &+ 8\alpha_2 = 8 \\ 8\alpha_1 &+ 10\alpha_2 = 4 \end{aligned}$$

gives $\alpha_1 = \frac{4}{3} \approx 1.3333$ and $\alpha_2 = -\frac{2}{3} \approx -0.6667$.

$\square$

Remarks:

(a) Lattice methods do not give exact solutions to the normal equations of §9.4 defined by (9.13).

(b) The Burg algorithm uses only available samples with no assumptions about the unavailable data, whereas Durbin's method uses only available samples but assumes that unavailable data samples are zero.

(c) For 'small' sets of data, the Durbin algorithm has larger mean square errors and larger prediction errors than the Burg algorithm. However, for 'large' data sets, where end effects contribute little to the final results, the performance between the two algorithms is similar (see Giordano and Hsu [7] and, also, Parsons [25]).

(d) A MATLAB function m-file for the Burg lattice algorithm is presented in Appendix B.3.

Details of fast algorithms, including lattice methods, for autoregressive spectral estimation can be found in the text by Marple [19].

9.8   THE CEPSTRUM

Often signals are combined using operations of multiplication and convolution. The system models, representing the above operations, are examples of a class of systems called **homomorphic systems**, which are discussed in the next section (§9.9). It is shown that the output signal from a homomorphic system, for which the response of the system is a convolution of input signals, can be deconvolved using a process that involves determination of a function known as the **cepstrum** (this terminology is derived from the word 'spectrum').

Let $x(k)$ denote the convolution of two real signals $x_1(k)$ and $x_2(k)$, that is $x(k) = (x_1 * x_2)(k)$. Then, provided the corresponding z-transforms exist,

$$X(z) = X_1(z)X_2(z) \ .$$

Remark: In this section and subsequent sections, all references to z-transforms implicitly imply the use of the bilateral z-transform.

Using the complex logarithm, the product term can be separated to give

$$\log_e(X(z)) = \log_e(X_1(z)) + \log_e(X_2(z)) \ ,$$

where $\log_e(X(z)) = \log_e|X(z)| + j \arg(X(z))$. If the corresponding inverse z-transforms exist, the resulting sequence takes the form:

$$\{c_x^{(c)}(k)\} \overset{\text{def}}{=} \mathcal{Z}^{-1}[\log_e(X(z))] \ ,$$

where $c_x^{(c)}(k) = c_{x_1}^{(c)}(k) + c_{x_2}^{(c)}(k)$ and the superscript $^{(c)}$ notation denotes that the *complex* logarithm has been used to generate the sequence. Here, $c_x^{(c)}(k)$ is known as the **complex cepstrum** of $x(k)$. If $\log_e(X(z))$ has a convergent power series representation, with region of convergence

$$\{z \in \mathbb{C} : r_1 < |z| < r_2, \ 0 < r_1 < 1 \text{ and } r_2 > 1\} \ ,$$

then it can be shown (for more details see Oppenheim and Schafer [21]) that

$$\{c_x^{(c)}(k)\} = \frac{1}{2\pi} \int_{-\pi}^{\pi} \log_e(X(e^{j\,\theta}))e^{j\,\theta k} \, d\theta \ , \tag{9.25}$$

which can be identified as an IDTFT as shown in §7.3, Equation (7.13b). With reference to (9.25), adding a multiple of 2π to $\arg X(z)$ does not produce a different sequence $\{c_x^{(c)}(k)\}$. However, if $\arg X(e^{j\,\theta})$ is not restricted to its principal values and it is assumed to be a continuous function of θ, then, for each value of θ, the sequence $\{c_x^{(c)}(k)\}$ is unique. If the DFT is used for the computation of the Fourier spectrum of $\{x(k)\}$, then sample values of $\arg X(e^{j\,\theta})$ contain only principal values (that is values between $-\pi$ and π). These are known as **wrapped** values. For continuity of $\arg X(e^{j\,\theta})$, the phase of $X(e^{j\,\theta})$ must be **unwrapped** before $\{c_x^{(c)}(k)\}$ can be computed. More details can be found in the texts by Oppenheim and Schafer [21] and Rabiner and Schafer [27].

Before presenting some properties of the complex cepstrum, we introduce the concept of a stable sequence and definitions of minimum-phase and maximum-phase sequences.

9.8.1 Stable sequences

Recall from Chapter 6, §6.2, that a system is (BIBO) stable if its impulse response sequence $\{y^\delta(k)\}$ decays to zero, which implies that its impulse response is absolutely summable, i.e. $\sum_{k=0}^{\infty} |y^\delta(k)| < \infty$. This motivates the definition of a **stable sequence** $\{x(k) : k \in \mathbb{Z}\}$ as *a sequence which is absolutely summable, that is*

$$\sum_{k=-\infty}^{\infty} |x(k)| < \infty.$$

Consider a sequence $\{x\}$ with z-transform $X(z)$. Depending on the poles of $X(z)$, there may be several regions of convergence, that is regions for which the power series defining $X(z)$ converges. For example, the three sequences

$$(i)\ x(k) = \begin{cases} 0, & k \geq 0 \\ \frac{2}{5}\left[(-2)^k - (\frac{1}{2})^k\right], & k < 0 \end{cases} \qquad (ii)\ x(k) = \begin{cases} \frac{2}{5}(\frac{1}{2})^k, & k \geq 0 \\ \frac{2}{5}(-2)^k, & k < 0 \end{cases}$$

$$(iii)\ x(k) = \begin{cases} \frac{2}{5}\left[(\frac{1}{2})^k - (-2)^k\right], & k \geq 0 \\ 0, & k < 0 \end{cases}$$

have z-transform $X(z) = \dfrac{z}{(z - \frac{1}{2})(z + 2)}$, but with different regions of convergence:

$(i)\ \{z \in \mathbb{C} : |z| < \frac{1}{2}\}$ $\qquad (ii)\ \{z \in \mathbb{C} : \frac{1}{2} < |z| < 2\}$ $\qquad (iii)\ \{z \in \mathbb{C} : |z| > 2\}$,

respectively. Since the absolute summability condition is equivalent to the condition

$$\sum_{k=-\infty}^{\infty} |x(k)z^{-k}| < \infty \quad \text{with } |z| = 1,$$

the sequence $\{x\}$ is stable if the region of convergence of $X(z)$ includes the unit circle. Thus, for the above example, with regions $(i)\ \{z \in \mathbb{C} : |z| < \frac{1}{2}\}$ and (iii) $\{z \in \mathbb{C} : |z| > 2\}$, the corresponding sequence is unstable. However, the sequence associated with $X(z)$ and region of convergence $\{z \in \mathbb{C} : \frac{1}{2} < |z| < 2\}$ is stable.

Note that a causal sequence $\{x\}$ always has a region of convergence of the form $\{z \in \mathbb{C} : |z| > r_p\}$, where $r_p \geq |z_p|$ and $|z_p|$ is the greatest magnitude of the poles of $X(z)$. Therefore, if a sequence $\{x\}$ is both causal and stable, then all the poles of $X(z)$ must lie inside the unit circle of the z-plane.

9.8.2 Minimum-phase and maximum-phase sequences

As stated in §9.8.1, if a real sequence $\{x\}$ is causal and stable, then all the poles of $X(z)$ must lie inside the unit circle. For some problems it may be desirable that the sequence with z-transform $[X(z)]^{-1}$ is also stable and causal. Since the poles of $[X(z)]^{-1}$ are the zeros of $X(z)$, this implies that the zeros of $X(z)$ must also lie inside the unit circle. A sequence with this property is known as a **minimum-phase sequence**, that is *a minimum-phase sequence is a causal sequence whose z-transform is such that its poles and zeros are inside the unit circle in the z-plane.* An explanation of the derivation of the name 'minimum-phase' is given in

Oppenheim and Schafer [21]. An example of a minimum-phase sequence is given by the unilateral z-transform $X(z) = (1 - \frac{1}{4}z^{-1})(1 + \frac{1}{2}z^{-1})$. In this case $X(z)$ has poles $z = 0$, 0 and zeros $z = \frac{1}{4}$, $-\frac{1}{2}$. The sequence $\{x\}$ is specified by $x(0) = 1$, $x(1) = \frac{1}{4}$, $x(2) = -\frac{1}{8}$, $x(k) = 0$ otherwise, which can be expressed in the form

$$x(k) = \delta(k) + \left(-\frac{1}{2}\right)^{k+1} [\zeta(k-1) - \zeta(k-3)] .$$

Clearly, the sequence is causal and stable.

A **maximum-phase sequence** is *a stable sequence whose z-transform is such that its poles and zeros are outside the unit circle in the z-plane.* Such a sequence is **anticausal**, that is a sequence $\{x\}$ with the property that $x(k) = 0$ for all $k > 0$. As an example, consider the sequence defined by

$$x(k) = 25\delta(k) + 4(4k+9)[\zeta(-k-1) - \zeta(-k-3)] ,$$

i.e. $x(-2) = 4$, $x(-1) = 20$, $x(0) = 25$, $x(k) = 0$ otherwise. This sequence is stable and, since $X(z) = (5 + 2z)^2$ has zeros $z = -\frac{5}{2}$, $-\frac{5}{2}$ and no poles, is a maximum-phase sequence.

Properties of the complex cepstrum

Consider stable sequences with rational z-transforms and no poles or zeros on the unit circle. For these sequences, the following properties hold (for a detailed analysis, see, for example, Oppenheim and Schafer [21]).

1. The complex cepstrum of a real sequence is a real sequence.

2. $\{c_x^{(c)}(k)\}$ is of infinite duration, even if $\{x(k)\}$ has finite duration.

3. $c_x^{(c)}(k)$ decays at least as fast as $\dfrac{1}{|k|}$.

4. $\{c_x^{(c)}(k)\}$ is a causal sequence if and only if $\{x(k)\}$ is minimum-phase.

5. $c_x^{(c)}(k) = 0$ for $k > 0$ if and only if $\{x(k)\}$ is a maximum-phase sequence.

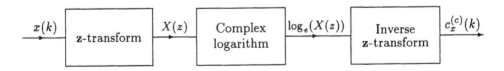

Figure 9.14: Process to obtain the complex cepstrum.

The processes required for the formation of $c_x^{(c)}(k)$ are illustrated diagrammatically in Figure 9.14 or, more simply as shown in Figure 9.15, where C_* is known as the **characteristic system.**

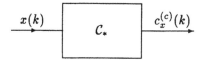

Figure 9.15: Characteristic system.

In some applications, the phase of $X(e^{j\theta})$ is unimportant, in which case the **real cepstrum**, $c_x^{(r)}(k)$, defined by

$$\{c_x^{(r)}(k)\} \stackrel{\text{def}}{=} \frac{1}{2\pi} \int_{-\pi}^{\pi} \log_e |X(e^{j\theta})| e^{j\theta k} \, d\theta \;,$$

is more useful. The relationship between $c_x^{(c)}(k)$ and $c_x^{(r)}(k)$ is given by

$$c_x^{(r)}(k) = \frac{1}{2}[c_x^{(c)}(k) + c_x^{(c)}(-k)] \;. \tag{9.26}$$

For real signals, $c_x^{(r)}(k)$ is the even part of $c_x^{(c)}(k)$ and, hence, $c_x^{(r)}(k)$ is an even function.

9.9 HOMOMORPHIC SYSTEMS AND DECONVOLUTION

Homomorphic systems are systems, with input-output transformation $\mathcal{T}$, that satisfy a generalized version of an application of the principle of superposition (see §1.4) in the form

$$\mathcal{T}\{u(k) \diamond v(k)\} = \mathcal{T}\{u(k)\} \circ \mathcal{T}\{v(k)\} \;,$$

where $\diamond$ and $\circ$ represent rules for combining inputs and outputs, respectively. It is known that such systems can be represented by algebraically linear (homomorphic) mappings between input and output signal spaces; for a lucid explanation of homomorphic systems, see Oppenheim and Schafer [21].

As discussed in the previous section, if the input to the characteristic system is the convolution $(x_1 * x_2)(k)$, then the output is $c_x^{(c)}(k) = c_{x_1}^{(c)}(k) + c_{x_2}^{(c)}(k)$. This is an example of a homomorphic system, where $\diamond$ represents the convolution operation and $\circ$ represents addition. This property of the characteristic system may be utilized in an application to **deconvolution** of a signal, namely deconvolving (or separating) the signal into its constitutive components.

If the sequences $\{c_{x_1}^{(c)}(k)\}$ and $\{c_{x_2}^{(c)}(k)\}$ have non-overlapping regions of support, that is $c_{x_1}^{(c)}(k) = 0$ for all k satisfying $c_{x_2}^{(c)}(k) \neq 0$, then a linear filter, $\{\ell(k)\}$, can be constructed so that

$$\ell(k)c_x^{(c)}(k) = c_{x_1}^{(c)}(k) \qquad [\text{ or, if desired, } \quad c_{x_2}^{(c)}(k)] \;.$$

Processing $c_{x_1}^{(c)}(k)$ (or $c_{x_2}^{(c)}(k)$) through the inverse characteristic system $\mathcal{C}_*^{-1}$, we are then able to recover $x_1(k)$ (or $x_2(k)$) from $x(k)$. Such filters are known as linear,

frequency-invariant filters. For example, if it is known that $x_1(k)$ is a minimum-phase sequence and $x_2(k)$ is a maximum-phase sequence with $x_2(0) = 0$, then a linear, frequency-invariant filter of the form:

$$\ell(k) = \zeta(-k-1) = \begin{cases} 1, & k < 0 \\ 0, & k \geq 0 \end{cases}$$

would remove $c_{x_1}^{(c)}(k)$ from $c_x^{(c)}(k)$, leaving $c_{x_2}^{(c)}(k)$. Hence, $x_2(k)$ can be recovered from $x(k)$. A fuller discussion of frequency invariant filters is given in Oppenheim and Schafer [21]. Thus, homomorphic deconvolution may be represented diagramatically by Figure 9.16. For the practical implementation of $\mathcal{C}_*$ or $\mathcal{C}_*^{-1}$, the operations of z-transform and inverse z-transform are replaced by DFT and inverse DFT, respectively.

A block diagram representation of $\mathcal{C}_*^{-1}$ is shown in Figure 9.17.

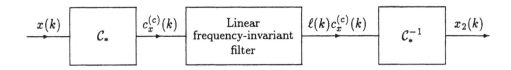

Figure 9.16: Homomorphic deconvolution.

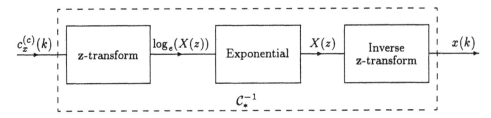

Figure 9.17: Inverse of characteristic system for homomorphic deconvolution.

9.10   CEPSTRAL ANALYSIS OF SPEECH SIGNALS

A simple model of the vocal tract system function is

$$V(z) = \frac{\displaystyle\sum_{n=0}^{N} \beta_n z^{-n}}{\displaystyle\sum_{m=0}^{M} \alpha_m z^{-m}}.$$

Since both numerator and denominator can be expressed as a product of linear factors, then, excluding poles and zeros on the unit circle, $V(z)$ may be rewritten in the form:

$$V(z) = \frac{Az^{-k_0} \prod\limits_{n=1}^{N_1} (1 - a_n z^{-1}) \prod\limits_{n=1}^{N_2} (1 - b_n z)}{\prod\limits_{m=1}^{M_1} (1 - c_m z^{-1}) \prod\limits_{m=1}^{M_2} (1 - d_m z)} ,$$

where $|a_n|$, $|b_n|$, $|c_m|$, $|d_m| < 1$ for all n, m. Hence, excluding $z = 0$ and $z = \infty$, there are M_1 poles and N_1 zeros inside the unit circle and M_2 poles and N_2 zeros outside the unit circle. It is assumed that the real constant A is positive. Now

$$\log_e(V(z)) = \log_e(A) + \log_e(z^{-k_0}) + \sum_{n=1}^{N_1} \log_e(1 - a_n z^{-1}) + \sum_{n=1}^{N_2} \log_e(1 - b_n z)$$
$$- \sum_{m=1}^{M_1} \log_e(1 - c_m z^{-1}) - \sum_{m=1}^{M_2} \log_e(1 - d_m z) .$$

The term z^{-k_0} represents an advance or delay of the sequence $\{v(k)\}$ and contributes only to the imaginary part of $\log_e(V(e^{j\theta}))$. This term may be removed by either shifting the time origin of $\{v(k)\}$ or by subtracting the contribution to $\log_e(V(e^{j\theta}))$. Thus, we consider

$$\log_e(V(z)) = \log_e(A) + \sum_{n=1}^{N_1} \log_e(1 - a_n z^{-1}) + \sum_{n=1}^{N_2} \log_e(1 - b_n z)$$
$$- \sum_{m=1}^{M_1} \log_e(1 - c_m z^{-1}) - \sum_{m=1}^{M_2} \log_e(1 - d_m z) ,$$

and assume that the region of convergence contains the unit circle. Hence, using the following two z-transform (bilateral) results:

$$\mathcal{Z}\left[\left\{-\frac{\alpha^k}{k}\zeta(k-1)\right\}\right] = \log_e(1 - \alpha z^{-1}) , \quad |z| > |\alpha|, \qquad (9.27)$$

and

$$\mathcal{Z}\left[\left\{\frac{\beta^{-k}}{k}\zeta(-k-1)\right\}\right] = \log_e(1 - \beta z) , \quad |z| < |\beta|^{-1}, \qquad (9.28)$$

and, since $\{c_v^{(c)}(k)\} = \mathcal{Z}^{-1}[\log_e(V(z))]$, $c_v^{(c)}(k)$ may be expressed as

$$c_v^{(c)}(k) = \log_e(A)\delta(k) - \frac{1}{k}\sum_{n=1}^{N_1}(a_n)^k \zeta(k-1) + \frac{1}{k}\sum_{n=1}^{N_2}(b_n)^{-k}\zeta(-k-1)$$
$$+ \frac{1}{k}\sum_{m=1}^{M_1}(c_m)^k \zeta(k-1) - \frac{1}{k}\sum_{m=1}^{M_2}(d_m)^{-k}\zeta(-k-1) ,$$

or

$$c_v^{(c)}(k) = \log_e(A)\delta(k) + \frac{1}{k}\left[\sum_{m=1}^{M_1}(c_m)^k - \sum_{n=1}^{N_1}(a_n)^k\right]\zeta(k-1)$$

$$+ \frac{1}{k}\left[\sum_{n=1}^{N_2}(b_n)^{-k} - \sum_{m=1}^{M_2}(d_m)^{-k}\right]\zeta(-k-1) \ .$$

Remarks:

(a) For $k \neq 0$, each term of $c_v^{(c)}(k)$ is a product of $\frac{1}{k}$ and a decaying exponential power and, hence, decays at least as fast as $\frac{1}{|k|}$ (see Property 3 of §9.8.2).

(b) If $\{v(k)\}$ is a minimum-phase sequence (namely $b_m = d_m = 0$ for all m), then its complex cepstrum is causal (see Property 4 of §9.8.2). For voiced speech, the contribution of the vocal tract will, in general, be non-mimimum-phase and thus the complex cepstrum will be nonzero for all k.

(c) The contribution of the vocal tract response to the complex cepstrum is mainly concentrated around the origin.

To model voiced speech, a window which tapers smoothly to zero at both ends, such as a Hamming window (see §8.6.2), is used to minimize end effects due to impulses that occur before the beginning of the analysis interval. We assume that the windowed speech segment, say $\{s(k)\}$, is modelled by

$$s(k) = (v * p_w)(k) \ ,$$

where $p_w(k) = w(k)p(k)$, $\{w(k)\}$ is a window sequence that tapers smoothly to zero at both ends, and $\{p(k)\}$ is the input sequence. For voiced speech, $p(k)$ is chosen to be a train of delayed and weighted impulses. The complex cepstrum of $\{s(k)\}$ is given by

$$c_s^{(c)}(k) = c_v^{(c)}(k) + c_{p_w}^{(c)}(k)$$

and, since the complex cepstrum of the vocal tract response is mainly concentrated around the origin, a frequency-invariant filter can be used to isolate $c_v^{(c)}(k)$ or $c_{p_w}^{(c)}(k)$.

Example 9.5
Using the results:

$$\sum_{r=0}^{n}q^r = \frac{1-q^{n+1}}{1-q} \ , \quad q \neq 1 \ , \tag{9.29}$$

$$\log_e(1-\alpha z^{-1}) = -\sum_{r=1}^{\infty}\frac{\alpha^r}{r}z^{-r} \ , \quad |z| > |\alpha| \ , \tag{9.30}$$

and given $p(k)$ has the form:

$$p(k) = \delta(k) + a\delta(k-\tau) + a^2\delta(k-2\tau) + a^3\delta(k-3\tau) \ ,$$

with $0 < a < 1$ and $\tau \in \mathbb{N}$, determine $c_p^{(c)}(k)$.

Now,

$$p(k) = \sum_{i=0}^{3} a^i \delta(k - i\tau) \ .$$

Thus, invoking (9.29),

$$P(z) = \sum_{i=0}^{3} a^i z^{-i\tau}$$

$$= \frac{1 - (az^{-\tau})^4}{1 - az^{-\tau}}$$

and, therefore, utilizing (9.30) we have

$$\log_e (P(z)) = \log_e \left(1 - a^4 z^{-4\tau}\right) - \log_e \left(1 - az^{-\tau}\right)$$

$$= \sum_{r=1}^{\infty} \left(-\frac{a^{4r}}{r} z^{-4\tau r} + \frac{a^r}{r} z^{-\tau r}\right) \ .$$

Hence,

$$c_p^{(c)}(k) = \sum_{r=1}^{\infty} \frac{1}{r} \left[a^r \delta(k - \tau r) - a^{4r} \delta(k - 4\tau r)\right] \ ,$$

that is a train of impulses at multiples of τ.

□

In the above example, in the context of a speech signal, τ would represent the pitch period. When $\{p(k)\}$ consists of an impulse train with impulses spaced at intervals of τ, Oppenheim and Schafer [21] (see §12.9.1, Chapter 12) show that impulses, also spaced at intervals of τ, appear in $\{c_{p_w}^{(c)}(k)\}$. As

$$c_s^{(c)}(k) = c_v^{(c)}(k) + c_{p_w}^{(c)}(k) \ ,$$

and since the vocal tract component of the complex cepstrum decays rapidly, $c_{p_w}^{(c)}(k)$ will be the dominant contribution to $c_s^{(c)}(k)$ for k sufficiently large. Thus, peaks in the sequence $\{c_{p_w}^{(c)}(k)\}$ will surface in the sequence $\{c_s^{(c)}(k)\}$. This suggests that the pitch period can be estimated by determining the difference between two consecutive peaks in $\{c_s^{(c)}(k)\}$ for sufficiently large k. This technique is illustrated using the real cepstrum of the data used in §9.3. The real cepstrum may be determined in MATLAB using the function m-file **rceps(x)**. If this function m-file is not available in your version of the MATLAB Signal Processing Toolbox, use the MATLAB command: **y= real(ifft(log(abs(fft(x)))))**. The graph of the real cepstrum sequence is illustrated in Figure 9.18. Note that the voiced speech signal is not truly periodic and it will include some noise. Also, the cepstrum decays rapidly, as indicated by property 3 of §9.8. For these reasons, it is difficult to identify peaks in the graph of the cepstrum. One way to partly overcome this problem is to apply a frequency-invariant filter to the cepstrum in order to remove some

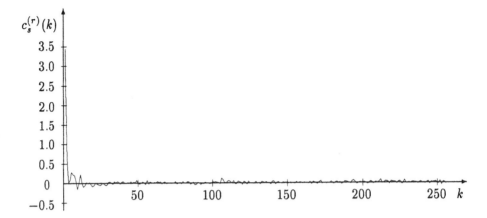

Figure 9.18: Real cepstrum incorporating vocal tract component.

of the vocal tract contribution with relatively large magnitude. Performing this technique with the frequency-invariant filter $\ell(k) = \zeta(k - 6)$, a plot of the resulting modified real cepstrum is shown in Figure 9.19. A peak occurring near 110 is more easily distinguished; however, a dominant peak in the approximate range 210-220 is still difficult to locate. As indicated above, a linear, frequency-invariant filter

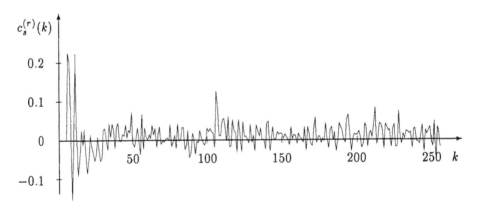

Figure 9.19: Real cepstrum with peaks corresponding to pitch period.

can be used to isolate either the vocal tract contribution or periodic component. The result is processed by the inverse characteristic system to recover the desired component.

9.11 FORMANT ESTIMATION USING CEPSTRAL ANALYSIS

A strongly periodic component in the log spectrum at a frequency T^{-1}, where T is the pitch period, become periodic peaks in the cepstrum. In cepstral analysis, the independent variable in the cepstrum has units of time and is referred to as **que-frency** (this terminology being derived from the word 'frequency'). For most voiced speech, a clear distinction can be made between the periodic excitation component in the cepstrum and the vocal tract contribution which appears as components at low values of quefrency.

A window function, that is a low quefrency **lifter (liftering** is the process of filtering in the quefrency domain), can be used to remove the high quefrency periodic excitations. The DFT of the resulting signal then gives a smoothed version of the log spectrum of the vocal tract filter. By investigating the peaks in the cepstrally smoothed log spectrum, the frequency, bandwidth and amplitude of the formants can be estimated from the spectral envelope. For more details on this technique see, for example, Deller *et al* [5] and Oppenheim and Schafer [21].

9.12 EXERCISES

1. Evaluate the autocorrelation function

$$R_{xx}(\tau) = \int_{-\infty}^{\infty} x(s)x(s+\tau)\,ds$$

for the deterministic, finite-energy signal $x(t) \stackrel{\text{def}}{=} e^{-t}\zeta(t)$.

2. Test data for non-nasal, voiced speech can be constructed by passing an impulse train through a stable filter. For example, the MATLAB commands:

```
e = eye(16,16);          % generates 16 by 16 identity matrix
x = e(5:254);            % forms impulse train
a = [1 -0.2 -0.23 -0.05 -0.12];
b = [1];
s = filter(b,a,x);       % output from the stable filter
```

produce such a signal, with 250 data values and 14 impulses every 16 units. To make the signal more realistic, additive 'noise' may be added with the MATLAB command: xs = s+k*randn(size(s)), with k suitably chosen (for example, $k = 1.2$). For the signal xs, use MATLAB to evaluate and plot xs and its autocorrelation function. From the graph, estimate the periodicity in the signal s.

3. Given the autocorrelation matrices:

$$R_1 = \begin{bmatrix} 3 & 1 \\ 1 & 3 \end{bmatrix}, \qquad R_2 = \begin{bmatrix} 1 & \frac{1}{2} & \frac{1}{4} \\ \frac{1}{2} & 1 & \frac{1}{2} \\ \frac{1}{4} & \frac{1}{2} & 1 \end{bmatrix}$$

and the equations

$$R_1\alpha = \begin{bmatrix} 5 \\ 3 \end{bmatrix}, \qquad R_2\beta = \begin{bmatrix} \frac{1}{2} \\ \frac{1}{4} \\ \frac{1}{8} \end{bmatrix},$$

show that the solutions obtained by an exact method and Durbin's algorithm are identical.

4. Given the data set $\{1\ 2\ 4\}$, determine the reflection coefficients k_1 and k_2 using Burg's algorithm and confirm that $|k_i| \leq 1$.

5. Determine the predictor coefficients for a linear predictor of order 2 given the normal equations:

$$\begin{bmatrix} 1 & 4 \\ 4 & 25 \end{bmatrix} \begin{bmatrix} \alpha_1 \\ \alpha_2 \end{bmatrix} = \begin{bmatrix} 5 \\ 2 \end{bmatrix} ,$$

using (a) the Cholesky decomposition algorithm
and   (b) an exact method.

6. Suppose an estimate of a signal is required based on past data $x(k)$. If the signal is modelled by the linear representation

$$\tilde{x}(k) = \sum_{i=1}^{3} a_i x(k - i) ,$$

use Durbin's algorithm to obtain the estimate $\tilde{x}(k)$ assuming that the elements of the autocorrelation matrix for the process $x(k)$ are given by

$$r(i, k) = r(|i - k|) = \alpha^{|i-k|} ,$$

where α is a real constant satisfying $|\alpha| < 1$.

7. Suppose a signal $x(k)$ is estimated using the AR model

$$\hat{x}(k) = \sum_{i=1}^{p} \alpha_i x(k - i) .$$

By applying Burg's method, determine the parameters α_i with given data $\{x(1)\ x(2)\ x(3)\} = \{0\ 1\ a\}$, where $a \neq 0$ is a real constant, for the case $p = 2$.

Write down the estimate $\hat{x}(k)$ when $p = 1$.

8. If the sequence $\{w(k)\}$ is related to the sequences $\{x(k)\}$ and $\{y(k)\}$ by the relationship
$$w(k) = 100(x * y)(k) ,$$

where $*$ denotes convolution, obtain an expression for the complex cepstrum of $\{y(k)\}$.

9. Using the z-transform property (9.27), determine (a) the complex cepstrum and (b) the real cepstrum of $\{x(k)\}$, given by

$$x(k) = 2\delta(k) - 2\delta(k - 1) + \frac{1}{2}\delta(k - 2) .$$

10. If $p(k)$ is the train of impulses

$$p(k) = \sum_{m=0}^{M-1} a^m \delta(k - mN), \qquad N > 0 \text{ and } M > 1,$$

find $c_p^{(c)}(k)$, the complex cepstrum of $p(k)$, using the results (9.29) and (9.30). If $M = 3$, $N = 10$ and $a = \frac{1}{2}$, determine whether the sequence $\{p(k)\}$ is minimum-phase and sketch $c_p^{(c)}(k)$ for $-80 \le k \le 80$.

11. Using the test data for non-nasal, voiced speech $\{s\}$ generated in Exercise 2 above, introduce noise in the signal by convolving the sequence with a random noise sequence. The MATLAB command: cs = conv(s,randn(size(s))) will achieve this objective. For the signal cs, use MATLAB to evaluate and plot its real cepstrum. Using an appropriate frequency-invariant filter, esti- mate the periodicity in the signal. Does the value, so obtained, agree with your expectations?

12. If $u(k) = K\alpha^k \zeta(k)$, where K, α are real constants satisfying $K > 0$ and $|\alpha| < 1$, show, using (9.27), that

$$c_u^{(c)}(k) = \frac{\alpha^k}{k}\zeta(k - 1) + \log_e(K)\delta(k) .$$

Suppose the impulse response of a linear, time-invariant filter satisfies

$$(h * u)(k) = \delta(k) + \beta\delta(k - 1) ,$$

where $\beta \in \mathbb{R}$ satisfies $0 < \beta < 1$. Find $c_h^{(c)}(k)$ and, hence, determine whether the filter is minimum-phase.

13. Show that the complex cepstrum, $c_x^{(c)}(k)$, of a maximum-phase signal is related to its real cepstrum, $c_x^{(r)}(k)$, by

$$c_x^{(c)}(k) = c_x^{(r)}(k)\ell(k) ,$$

where $\{\ell(k)\}$ is the frequency-invariant, linear filter given by

$$\ell(k) \stackrel{\text{def}}{=} 2\zeta(-k) - \delta(k) .$$

A

The complex exponential

A.1 MACLAURIN SERIES FOR EXP, COS AND SIN

Let $y = f(x)$ be a real-valued function of a real variable. Then, under certain assumptions, f may be represented by its **Maclaurin series expansion**:

$$f(x) = f(0) + f'(0)x + f''(0)\frac{x^2}{2!} + \cdots = \sum_{n=0}^{\infty} f^{(n)}(0)\frac{x^n}{n!}, \qquad (A.1)$$

where $f^{(n)}(x) \overset{\text{def}}{=} \dfrac{\mathrm{d}^n f}{\mathrm{d}x^n}(x)$, with $f^{(0)}(x) \overset{\text{def}}{=} f(x)$. The Maclaurin series expansions for exp, cos and sin are given by,

$$\exp(x) = 1 + x + \frac{x^2}{2!} + \cdots = \sum_{n=0}^{\infty} \frac{x^n}{n!}; \qquad (A.2)$$

$$\cos(x) = 1 - \frac{x^2}{2!} + \frac{x^4}{4!} - \cdots = \sum_{n=0}^{\infty} (-1)^n \frac{x^{2n}}{(2n)!}; \qquad (A.3)$$

$$\sin(x) = x - \frac{x^3}{3!} + \frac{x^5}{5!} - \cdots = \sum_{n=0}^{\infty} (-1)^n \frac{x^{2n+1}}{(2n+1)!}. \qquad (A.4)$$

Equation (A.2) is often considered to be the definition of $y = \exp(x)$. It can then be shown that $\exp(x) = e^x$, where $e \overset{\text{def}}{=} \exp(1)$. To show this, it is necessary to show that $\exp(x + y) = \exp(x)\exp(y)$, which is a little messy. Clearly, we have $\exp(0) = e^0 = 1$.

The expansions (A.2)–(A.4) may then be used to extend the definitions of exp, cos and sin from the real domain to the complex domain. In particular, for any complex number z, we can define e^z by

$$e^z \overset{\text{def}}{=} 1 + z + \frac{z^2}{2!} + \cdots = \sum_{n=0}^{\infty} \frac{z^n}{n!}.$$

The relation $e^{z+w} = e^z e^w$ can be shown to hold for general complex numbers z and w, and so, if we write $z = x + j\,y$, we have

$$e^z = e^{x+j\,y} = e^x e^{j\,y},$$

where

$$e^{j\,y} = 1 + j\,y + \frac{(j\,y)^2}{2!} + \cdots$$

$$= 1 + j\,y - \frac{y^2}{2!} - j\,\frac{y^3}{3!} + \frac{y^4}{4!} + j\,\frac{y^5}{5!} - \cdots$$

$$= \left(1 - \frac{y^2}{2!} + \frac{y^4}{4!} - \cdots\right) + j\left(y - \frac{y^3}{3!} + \frac{y^5}{5!} - \cdots\right). \qquad \text{(A.5)}$$

If we now compare (A.5) with (A.3) and (A.4), we obtain the remarkable equation,

$$e^{j\,y} = \cos(y) + j\,\sin(y) \qquad \textbf{Euler's Formula.} \qquad \text{(A.6)}$$

In conclusion, we see that, given any complex number $z = x + j\,y$, we can define e^z via

$$e^z = e^{x+j\,y} = e^x e^{j\,y} = e^x\Big(\cos(y) + j\,\sin(y)\Big).$$

Note: Henceforth, we use the term 'complex exponential' to refer to purely imaginary powers of e.

A.2 PROPERTIES OF THE COMPLEX EXPONENTIAL

Using Pythagoras' theorem, we see that

$$|\,e^{j\,y}\,| = \sqrt{\cos^2(y) + \sin^2(y)} = \sqrt{1} = 1.$$

Recall from basic trigonometry that, if we rotate a line of unit length in the (x, y)-plane, pivoted at the origin, anticlockwise through an angle of θ radians and lying initially along the positive x-axis, the coordinates of the end-point are given by $x = \cos(\theta)$, $y = \sin(\theta)$, (see Figure A.1). This is true for any real value of θ and provides a geometrical picture for the values of $\cos(\theta)$ and $\sin(\theta)$ in any quadrant.

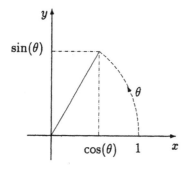

Figure A.1: Geometrical picture for $\cos(\theta)$ and $\sin(\theta)$.

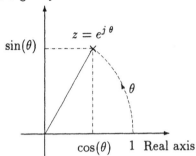

Figure A.2: Geometrical picture for $z = e^{j\,\theta}$.

If we now replace the (x, y)-plane with the complex plane, we see that the end-point is associated with the complex number, $z = \cos(\theta) + j\,\sin(\theta)$. But, from Euler's formula (A.6), this is simply $z = e^{j\,\theta}$, (see Figure A.2).

Example A.1
Simplify the following expressions.

1. $e^{j\,\frac{\pi}{2}}$.   2. $e^{j\,\pi}$.
3. $e^{j\,2\pi}$.   4. $e^{-j\,\frac{7\pi}{2}}$.
5. $e^{j\,2n\pi}$, n an integer.   6. $e^{j\,n\pi}$, n an integer.

1. Since $\theta = \dfrac{\pi}{2}$ represents a 90° anticlockwise rotation, we see that $e^{j\,\frac{\pi}{2}}$ lies one unit up the imaginary axis, i.e. $e^{j\,\frac{\pi}{2}} = j$.

2. $e^{j\,\pi} = -1$.

3. $e^{j\,2\pi} = 1$.

4. $e^{-j\,\frac{7\pi}{2}} = e^{j\,\frac{\pi}{2}} = j$.

5. Since $\theta = 2n\pi$ represents a rotation through an integer multiple of 2π, we see that $e^{j\,2n\pi} = 1$ for all n.

6. If n is even, we again see that $e^{j\,n\pi} = 1$. However, if n is odd, we obtain $e^{j\,n\pi} = -1$. This can be written more succinctly as

$$e^{j\,n\pi} = (-1)^n \qquad \text{for all integers, } n.$$

This same formula may also be seen by noting that $e^{j\,n\pi} = (e^{j\,\pi})^n$ and that $e^{j\,\pi} = -1$.

□

Example A.2

Use the complex exponential to find a formula for $\cos(5\theta)$ in terms of $c \overset{\text{def}}{=} \cos(\theta)$ and $s \overset{\text{def}}{=} \sin(\theta)$.

We have
$$e^{j\,5\theta} = \cos(5\theta) + j\,\sin(5\theta). \tag{A.7}$$

We also have
$$e^{j\,5\theta} = (e^{j\,\theta})^5 = (c + j\,s)^5$$
$$= c^5 + 5c^4(j\,s) + 10c^3(j\,s)^2 + 10c^2(j\,s)^3 + 5c(j\,s)^4 + (j\,s)^5$$
$$= c^5 + j\,5c^4 s - 10c^3 s^2 - j\,10c^2 s^3 + 5cs^4 + j\,s^5. \tag{A.8}$$

Equating real parts for (A.7) and (A.8) gives
$$\cos(5\theta) = c^5 - 10c^3 s^2 + 5cs^4.$$

If desired, this can be rewritten using $c^2 + s^2 = 1$. For example, we have
$$\cos(5\theta) = c^5 - 10c^3(1 - c^2) + 5c(1 - c^2)^2$$
$$= c^5 - 10c^3 + 10c^5 + 5c - 10c^3 + 5c^5$$
$$= 16c^5 - 20c^3 + 5c.$$

□

Example A.3

Write $\sin^3(\theta)$ as a linear combination of $\sin(\theta)$ and $\sin(3\theta)$.

Using $e^{j\,\theta} = \cos(\theta) + j\,\sin(\theta)$ $e^{-j\,\theta} = \cos(\theta) - j\,\sin(\theta)$
we see that
$$\cos(\theta) = \frac{1}{2}\left(e^{j\,\theta} + e^{-j\,\theta}\right) \tag{A.9}$$
$$\sin(\theta) = \frac{1}{2j}\left(e^{j\,\theta} - e^{-j\,\theta}\right). \tag{A.10}$$

In particular,
$$\sin^3(\theta) = \left[\frac{1}{2j}\left(e^{j\,\theta} - e^{-j\,\theta}\right)\right]^3$$
$$= \frac{1}{8j^3}\left(e^{j\,3\theta} - 3e^{j\,\theta} + 3e^{-j\,\theta} - e^{-j\,3\theta}\right)$$
$$= \frac{-1}{8j}\left(e^{j\,3\theta} - e^{-j\,3\theta}\right) + \frac{3}{8j}\left(e^{j\,\theta} - e^{-j\,\theta}\right)$$
$$= \frac{3}{4}\sin(\theta) - \frac{1}{4}\sin(3\theta).$$

□

A.3　THE COMPLEX EXPONENTIAL FORM

Let z be a complex number with **modulus**, $|z| = r$, and **argument**, $\arg z = \theta$ radians, see Figure A.3.

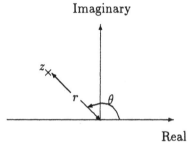

Imaginary

Real

Figure A.3: Modulus and argument of a complex number z.

If $r = 1$, then, as we have seen, $z = e^{j\theta}$. In general, we have

$$z = re^{j\theta}.$$

Usually θ is chosen in the range $-\pi < \theta \le \pi$. (Aside: If $z = 0$ then $r = 0$ and θ is unspecified or, alternatively, may be chosen arbitrarily.) The form $z = re^{j\theta}$ is referred to as **the complex exponential form** for z. Conversion between $a + jb$ form and $re^{j\theta}$ form is identical to conversion between rectangular and polar coordinates, with the restriction that θ be given in radians. Most scientific calculators have such a facility.

Example A.4

1. Convert the following to complex exponential form.

    (a) $-1 + j$　(b) $-1 - j$　(c) $-j$　(d) $2 - j$　(e) $-2 + j$.

2. Convert the following to $a + jb$ form.

    (a) $2e^{-j\frac{\pi}{3}}$　　(b) $5e^{j\frac{3\pi}{4}}$　　(c) $-2e^{j\frac{\pi}{4}}$.

1. (a) We have $|-1 + j| = \sqrt{1^2 + 1^2} = \sqrt{2}$ and $\arg(-1 + j) = \dfrac{3\pi}{4}$. So,

    $$-1 + j = \sqrt{2}e^{j\frac{3\pi}{4}} \approx 1.414e^{j\,2.356}.$$

    This example can be performed readily by hand. The reader may wish to discover how to use rectangular to polar conversion with a calculator to reproduce this result.

(b) We have $|-1-j| = \sqrt{2}$ and $\arg(-1-j) = -\dfrac{3\pi}{4}$ and so

$$-1-j = \sqrt{2}e^{-j\frac{3\pi}{4}} \approx 1.414e^{-j\,2.356}.$$

We note that the above two solutions illustrate the general result that the complex conjugate of $z = re^{j\theta}$ is $\bar{z} = re^{-j\theta}$.

(c) We have $|-j| = 1$ and $\arg(-j) = -\dfrac{\pi}{2}$ and so

$$-j = e^{-j\frac{\pi}{2}}.$$

(d) We have $|2-j| = \sqrt{5}$ and $\arg(2-j) = -\tan^{-1}\left(\dfrac{1}{2}\right)$ and so

$$2-j = \sqrt{5}e^{-j\,\tan^{-1}(0.5)} \approx 2.236e^{-j\,0.464}.$$

(e) Finally, we have $|-2+j| = \sqrt{5}$ and $\arg(-2+j) = \pi - \tan^{-1}\left(\dfrac{1}{2}\right)$ which gives

$$-2+j = \sqrt{5}e^{j\,(\pi-\tan^{-1}(0.5))} \approx 2.236e^{j\,2.678}.$$

It is important to note that the oft-quoted formula, $\arg(a+jb) = \tan^{-1}(b/a)$ would not give the correct result here. Rather than use such a formula, think carefully about the quadrant in which the complex number lies and then use basic trigonometry to find the argument.

2. Using Euler's formula, we have

(a) $2e^{-j\frac{\pi}{3}} = 2\left(\cos\left(\dfrac{\pi}{3}\right) - j\sin\left(\dfrac{\pi}{3}\right)\right)$

$$= 2\left(\dfrac{1}{2} - j\dfrac{\sqrt{3}}{2}\right) = 1 - j\sqrt{3}.$$

(b) $5e^{j\frac{3\pi}{4}} = 5\left(\cos\left(\dfrac{3\pi}{4}\right) + j\sin\left(\dfrac{3\pi}{4}\right)\right)$

$$= 5\left(-\dfrac{1}{\sqrt{2}} + j\dfrac{1}{\sqrt{2}}\right) = \dfrac{5}{\sqrt{2}}(-1+j).$$

(c) $-2e^{j\frac{\pi}{4}} = -2\left(\cos\left(\dfrac{\pi}{4}\right) + j\sin\left(\dfrac{\pi}{4}\right)\right)$

$$= -2\left(\dfrac{1}{\sqrt{2}} + j\dfrac{1}{\sqrt{2}}\right) = -\sqrt{2}(1+j).$$

Note that $-2e^{j\frac{\pi}{4}}$ is not strictly speaking in complex exponential form, since $r = -2$ is negative and, hence, not a modulus. By a graphical method, or otherwise, it can be seen that the correct exponential form for $-2e^{j\frac{\pi}{4}}$ is given by

$$-2e^{j\frac{\pi}{4}} = 2e^{-j\frac{3\pi}{4}}.$$

□

As a general rule, $a + jb$ form is the most convenient form for performing additions and subtractions and $re^{j\theta}$ form is preferable for multiplications and divisions, including powers and roots.

Example A.5

1. Rewrite $\dfrac{(-1+j)^5(\sqrt{3}+j)^2}{(1-j\sqrt{3})^6}$ in $a + jb$ form.

2. Rewrite $e^{j\frac{\pi}{4}} - e^{j\frac{\pi}{3}}$ in $re^{j\theta}$ form.

3. Find all complex numbers z satisfying $z^3 = j$.

1. Since the expression involves powers, multiplications and divisions, we first convert the numbers $-1+j$, $\sqrt{3}+j$, $1-j\sqrt{3}$ to complex exponential form. We have

$$-1+j = \sqrt{2}e^{j\frac{3\pi}{4}}, \quad \sqrt{3}+j = 2e^{j\frac{\pi}{6}} \quad \text{and} \quad 1-j\sqrt{3} = 2e^{-j\frac{\pi}{3}}.$$

The desired complex number is therefore given by,

$$\frac{(\sqrt{2}e^{j\frac{3\pi}{4}})^5(2e^{j\frac{\pi}{6}})^2}{(2e^{-j\frac{\pi}{3}})^6} = 2^{(\frac{5}{2}+2-6)}e^{j(\frac{15\pi}{4}+\frac{\pi}{3}+2\pi)}$$

$$= 2^{-\frac{3}{2}}e^{j(-\frac{\pi}{4}+\frac{\pi}{3})} = 2^{-\frac{3}{2}}e^{j\frac{\pi}{12}}.$$

Finally, conversion back to $a + jb$ form gives

$$2^{-\frac{3}{2}}\left(\cos\left(\frac{\pi}{12}\right) + j\sin\left(\frac{\pi}{12}\right)\right) \approx 0.342 + j\,0.915.$$

2. Since this problem involves subtraction, we first convert the two complex exponentials to $a + jb$ form. We obtain

$$e^{j\frac{\pi}{4}} - e^{j\frac{\pi}{3}} = \frac{1}{\sqrt{2}}(1+j) - \frac{1}{2}(1+j\sqrt{3})$$

$$= \left(\frac{\sqrt{2}-1}{2}\right) - j\left(\frac{\sqrt{3}-\sqrt{2}}{2}\right).$$

Converting this to complex exponential form leads to the approximation

$$0.261 e^{-j\,0.654}.$$

Another approach to this problem, which has useful applications in Fourier analysis, is to first find the average value of the exponents, i.e.

$$\frac{1}{2}\left(\frac{j\pi}{4} + \frac{j\pi}{3}\right) = \frac{j\,7\pi}{24}.$$

We can then rewrite the difference between the two given exponentials in the form

$$e^{j\frac{\pi}{4}} - e^{j\frac{\pi}{3}} = e^{j\frac{7\pi}{24}}\left(e^{-j\frac{\pi}{24}} - e^{j\frac{\pi}{24}}\right).$$

Now, using (A.10), we obtain

$$e^{j\frac{7\pi}{24}}\left(-2j\,\sin\left(\frac{\pi}{24}\right)\right).$$

From this, we can easily read off the modulus as $2\sin\left(\dfrac{\pi}{24}\right) \approx 0.261$, as obtained before. Using the fact that $\arg(-j) = -\dfrac{\pi}{2}$, we deduce that the argument is given by $\dfrac{7\pi}{24} - \dfrac{\pi}{2} = -\dfrac{5\pi}{24} \approx 0.654$, also as obtained before.

3. This problem is concerned with finding cube roots and so the complex exponential form is best suited. If $z = re^{j\theta}$ then $z^3 = r^3 e^{j\,3\theta}$. We want this to equal j, which in complex exponential form is $e^{j\frac{\pi}{2}}$. This means that r^3 must equal 1 and 3θ must be equivalent to $\dfrac{\pi}{2}$. It follows immediately that r must equal 1. Since 3θ is equivalent to $\dfrac{\pi}{2}$, we have

$$3\theta = \frac{\pi}{2} + 2n\pi \qquad n = 0, \pm 1, \pm 2, \ldots$$

$$\Longleftrightarrow \qquad \theta = \frac{\pi}{6} + \frac{2n\pi}{3} \qquad n = 0, \pm 1, \pm 2, \ldots.$$

Restricting θ to lie in the range $-\pi < \theta \le \pi$, gives the three solutions: $\theta = -\dfrac{\pi}{2}, \dfrac{\pi}{6}$ and $\dfrac{5\pi}{6}$. So, in complex exponential form, the three solutions to $z^3 = j$ are given by

$$z_1 = e^{-j\frac{\pi}{2}}, \qquad z_2 = e^{j\frac{\pi}{6}}, \qquad z_3 = e^{j\frac{5\pi}{6}}.$$

In $a + jb$ form, these become,

$$z_1 = -j, \qquad z_2 = \frac{1}{2}(\sqrt{3} + j), \qquad z_3 = \frac{1}{2}(-\sqrt{3} + j).$$

$\square$

This concludes our brief introduction to the complex exponential. Further results and applications can be found in the main body of the text.

B

Linear predictive coding algorithms

The method of linear prediction or *linear predictive coding*, as it is sometimes known, has a wide range of applicability in signal processing. This method has been applied to the problem of speech processing, as discussed in Chapter 9, and three particular algorithms (namely, Durbin, Cholesky decomposition, and Burg) were presented. These algorithms have been coded using the MATLAB programming language and the set of commands are prescribed in the following three sections.

B.1   Durbin-Levinson algorithm

A MATLAB function m-file for the Durbin-Levinson algorithm is illustrated below.

```
%
%    Autocorrelation method for determination of the predictor
%
%    parameters using the Durbin-Levinson algorithm
%
%
function[alpha,k]=Durbin(x,p)
%
%    x - speech data segment
%    p - order of predictor
%    k - PARCOR coefficients
%    alpha - predictor coefficients (row vector)
%
x=x(:)';
M=length(x);
if M<=1,
error('Data must have more than one entry'),
end
if p>M-1,
error('Predictor order must be less than size of data vector'),
end
%
```

```
%     Construction of the elements defining the Toeplitz matrix
%
r=zeros(1,p+1);
for i=1:p+1
    r(i)=x(1:M-i+1)*x(i:M)';
end
%
%     Initialization
%
k=zeros(1,p);
e=k;
a=zeros(p,p);
k(1)=r(2)/r(1);
a(1,1)=k(1);
e(1)=(1-k(1)*k(1))*r(1);
%
%     DURBIN-LEVINSON ALGORITHM
%
for i=2:p
    k(i)=(r(i+1)-r(i:-1:2)*a(1:i-1,i-1))/e(i-1);
    a(i,i)=k(i);
    for n=1:i-1
        a(n,i)=a(n,i-1)-k(i)*a(i-n,i-1);
    end
    e(i)=(1-k(i)*k(i))*e(i-1);
end
alpha=a(:,p)';
```

B.2   Cholesky decomposition algorithm

A MATLAB function m-file for the Cholesky decomposition algorithm follows.

```
%
%     Covariance method for determination of the predictor
%
%     parameters using the Cholesky decomposition algorithm
%
function[alpha]=Chol(x,p)
%
%     x - speech data segment
%     p - order of predictor
%     alpha - predictor coefficients (row vector)
%
x=x(:)';
M=length(x);
if M<=1,
error('Data must have more than one entry'),
```

```
end
if p>(M-1)/2,
error('Twice predictor order must be less than size of data vector'),
end
%
%    Construction of the covariance matrix
%
rho=zeros(p,p+1);
w=rho;
for i=1:p
    for n=1:p
    rho(i,n)=x(p+1-i:M-i)*x(p+1-n:M-n)';
    end
end
%
%    CHOLESKY DECOMPOSITION ALGORITHM
%
%
%    Determination of the triangular matrix W
%
for k=1:p-1
    w(k,k)=sqrt(rho(k,k));
    w(k,k+1:p)=rho(k,k+1:p)/w(k,k);
    for n=k+1:p
        rho(n,n:p)=rho(n,n:p)-w(k,n:p)*w(k,n);
    end
end
w(p,p)=sqrt(rho(p,p));
%
%    Evaluation of elements on the RHS of the Normal equations
%
for n=1:p
    rho(n,p+1)=x(p+1:M)*x(p+1-n:M-n)';
end
%
%    Evaluation of the predictor coefficients
%
a=zeros(1,p);
a(1)=rho(1,p+1)/w(1,1);
for k=2:p
    a(k)=(rho(k,p+1)-a(1:k-1)*w(1:k-1,k))/w(k,k);
end
alpha=zeros(1,p);
alpha(p)=a(p)/w(p,p);
for k=p-1:-1:1
    alpha(k)=(a(k)-alpha(k+1:p)*w(k,k+1:p)')/w(k,k);
end
```

B.3   Burg algorithm

A MATLAB function m-file for the Burg algorithm is presented.

```
%
%      Lattice method for determination of the predictor parameters
%
%
function[alpha,k]=Burg(x,p)
%
%      x - speech data segment
%      p - order of predictor
%      k - PARCOR coefficients
%      alpha - predictor coefficients (row vector)
%
x=x(:)';
M=length(x);
if M<=1,
error('Data must have more than one entry'),
end
if p>M-1,
error('Predictor order must be less than size of data vector'),
end
%
%      Initialization
%
fp=x(2:M);
bp=x(1:M-1);
%
%      Determination of PARCOR coefficient k(1) and
%      sum of mean-square forward and backward prediction
%      errors
%
k=zeros(1,p);
a=zeros(p,p);
e=fp*fp'+bp*bp';
k(1)=2*fp*bp'/e;
a(1,1)=k(1);
%
%      Lattice filter
%
for i=2:p
    m=M-i+1;
    f=fp(1:m)-k(i-1)*bp(1:m);
    b=bp(1:m)-k(i-1)*fp(1:m);
%
%      Determination of PARCOR coefficients k(i), i>1
%      and sum of mean-square prediction errors
```

```
%
    fp=f(2:m);
    bp=b(1:m-1);
    e=fp*fp'+bp*bp';
    k(i)=2*fp*bp'/e;
%
%      Evaluation of predictor coefficients
%
    a(i,i)=k(i);
    for n=1:i-1
        a(n,i)=a(n,i-1)-k(i)*a(i-n,i-1);
    end
end
alpha=a(:,p)';
```

C

Answers to the exercises

Exercises §1.11

1. $\dfrac{dv_L}{dt}(t) + \dfrac{R}{L}v_L(t) = \dfrac{de}{dt}(t)$

2. (a) $\dfrac{5}{s} - \dfrac{2s}{2s^2 + 1}$ (b) $\dfrac{3}{s+2} - \dfrac{6}{s^4}$ (c) $\dfrac{2}{s^2 + 2s + 2} - \dfrac{s^2 - 1}{(s^2 + 1)^2}$

   (d) $\dfrac{1}{s^2 - 1} + \dfrac{2(s+1)e^{-s}}{s^2}$

4. (a) $\dfrac{1}{3}(e^{-2t} - e^{-5t})$, $t \geq 0$ (b) $\dfrac{1}{2}(e^t - \cos(t) + \sin(t))$, $t \geq 0$

   (c) $\dfrac{1}{16}(e^{2t} - e^{-2t} - 4te^{-2t})$ or $\dfrac{1}{8}\sinh(2t) - \dfrac{1}{4}te^{-2t}$, $t \geq 0$

5. (i) $\delta(t) - 3e^{-3t}\zeta(t)$ (ii) $a\sin(at)\zeta(t)$

7. $\dfrac{1}{s^2 + 5s + 6}$

8. $\dfrac{1}{(s-1)^2}$

9. (a) $e^t - t - 1$ (b) e^t

10. $\dfrac{s+2}{s^2 + s + 2}$

11. $\mathbf{A} = \begin{bmatrix} -\dfrac{1}{RC_1} & 0 & -\dfrac{1}{C_1} \\ 0 & 0 & \dfrac{1}{C_2} \\ \dfrac{1}{L} & -\dfrac{1}{L} & 0 \end{bmatrix}$ $\mathbf{B} = \begin{bmatrix} \dfrac{1}{RC_1} \\ 0 \\ 0 \end{bmatrix}$ $\mathbf{C} = [0 \ 1 \ 0]$

$\det(\lambda\mathbf{I} - \mathbf{A}) = \lambda^3 + \dfrac{1}{RC_1}\lambda^2 + \dfrac{1}{L}\left(\dfrac{1}{C_1} + \dfrac{1}{C_2}\right)\lambda + \dfrac{1}{RLC_1C_2}$

15. (i) $y(t) = ae^{-3t} + be^{-4t} + (11\sin(t) - 7\cos(t))/170$
     As $t \to \infty$, $y(t) \to (11\sin(t) - 7\cos(t))/170$ which is independent of the
     initial conditions, since the initial conditions only determine the con-
     stants a and b.

     (ii) $y(t) = ae^{-3t} + be^{-4t} + (4\sin(2t) - 7\cos(2t))/130$
     As $t \to \infty$, $y(t) \to (4\sin(2t) - 7\cos(2t))/130$ which is independent of the
     initial conditions.

     When $u(t) = \sin(t) + \sin(2t)$,
     $y(t) \to (11\sin(t) - 7\cos(t))/170 + (4\sin(2t) - 7\cos(2t))/130$ as $t \to \infty$.

Exercises §2.7

1. (a) $\dfrac{1}{s^2 + 5s + 4}$; $\frac{1}{3}\left(e^{-t} - e^{-4t}\right)$, $t \geq 0$

    (b) $\dfrac{1}{s^2 + 2s + 2}$; $e^{-t}\sin(t)$, $t \geq 0$

    (c) $\dfrac{1}{8s^2 + 6s + 1}$; $\frac{1}{2}\left(e^{-\frac{1}{4}t} - e^{-\frac{1}{2}t}\right)$, $t \geq 0$

    (d) $\dfrac{s+2}{3s^2 + 2s + 2}$; $\frac{1}{3}e^{-\frac{1}{3}t}\left[\cos\left(\dfrac{\sqrt{5}}{3}t\right) + \sqrt{5}\sin\left(\dfrac{\sqrt{5}}{3}t\right)\right]$, $t \geq 0$

2. For $t \geq 0$,

$$y(t) = \begin{cases} 2e^{-t} + (t-2)e^{-\frac{1}{2}t} , & \text{if } a = \frac{1}{2} , \\ 2e^{-\frac{1}{2}t} - (t+2)e^{-t} , & \text{if } a = 1 , \\ \dfrac{2}{2a-1}e^{-\frac{1}{2}t} + \dfrac{1}{1-a}e^{-t} + \dfrac{1}{(1-a)(1-2a)}e^{-at} , & \text{otherwise} . \end{cases}$$

3. (a) Stable   (b) Stable   (c) Unstable   (d) Marginally stable   (e) Unstable

4. $y_c(t) = \frac{1}{10}\left(2 - e^{-\frac{1}{2}t}[2\cos(t) + \sin(t)]\right)\zeta(t)$, $y_\delta(t) = \frac{1}{4}e^{-\frac{1}{2}t}\sin(t)\zeta(t)$

5. $|H(j\omega)| = \dfrac{|\omega|}{\sqrt{\omega^6 + 1}}$

$$\arg H(j\omega) = \begin{cases} \frac{\pi}{2} - \tan^{-1}\left(\dfrac{\omega(\omega^2 - 2)}{2\omega^2 - 1}\right), & \omega < -\sqrt{2} \\[2mm] \frac{\pi}{2}, & \omega = -\sqrt{2} \\[2mm] \frac{\pi}{2} + \tan^{-1}\left(\dfrac{\omega(2 - \omega^2)}{2\omega^2 - 1}\right), & -\sqrt{2} < \omega < -\frac{1}{\sqrt{2}} \\[2mm] 0, & \omega = -\frac{1}{\sqrt{2}} \\[2mm] -\frac{\pi}{2} - \tan^{-1}\left(\dfrac{\omega(2 - \omega^2)}{1 - 2\omega^2}\right), & -\frac{1}{\sqrt{2}} < \omega < 0 \\[2mm] \text{undefined}, & \omega = 0 \\[2mm] \frac{\pi}{2} - \tan^{-1}\left(\dfrac{\omega(2 - \omega^2)}{1 - 2\omega^2}\right), & 0 < \omega < \frac{1}{\sqrt{2}} \\[2mm] 0, & \omega = \frac{1}{\sqrt{2}} \\[2mm] -\frac{\pi}{2} + \tan^{-1}\left(\dfrac{\omega(2 - \omega^2)}{2\omega^2 - 1}\right), & \frac{1}{\sqrt{2}} < \omega < \sqrt{2} \\[2mm] -\frac{\pi}{2}, & \omega = \sqrt{2} \\[2mm] -\frac{\pi}{2} - \tan^{-1}\left(\dfrac{\omega(\omega^2 - 2)}{2\omega^2 - 1}\right), & \omega > \sqrt{2} \end{cases}$$

6. $y(t) = \dfrac{1}{4}e^{-t}[1 + 2\sin(2t) - \cos(2t)], \quad t \geq 0$.

7. (a) $\dfrac{1}{\sqrt{2}}\cos\left(2t - \dfrac{\pi}{4}\right) \approx 0.7071\cos(2t - 0.7854)$

   (b) $\dfrac{1}{\sqrt{5}}\cos\left(2t + \tan^{-1}\left(\dfrac{1}{2}\right)\right) \approx 0.4472\cos(2t + 0.4636)$

   (c) System is unstable.

   (d) $\dfrac{1}{2\sqrt{2}}\cos\left(2t + \dfrac{3\pi}{4}\right) \approx 0.3536\cos(2t + 2.3562)$

Exercises §3.10

2. $F_n = \dfrac{1}{2}\text{sinc}\left(\dfrac{n\pi}{2}\right) \quad$ for all n.

4. $\mathcal{F}\{\cos(bt)/(1 + t^2)\} = \dfrac{\pi}{2}\left(e^{-|\omega + b|} + e^{-|\omega - b|}\right)$

   $\mathcal{F}\{\sin(bt)/(1 + t^2)\} = j\dfrac{\pi}{2}\left(e^{-|\omega + b|} - e^{-|\omega - b|}\right)$

5. $\mathcal{F}\left\{\dfrac{\omega_b}{\pi}\text{sinc}(\omega_b t)\right\} = \begin{cases} 1, & |\omega| \leq \omega_b \\ 0, & |\omega| > \omega_b \end{cases}$.

6. $\mathcal{F}\{A\cos(\omega_0 t)\zeta(t)\} = \dfrac{A}{2}\left[\pi\big(\delta(\omega + \omega_0) + \delta(\omega - \omega_0)\big) - j\left(\dfrac{1}{\omega + \omega_0} + \dfrac{1}{\omega - \omega_0}\right)\right]$

$$\mathcal{F}\{A\sin(\omega_0 t)\zeta(t)\} = \frac{A}{2}\left[j\pi\Big(\delta(\omega+\omega_0) - \delta(\omega-\omega_0)\Big) + \left(\frac{1}{\omega+\omega_0} - \frac{1}{\omega-\omega_0}\right)\right]$$

7. $|H(j\omega)| = \dfrac{1}{\sqrt{\omega^4 + 1}}$

Amplitude spectrum of the output signal is

$$\frac{1}{\sqrt{\omega^4+1}}\left|\mathrm{sinc}\left(\frac{\omega}{2}\right) + \mathrm{sinc}\left(\frac{\omega}{2} - 5\right) + \mathrm{sinc}\left(\frac{\omega}{2} + 5\right)\right|.$$

The linear system attenuates the copy of the pulse modulating the carrier signal.

Exercises §4.7

1. Poles : $-\omega_c,\ \omega_c e^{\pm j\frac{3\pi}{5}},\ \omega_c e^{\pm j\frac{4\pi}{5}}$;

$$H_5(s) = \frac{\omega_c^5}{s^5 + 3.2361\omega_c s^4 + 5.2361\omega_c^2 s^3 + 5.2361\omega_c^3 s^2 + 3.2361\omega_c^4 s + \omega_c^5}$$

2. $H_3(s) = \dfrac{64 \times 10^9 \pi^3}{s^3 + 8\times 10^3\pi s^2 + 32\times 10^6\pi^2 s + 64\times 10^9\pi^3}$;   5th order filter

3. $H(s) = \dfrac{50\pi s}{s^2 + 50\pi s + \pi^2 \times 10^6}$

4. $H(s) = \dfrac{2500\pi^2 s^2}{s^4 + 50\sqrt{2}\pi s^3 + 2.0025\times 10^6\pi^2 s^2 + 50\sqrt{2}\times 10^6\pi^3 s + \pi^4\times 10^{12}}$

5. $L = 0.3827\dfrac{R}{\omega_c},\quad C_1 = \dfrac{1.0824}{R\omega_c},\quad L_1 = 1.5772\dfrac{R}{\omega_c},\quad C_2 = \dfrac{1.5307}{R\omega_c}$

6. $H(s) = \dfrac{s^2}{s^2 + 200\sqrt{2}\pi s + 40000\pi^2}$

7. $H(s) = \dfrac{s^4 + 2\times 10^4 s^2 + 10^8}{s^4 + 10\sqrt{2}s^3 + 2.01\times 10^4 s^2 + \sqrt{2}\times 10^5 s + 10^8}$

8. $H(s) = 0.2713\times 10^{30}/(s^6 + 1.166\times 10^5 s^5 + 3.049\times 10^{10} s^4 + 2.386\times 10^{15} s^3 + 2.342\times 10^{20} s^2 + 0.9623\times 10^{25} s + 0.2173\times 10^{30})$

9. $H(s) = \dfrac{0.1741\times 10^5 s^2}{s^4 + 1.379\times 10^2 s^3 + 1.794\times 10^6 s^2 + 1.225\times 10^8 s + 0.7890\times 10^{12}}$;

30.88 dB,   19.93 dB,   0.27 dB

Exercises §5.10

1. (a) $\dfrac{1}{z^2} + \dfrac{2}{z^3} - \dfrac{2}{z^4} - \dfrac{1}{z^5} + \dfrac{2}{z^7}$

(b) $\dfrac{z^6 - 1}{z^5(z-1)}$ or $1 + \dfrac{1}{z} + \dfrac{1}{z^2} + \dfrac{1}{z^3} + \dfrac{1}{z^4} + \dfrac{1}{z^5}$

(c) $\dfrac{z}{z-5} + \dfrac{3z}{z+\frac{1}{4}}$

(d) $\dfrac{8z}{z-\frac{1}{2}}$

(e) $\dfrac{5z(z+1)}{(z-1)^3}$

3. (a) $\left\{(1+2^k)\zeta_k\right\}$ (b) $\left\{\left[3(4)^{k-1} - (3+k)(2)^{k-2}\right]\zeta_k\right\}$ (c) $\left\{(2-2^{k-1})\zeta_{k-1}\right\}$

4. (a) $y_k = \frac{1}{3}\left[\left(\frac{1}{2}\right)^{k-2} - 2^k\right]$; $y_2 = -1$, $y_3 = -\frac{5}{2}$, $y_4 = -\frac{21}{4}$

(b) $y_k = 1 + \frac{1}{7}(-1)^{k+1} - \frac{1}{7}\left(\frac{1}{6}\right)^{k-1}$; $y_2 = \frac{5}{6}$, $y_3 = \frac{41}{36}$, $y_4 = \frac{185}{216}$

(c) $y_k = (-1)^k + \left(\frac{1}{3}\right)^{k-1} - \left(\frac{1}{4}\right)^{k-1}$; $y_2 = \frac{13}{12}$, $y_3 = -\frac{137}{144}$, $y_4 = \frac{1765}{1728}$

(d) $y(k) = (\sqrt{2})^k \cos\left(\dfrac{k\pi}{2}\right) + \left[1 + (\sqrt{2})^{k-1}\right]\sin\left(\dfrac{k\pi}{2}\right)$;

$y(2) = -2$, $y(3) = -3$, $y(4) = 4$

5. $D(z) = \dfrac{z(z+1)}{z^3 + z^2 + 2}$

Exercises §6.7

1. $\{y_k^\delta\}$, where $y_k^\delta = \begin{cases} 0, & k \le 1, \\ \left(\frac{1}{2}\right)^k - 2\left(\frac{1}{4}\right)^k, & k \ge 2, \end{cases} = \left[\left(\frac{1}{2}\right)^k - 2\left(\frac{1}{4}\right)^k\right]\zeta(k-2)$

System is stable since the poles $z = \frac{1}{4}, \frac{1}{2}$ lie in the open unit disk.

2. $y_k^\delta = \left[7\left(\frac{2}{5}\right)^{k-1} - \frac{13}{2}\left(\frac{3}{10}\right)^{k-1}\right]\zeta(k-1)$

$y_k^\varsigma = \left[\frac{50}{21} - \frac{35}{3}\left(\frac{2}{5}\right)^k + \frac{65}{7}\left(\frac{3}{10}\right)^k\right]\zeta(k)$

3. $D\left(e^{j\theta}\right) = \dfrac{1}{2e^{j\theta} + 1}$

4. $D(z) = \dfrac{2z+1}{(4z-1)(2z+1)}$

System is stable since the poles $z = \frac{1}{4}, -\frac{1}{2}$ lie in the unit open disk.

6. $D\left(e^{j\theta}\right) = \dfrac{1 - a\cos(\theta) - ja\sin(\theta)}{1 - 2a\cos(\theta) + a^2}$; $y_k = \dfrac{\cos(k\theta) - a\cos((k+1)\theta)}{1 - 2a\cos(\theta) + a^2}$

7. (a) $\dfrac{(z-1)(z+2)}{(z-\frac{1}{2})(z+\frac{3}{4})}$; Poles at $z = \frac{1}{2}, -\frac{3}{4}$ and zeros at $z = 1, -2$.

(b) $\frac{16}{3}\delta(k) - \left(\frac{1}{2}\right)^{k-1} - \frac{7}{3}\left(-\frac{3}{4}\right)^k$

(c) Poles $z = \frac{1}{2}, -\frac{3}{4}$ lie in the open unit disk.

(d) $y(k+2) + \frac{1}{4}y(k+1) - \frac{3}{8}y(k) = u(k+2) + u(k+1) - 2u(k)$

8. BIBO stable : $-1 < \alpha < \frac{1}{2}$

Exercises §7.8

1. $e^{-j\,3\theta}(1 + 4\cos(\theta) + 2\cos(3\theta))$

2. $\dfrac{1 - e^{-(3+j\,\frac{5\pi}{8}n)}}{1 - e^{-(0.3+j\,\frac{\pi}{16}n)}}$; $n = 0, 1, \ldots, 31$

3. (a) 8000 Hz   (b) 0.1 sec.   (c) 1024

4. $U_0 = 2$ $U_1 = \frac{1}{2}(3 + j)$ $U_2 = -1$ $U_3 = \overline{U}_1 = \frac{1}{2}(3 - j)$

5. (a) $W(n) = \{0.5N, -0.25N, 0, 0, \ldots, 0, -0.25N\}$
   (b) $\{1.5, -2.0 - j0.5, 2.5, -2.0 + j0.5\}$

6. $\{-j, 3 - j2, 2 - j\}$

7. $\{13, -2 + \frac{1}{\sqrt{2}} + j\frac{1}{\sqrt{2}}, 2 - j, -2 - \frac{1}{\sqrt{2}} + j\frac{1}{\sqrt{2}}, -1, -2 - \frac{1}{\sqrt{2}} - j\frac{1}{\sqrt{2}}, 2 + j,$
   $-2 + \frac{1}{\sqrt{2}} - j\frac{1}{\sqrt{2}}\}$

8. $\{1, 2, -1, 3\}$

9. $A_0 = \dfrac{1}{\pi}$, $A_{-n} = A_n = \dfrac{-1}{(n^2 - 1)\pi}$, n even, $n \geq 2$

   $A_1 = -\dfrac{1}{4}j$, $A_{-1} = \dfrac{1}{4}j$, $A_{-n} = A_n = 0$, n odd, $n \geq 3$

   $\{A_n; n = 0, 1, \ldots, 8\} =$
   $\{0.3183, -0.25j, -0.1061, 0, -0.0212, 0, -0.0091, 0, -0.0051\}$

Exercises §8.7

3. $D(z) = \dfrac{0.0025e^{-0.1}z}{(z - e^{-0.1})^2}$

   $H(j\omega) = \dfrac{1}{(j\omega + 2)^2}$ or $\dfrac{4 - \omega^2 - j\,4\omega}{(\omega^2 + 4)^2}$ and $D(e^{j\,\omega T}) = \dfrac{0.0025e^{-0.1 + j\,0.05\omega}}{(e^{j\,0.05\omega} - e^{-0.1})^2}$

   $H(0) = 0.25$ $D(1) = 0.2498$ (4D); dc gains differ due to aliasing.

4. $y([k + 2]T) - 2e^{-aT}\cos(aT)y([k + 1]T) + e^{-2aT}y(kT) =$
   $2aTe^{-aT}\sin(aT)u([k + 1]T)$

5. $y([k + 2]T) - 2e^{-5\sqrt{2}T}\cos(5\sqrt{2}T)y([k + 1]T) + e^{-10\sqrt{2}T}y(kT)$
   $= 10\sqrt{2}e^{-5\sqrt{2}T}\sin(5\sqrt{2}T)u([k + 1]T)$,   where $T = \frac{\pi}{25} \approx 0.1257$ sec.

6. $y([k + 2]T) - 2e^{-5\sqrt{2}T}\cos(5\sqrt{2}T)y([k + 1]T) + e^{-10\sqrt{2}T}y(kT)$
   $= \left[1 - e^{-5\sqrt{2}T}(\cos(5\sqrt{2}T) + \sin(5\sqrt{2}T))\right]u([k + 1]T) +$
   $e^{-5\sqrt{2}T}\left[e^{-5\sqrt{2}T} - \cos(5\sqrt{2}T) + \sin(5\sqrt{2}T)\right]u(kT)$,   where $T = \frac{\pi}{25}$ sec.

7. $D_1(z) = Tz \left(\dfrac{3}{z - e^{-2T}} - \dfrac{1}{z - e^{-T}} \right) = \dfrac{Tz(2z - 3e^{-T} + e^{-2T})}{(z - e^{-T})(z - e^{-2T})}$

$H(0) = \dfrac{1}{2}$ $\qquad$ $D_1(1) = \dfrac{T(2 - e^{-T})}{1 - e^{-2T}}$

$D_2(z) = \dfrac{(1 - e^{-2T})z(2z - 3e^{-T} + e^{-2T})}{2(2 - e^{-T})(z - e^{-T})(z - e^{-2T})}$

9. $D(z) = \dfrac{z^3 + 3z^2 + 3z + 1}{az^3 + bz^2 + cz + d}$,

where $a = 1 + 2\alpha + 2\alpha^2 + \alpha^3$, $b = 3 + 2\alpha - 2\alpha^2 - 3\alpha^3$, $c = 3 - 2\alpha - 2\alpha^2 + 3\alpha^3$,

$d = 1 - 2\alpha + 2\alpha^2 - \alpha^3$ and $\alpha = \cot\left(\dfrac{\pi \omega_d}{\omega_s}\right)$;

T	ω_s	ω_d
1	2π	0.60879
0.1	20π	6.0879
0.0001	20000π	6087.9

10. (a) $D(z) = \dfrac{z^2 + 2z + 1}{(4 + \sqrt{6})z^2 - 4z + (4 - \sqrt{6})}$

$D(1) = 1,$ $\qquad$ $H(0) = 1$

(b) dc gain ≈ 0.910

Suitable MATLAB commands are:

```
th = pi/256*[0:255];
thc = pi/3;
num = thc^2;
den = [1 sqrt(2)*thc thc^2];
H = freqs(num,den,th);
gain = abs(H);
numD = [1 2 1];
denD = [(4 + sqrt(6)) -4 (4 - sqrt(6))];
D = freqz(numD,denD,256);
gainD = abs(D);
k = thc/sqrt(2);
numD1 = [0 2*k*exp(-k)*sin(k)];
denD1 = [1 -2*exp(-k)*cos(k) exp(-2*k)];
D1 = freqz(numD1,denD1,256);
gainD1 = abs(D1);
gains = [gain(:) gainD(:) gainD1(:)];
plot(th,gains)
```

If the frequency range is increased to $0 \le \theta < 2\pi$, both the digital gain spectra will show symmetry about $\theta = \pi$ and, therefore, will rise back towards their dc values. In contrast, the analogue gain will continue to decrease towards zero.

The MATLAB commands for the last part of this question are similar to those above and, hence, are omitted.

11. $D(z) = H(s)\big|_{s \mapsto \frac{3}{T}\left(\frac{z^2-1}{z^2+4z+1}\right)}$

When $H(s) = \dfrac{\omega_c}{s + \omega_c}$, $\quad D(z) = \dfrac{\omega_c T(z^2 + 4z + 1)}{(3 + \omega_c T)z^2 + 4\omega_c T z + (\omega_c T - 3)}$.

12. $\{0.06366 \ 0 \ -0.10610 \ 0 \ 0.31831 \ 0.5 \ 0.31831 \ 0 \ -0.10610 \ 0 \ 0.06366\}$,
$D(z) = 0.06366(1 + z^{-10}) - 0.10610(z^{-2} + z^{-8}) + 0.31831(z^{-4} + z^{-6}) + 0.5z^{-5}$

13. $\{0.00509 \ 0 \ -0.04221 \ 0 \ 0.29035 \ 0.5 \ 0.29035 \ 0 \ -0.04221 \ 0 \ 0.00509\}$,
$D(z) = 0.00509(1 + z^{-10}) - 0.04221(z^{-2} + z^{-8}) + 0.29035(z^{-4} + z^{-6}) + 0.5z^{-5}$

14. $14.8246y([k + 2]T) - 16.9443y([k + 1]T) + 6.1196y(kT) = u([k + 2]T) + 2u([k + 1]T) + u(kT)$

Exercises §9.12

1. $\frac{1}{2}e^{-|\tau|}$

3. $\alpha = \begin{bmatrix} 1.5 \\ 0.5 \end{bmatrix} \quad \beta = \begin{bmatrix} \frac{1}{2} \\ 0 \\ 0 \end{bmatrix}$

4. $k_1 = \dfrac{4}{5} \quad k_2 = -\dfrac{8}{17}$

5. $\alpha_1 = 13 \quad \alpha_2 = -2$

6. $\tilde{x}(k) = \alpha x(k - 1)$

7. $\alpha_1 = \dfrac{2a(a^2 + 2)}{a^4 + 4}, \quad \alpha_2 = \dfrac{-4a^2}{a^4 + 4}; \quad \hat{x}(k) = \dfrac{2a}{a^2 + 2}x(k - 1)$

8. $c_w^{(c)}(k) - c_x^{(c)}(k) - \log_e(100)\delta(k)$ or $c_w^{(c)}(k) - c_x^{(c)}(k) - 2\log_e(10)\delta(k)$

9. (a) $\log_e(2)\delta(k) - \dfrac{1}{k}(\frac{1}{2})^{k-1}\zeta(k - 1)$

(b) $\log_e(2)\delta(k) + \dfrac{1}{k}[(2)^k\zeta(-k - 1) - (\frac{1}{2})^k\zeta(k - 1)]$

10. $c_p^{(c)}(k) = \displaystyle\sum_{r=1}^{\infty} \dfrac{1}{r}[a^r\delta(k - Nr) - a^{Mr}\delta(k - MNr)];$
yes, since the complex cepstrum is causal.

11. A possible sequence of MATLAB commands is:

```
e = eye(16,16);                  % generates 16 by 16 identity matrix
x = e(5:254);                    % forms impulse train
a = [1 -0.2 -0.23 -0.05 -0.12];
b = [1];
s = filter(b,a,x);               % output from the stable filter
k=1.2; n=k*randn(size(s));       % generation of noise
cs = conv(s,n);                  % convolving the signals
rc = rceps(cs);                  % real cepstrum evaluation
M = length(rc);
plot(rc(1:M/2)),title('Real Cepstrum');
pause;
%
% filtered cepstrum with rc(k)=0 for k<10
%
frc=[zeros(1,9) rc(10:M/2)];
plot(frc),grid,title('Filtered Cepstrum');
pause;
```

12. $c_h^{(c)}(k) = \log_e \left(\frac{1}{K} \right) \delta(k) - \frac{1}{k} \left[\alpha^k + (-\beta)^k \right] \zeta(k-1)$

$\{h(k)\}$ is a minimum-phase sequence, since $\{c_h^{(c)}(k)\}$ is causal.

References

[1] P.R. Adby. *Applied Circuit Theory*. Ellis Horwood, Chichester, U.K., 1980.

[2] E.O. Brigham. *The Fast Fourier Transform*. Prentice-Hall, Englewoods Cliffs, New Jersey, U.S.A., 1974.

[3] D.N. Burghes and A. Graham. *Introduction to Control Theory, including Optimal Control*. Ellis Horwood, Chichester, U.K., 1986.

[4] J.W. Cooley and J.W. Tukey. An algorithm for machine calculation of complex Fourier series. *Math. Computation*, 19:297–301, 1965.

[5] J.R. Deller Jr., J.G. Proakis, and J.H.L. Hansen. *Discrete-time Processing of Speech Signals*. Macmillan, New York, U.S.A., 1993.

[6] R.A. Gabel and R.A. Roberts. *Signals and Linear Systems*. Wiley, Singapore, 1987.

[7] A.A. Giordano and F.M. Hsu. *Least Square Estimation with Applications to Digital Signal Processing*. Wiley, New York, U.S.A., 1985.

[8] R.A. Haddad and T.W. Parsons. *Digital Signal Processing: Theory, Applications, and Hardware*. Computer Science Press, New York, U.S.A., 1991.

[9] R.F. Hoskins. *Generalised Functions*. Ellis Horwood, Chichester, U.K., 1979.

[10] E.C. Ifeachor and B.W. Jervis. *Digital Signal Processing: A Practical Approach*. Addison-Wesley, Wokingham, U.K., 1993.

[11] L.B. Jackson. *Digital Filters and Signal Processing*. Kluwer Academic Publishers, Boston, MA, U.S.A., 1986.

[12] M.T. Jong. *Methods of Discrete Signals and Systems Analysis*. McGraw-Hill, New York, U.S.A., 1982.

[13] R.P. Kanwal. *Generalized Functions: Theory and Technique*. Academic Press, London, U.K., 1983.

[14] F.F. Kuo. *Network Analysis and Synthesis*. Wiley, New York, U.S.A., 1966.

[15] M.J. Lighthill. *Introduction to Fourier Analysis and Generalised Functions*. Cambridge University Press, Cambridge, U.K., 1964.

[16] L. Ljung. *System Identification: Theory for the User.* Prentice-Hall, Englewood Cliffs, New Jersey, U.S.A., 1987.

[17] J. Makhoul. Linear prediction: a tutorial review. *Proc. IEEE*, 63:561–580, 1975.

[18] J.D. Markel and A.H. Gray Jr. *Linear Prediction of Speech.* Springer-Verlag, Berlin, Germany, 1976.

[19] S.L. Marple Jr. *Digital Spectral analysis with Applications.* Prentice-Hall, Englewood Cliffs, New Jersey, U.S.A., 1987.

[20] P.V. O'Neil. *Advanced Engineering Mathematics.* Wadsworth, Belmont, U.S.A., 1983.

[21] A.V. Oppenheim and R.W. Schafer. *Discrete-time Signal Processing.* Prentice-Hall, Englewood Cliffs, New Jersey, U.S.A., 1989.

[22] A.V. Oppenheim, A.S. Willsky, and I.T. Young. *Signals and Systems.* Prentice-Hall, Englewood Cliffs, New Jersey, U.S.A., 1983.

[23] A. Page. *Mathematical Analysis and Techniques.* Oxford University Press, London, U.K., 1976.

[24] A. Papoulis. *Circuits and Systems: a Modern Approach.* Holt, Rinehart and Winston, New York, U.S.A., 1980.

[25] T.W. Parsons. *Voice and Speech Processing.* McGraw-Hill, New York, U.S.A., 1987.

[26] A.D. Poularikas and S. Seely. *Elements of Signals and Systems.* PWS-Kent, Boston, U.S.A., 1988.

[27] L.R. Rabiner and R.W. Schafer. *Digital Processing of Speech Signals.* Prentice-Hall, Englewood Cliffs, New Jersey, U.S.A., 1978.

[28] S. Saito and K. Nakata. *Fundamentals of Speech Signal Processing.* Academic Press, Orlando, Florida, U.S.A., 1985.

[29] S.D. Stearns and D.R. Hush. *Digital Signal Analysis.* Prentice-Hall, Englewood Cliffs, New Jersey, U.S.A., 1990.

[30] L. Weinberg. *Network Analysis and Synthesis.* McGraw-Hill, New York, U.S.A., 1962.

[31] R.E. Ziemer, W.H. Tranter, and D.R. Fannin. *Signals and Systems.* Macmillan, New York, U.S.A., 1983.

Index

Printed and bound by CPI Group (UK) Ltd, Croydon, CR0 4YY

03/10/2024

01040342-0003